# The MILLENNIAL PROJECT

R

# The MILLENNIAL PROJECT

## *Colonizing the Galaxy in Eight Easy Steps*

*MARSHALL T. SAVAGE*

*With an Introduction by Arthur C. Clarke*

Illustrations by Keith Spangle

*Little, Brown and Company*

Boston New York Toronto London

Library of Congress Cataloging-in-Publication Data
Savage, Marshall T. (Marshall Thomas).
The millennial project : colonizing the galaxy in eight easy steps / by Marshall T. Savage with an introduction by Arthur C. Clarke ; illustrations by Keith Spangle.
p. cm.
Originally published: Denver : Empyrean Pub., c1992.
Includes index
ISBN 0-316-77165-1 (hc)
ISBN 0-316-77163-5 (pb)
1. Space colonies. I. Title.
[TL795.7.S28 1994]
629.4 — dc20 94-15965

HC: 10 9 8 7 6 5 4 3 2 1

PB: 10 9 8 7 6 5 4 3 2 1

RRD-VA

Published simultaneously in Canada by
Little, Brown & Company (Canada) Limited

Printed in the United States of America

*To my mother,*
*whose love and wisdom*
*illuminate my life*

*It's a kind of magic.*
*One dream.*
*One soul.*
*One prize.*
*One goal.*
*One golden glance*
*of what should be.*
*It's a kind of magic . . .*

— *QUEEN*

# CONTENTS

***It would not be too much to say that myth is the secret opening through which the inexhaustible energies of the cosmos pour into human cultural manifestation.***

**Joseph Campbell**

## LIST OF TABLES

**LIST OF FIGURES**

## APPENDICES

# *INTRODUCTION*
## *by Arthur C. Clarke*

*THE MILLENNIAL PROJECT* is a book I wish I'd written: correction—it's a book I wish I *could* have written. I am completely awed, and I don't awe easily, by the author's command of a dozen engineering disciplines and his amazing knowledge of scientific and technical literature. Yet I note with approval that his interests also range from rock music through Emily Dickinson to Malory's *Morte d'Arthur*. On top of all this, he is an inspiring writer with an engaging sense of humour ("I'm just a simple home-boy, and take no great interest in anything beyond the Magellanic Clouds").

Do I sound a little envious? Well, let's say that if I was 50 years younger I might have considered terminating Mr.Savage with extreme prejudice. Now I'll merely warn my fellow science-fiction writers that if he decides to invade their turf they may be in deep trouble.

*The Millennial Project* has the modest subtitle "Colonizing the Galaxy in Eight Easy Steps," and that's what it's all about—though some readers may challenge the author's definition of "easy." However, even those who have no extraterrestrial ambitions will be fascinated by his projects for sea-farming, floating cities and underwater habitats. Like most of the book's concepts, these are illustrated by dramatic and often strikingly beautiful color plates.

I may be unable to judge *The Millennial Project* dispassionately as it touches on an extraordinary—indeed, uncanny—

number of my own interests; sometimes it seems that the author has been reading my mind, and plagiarising my books before they are published. His description of form-fitting, leotard-type spacesuits is a straight steal from *The Hammer of God*. The illustration of terraformed Mars has undoubtedly been downloaded from the *Snows of Olympus* file on my Amiga's hard disk. (Damn clever, as it doesn't have a modem.)

I could quote at least a dozen more resonances in this extraordinary book, but as they are of purely personal interest I'll mention only one. *The Millennial Project* has 727 references, many of them fascinating reading in their own right. Number 303 is my 1945 paper on communications satellites, written when I was working on Ground Controlled Approach radar. And a couple of inches away (number 308) is a name I'd almost forgotten—the RAF group captain (Edward Fennessy) who, more than half a century ago, completely changed my life by selecting me for that job. . . .

It's goose-pimple time, so I'll get out of the way and let Marshall start stretching your mind.

ARTHUR C. CLARKE
*Colombo, May 1993*

# Author's Introduction

***...our expansion into the universe is not just an expansion of men and machines. It is an expansion of all life, making use of man's brain for her own purposes.***
**Freeman Dyson**

Now is the watershed of Cosmic history. We stand at the threshold of the New Millennium. Behind us yawn the chasms of the primordial past, when this universe was a dead and silent place; before us rise the broad sunlit uplands of a living cosmos. In the next few galactic seconds, the fate of the universe will be decided. Life—the ultimate experiment—will either explode into space and engulf the star-clouds in a fire storm of children, trees, and butterfly wings; or Life will fail, fizzle, and gutter out, leaving the universe shrouded forever in impenetrable blankness, devoid of hope.

Teetering here on the fulcrum of destiny stands our own bemused species. The future of the universe hinges on what we do next. If we take up the sacred fire, and stride forth into space as the torchbearers of Life, this universe will be aborning. If we carry the green fire-brand from star to star, and ignite around each a conflagration of vitality, we can trigger a Universal metamorphosis. Because of us, the barren dusts of a million billion worlds will coil up into the pulsing magic forms of animate matter. Because of us, landscapes of radiation blasted waste, will be miraculously transmuted: Slag will become soil, grass will sprout, flowers will bloom, and forests will spring up in once sterile places. Ice, hard as iron, will melt and trickle into pools where starfish, anemones, and seashells dwell—a whole frozen universe will thaw and transmogrify, from howling desolation to blossoming paradise. Dust into Life; the very alchemy of God.

If we deny our awesome challenge; turn our backs on the living universe, and forsake our cosmic destiny, we will commit a crime of unutterable magnitude. Mankind alone has the power to carry out this fundamental change in the universe. Our failure would lead to consequences unthinkable. This is perhaps the first and only chance the universe will ever have to awaken from its long night and live. We are the caretakers of this delicate spark of Life. To let it flicker and die through ignorance, neglect, or sheer lack of imagination is a horror too great to contemplate.

**Prometheus Unchained**

The stars are our destiny. They are our legacy. Strewn like diamonds on a field of black velvet, they lie waiting for the hand of man to pluck them up. The gulf of space is like an infinite version of Ali Baba's cave, crammed with jewels and riches beyond counting.

All these treasures strewn before us are free for the taking. There is no guardian genie. There are no alien owners to be bargained with, no evil empires to be vanquished, not even a galactic bureaucracy to demand emigration forms in triplicate. The galaxy is free and open now in a way it never will be again. Our species can skate across the glassy spaces, sliding unfettered through the blizzard of stars, skimming down the frosty spiral arms to the snowy banks of the galactic nucleus.

For better or worse, Life has evolved *Homo sapiens* as the active agent of her purpose. We are the sentient tool-users. Perhaps Life should have bet on the dolphins. But, she put her money on us, and there is no time left for second guesses. Life has endowed us with the power to conquer the galaxy, and our destiny awaits us there, among the powdery star-fields of deep space. Now we must spring from our home planet and carry the living flame into the sterile wastes. It is time to return the gift of Prometheus to the heavens.

To fulfill our cosmic destiny and carry Life to the stars, we must act quickly. The same unleashed powers that enable us to enliven the universe are now, ironically, causing us to destroy the Earth. The longer we delay, the further we may slip into a pit of our own digging. If we wait too long, we will be swept into a world so poisoned by pollution, so overrun by masses of starving people, so stripped of surplus resources, that there will be no chance to ever leave this planet. Thus far, we have failed to use our new powers for the ends they were intended. The result is an accelerating slide toward disaster.

The litany of eco-crisis is numbingly familiar—like a Gregorian chant of doom: the ozone hole, the greenhouse effect, deforestation, desertification, overpopulation. Woe, lamentation, and gnashing of teeth. If you are still unaware of the emergency, you must already live on Mars.

The crisis is driven by the exponential explosion of human numbers. A hundred million new people enter the world each year. A new population the size of Iran every five months. Where will all of these new people live? What will they eat? What prospect for the future do they have?

There is no way, short of nuclear war, plague, or famine, to prevent human numbers from doubling. The parents of tomorrow have already been born, and when they bear children of their own, the global population will surge.

Our situation is analogous to yeast in a bottle. The yeast cells will double their number every day until the bottle is full—then they will all die. If the yeast die on the 30th day, then on what day is the bottle half full? The 29th day. We are in the 29th day of our history on Earth. We must do something now, or face extinction.

The obvious answer is to blow the lid off this bottle! We need to rupture the barriers that confine us to the land mass of a single planet. By breaking out, we can assure our survival and the continuation of Life.

Space beckons us. It is the clarion call of destiny. We are still evolving as a species and Life is still evolving as a force of nature. Only by leaving the womb of Mother Earth can man and Life survive and mature.

Within a thousand years, we will break forever the bonds of gravity and soar freely among the stars. This Great Divide in the topography of time coincides with the dawn of the Third Millennium. The coming Millennium is the Age of Aquarius—prelude to the endless emerald epoch of Life's galactic empire.

Life is too precious a thing in the Cosmos not to be preserved at all costs. It is entirely possible that ours is the only living planet in the universe. Throughout the star clouds of the Milky Way, planets probably teem by the hundreds of millions. But every one may be as dead and sterile as our own moon. Those myriad empty worlds could be just so many particles of barren galactic dust.

Yet, out on the margin of this vast slag-heap of stellar debris, there glows a single magic scintilla of blue-green living light. Like a lone incandescent spark in an endless landscape of cinders—this is Gaia. Earth, a single tiny glimmer of Life, utterly and eternally alone. And yet, for all our microscopic insignificance, we have the potential to suffuse our green fire through every granule of the whole lifeless pile. What is such a spark worth do you suppose? How many of the lifeless worlds would you give in exchange for the one living one? It is like asking how much coal ash you would trade for the Hope Diamond.

Consider for a moment the implications if we are alone: Then the entire responsibility for Life in the Cosmos is ours to bear. Compared to this duty, the burden of Atlas was nothing. As the sole caretakers of Life it is our sacred duty not only to preserve Life here on Earth, but also to disseminate the magic among the stars.

The universe may teem with Life. We don't know that it doesn't. Conversely, we don't know that it does. The bottom line is that as far as we know, this planet holds the only reservoir of Life Force in the Cosmos. Until we find out otherwise, it is incumbent upon us to act as if the Earth is the lone spark in ten billion parsecs of frozen desolation.

**Let's Go!**

This book presents a plan of action for building a stairway to the stars. Step by step we can pile up the cyclopean stones; like the mythic Titans we will heap Ossa on Pelion on Olympus, erecting a pyramid to touch the heavens.

The message of this book is a simple one: the stars are within our reach. We now have the capacity, economically and technically, to leave this planet and begin the infinite task of enlivening the universe. We can accomplish our ends in eight easy steps: First, we will lay the Foundation, uniting ourselves around the green banner of Cosmic destiny. Then we will grow a crystalline city, floating on the waves of the sea. With power from the ocean, we will launch ourselves into space, propelled aloft by a rainbow-hued array of lasers. In orbit above the Earth, we will inflate gleaming golden bubbles to shelter our new generation of space dwelling people. On the face of the Moon, we will cap the craters with glistening domes, each sheltering a green oasis of life. Mars will be transformed into a glorious gem of blue oceans and swirling white clouds, vibrant and alive as Gaia herself. Among the asteroids we will strew a spreading ring-cloud of billions of billions of bubbles of life, shimmering like a galaxy of golden sparks. Finally, in the latter half of the Millennium, space arks will carry human colonists across the interstellar gulfs to inseminate new worlds with the chartreuse elixir of Life. By Millennium's end the night sky will twinkle with a handful of emerald stars—the initial scattering of our celestial seeds. From this first planting will spring a growing forest of living solar systems. Life will explode through the star clouds like beryllian fire through flash powder. Within a thousand millennia, the whole majestic pinwheel of the Milky Way, will be saturated with the lush aquamarine light of a hundred billion living suns. We will have created a living galaxy—seed of a living universe. Then the animate flame will leap the firebreak between galaxies and ignite new blazes among the great star clusters in the outer universe. The process will continue, unremitting, for the eternal lifetime of the Cosmos. (But of this I do not speculate. I am just a simple home-boy, and take no great interest in anything much beyond the Magellanic Clouds.)

**Manifest Density**

A million years from now, our descendants will populate this galaxy. From the red dwarves of the globular clusters to the blue giants of the galactic nucleus, a hundred billion stars will shine on the homes of a trillion trillion human beings. Their civilizations will span the heavens with powers transcending the feeble reach of our imaginations. Yet, each person of that countless multitude will look back in space and time to a tiny yellow star out on the rim of the Orion Arm. In their grandeur and their glory these demigods of a future time will remember us, and think of how it all began.

# CHAPTER 1

# AQUARIUS

*This is the Dawning of the Age...*
**Hair**

Aquarius - Zodiacal sign of the water bearer. The wisdom of the ancients decreed three ages for the reign of man. With the coming of the Third Millennium dawns the age of Aquarius.

It is our destiny to colonize space. Eventually we will spread our civilization among the stars, but our first step is to build space colonies on Earth.

At first glance the Earth may seem a little over-crowded for colonization. But really, three-quarters of this planet's surface—the oceans—are virtually uninhabited. Colonizing the oceans will be like discovering three new planets the size of Earth.

Our first space colonies will be floating islands, grown organically from the lambent waters of the tropical seas. There are four principal reasons why our first step toward space should take us to sea:

① If we are going to colonize space, it is best to colonize the easiest space first. The most accommodating space in this universe is right here on Earth. The tropical oceans are womb-like: warm, hospitable, nourishing, and wet. We will never find a better place to gestate our embryonic pan-galactic empire than right here on the mellow seas of Earth.

② Living in colonies at sea will teach us many crucial lessons about life in space. The isolation, self-sufficiency, and political autonomy of sea colonies are the same as those of space colonies. Both types will impose many of the same requirements on their inhabitants. While the external environments of sea and space colonies are as different as tropical islands from lunar craters, the internal social and personal environments are identical. Space colonization's hardware problems—questions of tool design—are easy to solve; the software problems—questions of social and individual evolution—are much tougher. We need to learn to live together in a colony environment long before we need to worry about how to live in the space environment. The Moon is a harsh mistress; we would be wise to learn these early lessons while still in Earth's gentle lap.

③ Before we go gallivanting off to populate the galaxy, we had better save the planet we're already on. The sea colonies can go far toward rescuing the Earth, producing enough food and energy to meet the needs of billions, without damaging the planetary ecosphere. The sea colonies can even repair some of the damage already done.

④ Getting into space requires enormous power; both physical power that flares out of a rocket, and financial power that flares out of a bank account. The sea colonies will produce both kinds in abundance: enough raw electrical power to blast us into space, and enough raw financial power to pay the fare.

**Malthusian Blues**

Our direst problems face us right here on Earth; and it is here we must solve them. Our future lies in space, but the Earth is the womb of life, and it will be a long time before we can cut our umbilical cord. The new worlds we wish to create can survive their infancy only if the Mother of Life (Gaia) is here to nourish them. If we are to fulfill our Cosmic destiny as the harbingers of Life, we must first insure survival of the home planet.

The world's immediate problem is that there are too many demands on too few resources. There is simply not enough 'stuff' to go around. This creates many attendant problems: Subsistence farmers, in quest of land, slash and burn tropical forests, ravaging the lungs of the world. The poor, in search of jobs, flood urban areas, engorging these already bloated tumors. The rich, in pursuit of 'the good life,' suck up the last vestiges of vanishing resources, spewing out mountains of garbage and rivers of toxins in exchange. Our rapacious demands are overtaxing the ability of Gaia to regenerate herself. The result is a dying planet.

We must find a way to avert this catastrophe. From Gaia's perspective, the answer to this disaster is a species-specific plague to wipe us out—AIDS perhaps, or maybe something even worse. While this might save the Earth, it is hardly an agreeable solution from our point of view. A viable answer must meet the needs of both the planet and the people: it must reduce or halt the destruction of Gaia's ecological tissues; it must decrease or eliminate the production of pollutants; it must be implemented without depleting scarce resources; and, at the same time, it must satisfy the food and energy needs of ten billion hungry humans. A solution which can fulfill all these requirements might seem impossible, but the answer is at hand—Aquarius, and her thousand sister sea colonies.

The primary commodities that support mankind are food and energy. So far, we have always obtained these staples at the expense of the environment. Originally, we got food by hunting animals and gathering plants directly from the food chain. This was so destructive of resources, however, that we could sustain ourselves only through nomadic wandering. Our numbers eventually grew so large that we could no longer wander freely enough to allow nature time to heal. We settled in fixed places and used agriculture to increase food production enough to keep pace with our expanding population. But agriculture demanded that we bend the environment to our will. We cleared forests and put grasslands to the plow. We appropriated the grazing for our herds and we exterminated the predators that preyed on them. We dammed rivers, flooded valleys, and ran irrigation networks across the landscape. In so doing, we changed utterly and forever the face of the world.

Energy is much like food. We have always supplied most of our growing needs for energy by burning organic fuels. (Renewable hydro-power and nuclear energy have made contributions in the past, but in the future these will be relatively small, and not without cost to the environment. Hydro dams dramatically alter the landscape, and nuclear power produces poisons of such lethality and longevity that they will still be deadly 20,000 years from now.) Wood is a historic mainstay of human energy consumption, but the effect of strip-mining this resource beyond its sustainable yield is fatal. The biggest problem, of course, is with fossil fuels. The industrial exhalations of acids and carbon dioxide produced by burning these fuels have already done substantial harm to the world's environment. Acid rain is increasingly killing the forests of the North, and accumulating $CO_2$ may be pushing us into an uncharted realm of higher global temperatures. If we try to supply the energy needs of 10 billion people—all desiring comfort, mobility, and sustenance—by burning the remaining stocks of fossil fuels, we surely face an environmental catastrophe. The planet simply can't stand 10 billion people all burning coal and gasoline like Americans.

We are perilously close to toppling the delicate balance of life already. If we destroy what little remains of the natural biosphere to support ourselves, we will surely push this planet over the brink. If we continue to rip resources from the Earth at the expense of the biosphere—essentially tearing them out of Mother Earth's hide—then the rise in our numbers and living standards will inevitably destroy the planet's viability as a human habitat.

So this is our quandary: the food and energy we need must come from somewhere or we will die, but if we rip these things from the body of Gaia, as we have done in the past, then the planetary ecosystem will collapse and we will die. Either way, we seem doomed. Doom is an unacceptable fate. We humans are a resourceful, resilient, adaptable, and above all stubborn species. We will never surrender. There must be a solution—we just have to find it.

The problem is that we need an abundance of resources, but we can no longer afford to take them from the biosphere. The solution is to find some source outside the biosphere. Eventually that source will be outer space itself. In the interim, these resources must come from some place which is on Earth but is not part of the existing biosphere. This may seem a contradiction, but such a source does exist.

There is a source from which we can get both the food and the energy we need to survive, without devouring or poisoning the planet. It is no coincidence that the original source of all life should in the end be our salvation. Just as she gave us birth in the beginning, now she will save us in the end. She is our original mother and our ultimate savior—the sea.

The global ocean can provide enough energy and nutrients for us to survive detonation of the population bomb. The warm surface waters of the sea hold an inexhaustible charge of solar energy. The oceans of the world function as gigantic solar collectors. The sun transmits to earth 18,000 times as much energy as mankind uses.[1] An enormous amount of this radiant flux is stored in the surface of the oceans. Each ton of sea water contains as much energy as two pounds of gasoline.[2] The energy contained in the world's sea water is equivalent to filling the ocean basins twenty feet deep in high-octane fuel.[3] Altogether, the world's oceans contain $5 \times 10^{21}$ BTU of potential energy—an amount equal to a million billion barrels of oil.[4] There is enough latent energy in the oceans to supply the entire world power demand for 25,000 years. And it is renewable.

The world's oceans contain 550 billion metric tons of nitrates.[5] This is 36 times more nitrogen than is held in the planet's entire biomass.[6] In 1986 the U.S. used 20 million tons of fertilizers; the nitrogen in the oceans could supply this demand for 27,000 years. These reserves of oceanic nutrients are the yolk of our planetary egg. To survive this embryonic phase of our species' development we need only tap the oceanic yolk sac.

**Healing Gaia**

The resources we need can be produced at virtually no cost to the Earth's failing ecosystem. The sea colonies can help solve the world's energy and food crises without exacerbating its environmental crisis. The sea colonies can double the world's supply of energy, and do it without increasing carbon dioxide or acid rain, without disturbing an acre of ground, and without depleting any limited resources. Marine colonies, will, like all space colonies, make use of space which is now ecologically barren. The open oceans are largely lifeless due to a lack of nutrients. The marine colonies will therefore displace no pre-existing ecosystems. (See Plate No. 1)

The total energy contribution of the marine colonies to the world will be enormous. Every colony will produce three principal products, each with an energy value:

**Table 1.1**
**Daily Production of a Millennial Marine Colony**

| Commodity | Daily Production | Energy Value |
|---|---|---|
| Hydrogen | $66 \times 10^6$ ft$^3$ | $16 \times 10^9$ BTU |
| Distilled Water | $175 \times 10^6$ gal. | $100 \times 10^9$ BTU |
| Protein | 410 tons | $465 \times 10^9$ BTU |
| **TOTAL** | | $581 \times 10^9$ BTU |

Every marine colony will annually generate commodities which would otherwise require 50 million barrels of oil to produce. The world presently consumes the energy equivalent of 60 billion barrels of oil per year.[7] Twelve hundred marine colonies could produce an equivalent flow of energy in the forms of electricity, hydrogen, distilled water, and food.

A single marine colony will produce 300 million lbs. of protein annually, saving vast amounts of fossil energy. To supply the same protein from feedlot beef would require 28 million barrels of oil.[8] If the same protein were extracted from the sea in the form of commercially netted fish, it would require 82 million barrels of oil.[9] If this protein were produced by Zebu cattle, the only other protein producers which approach phytoplankton in energy efficiency, it would require 440 million acres of African grazing land—an area three times the size of Kenya.[10] If the protein is produced by a Millennial colony floating in the open ocean, instead of by vast herds of cattle on the African plains, 700,000 square miles of the Earth's surface can be spared the ravages of overgrazing. When a thousand marine colonies are operating, they will produce as much protein as could be gleaned from 240 million square miles of prime range land—an area four times the land surface of the Earth.[11] A protein supply of this magnitude could relieve many of the terrible burdens man places on the land.

What would this mean for the world's ecology? Every barrel of oil burned produces 800 pounds of carbon dioxide. Of all the Earth's present maladies, accumulating atmospheric $CO_2$ is probably the most serious. Mankind is pumping 5 billion tons of $CO_2$ into the atmosphere every year by burning fossil fuels.[12] The marine colonies could rapidly reduce those emissions to under a billion tons a year.

**Table 1.2**
**World Wide Energy Usage**
**in $10^6$ Barrels of Oil Equivalent[13]**

| Source | Annual Total | bbl/day |
|---|---|---|
| Oil | 24,309 | 66.6 |
| Coal | 15,308 | 41.9 |
| Natural Gas | 12,550 | 34.4 |
| Nuclear & Other | 3,538 | 9.7 |
| Hydropower | 3,715 | 10.2 |
| **TOTALS** | **59,420** | **162.8** |

Mankind uses a lot of energy; the equivalent of around 166 million barrels of oil per day. A barrel of oil is the energy equivalent of 700 kilowatt hours of electricity.[14] A single sea colony will produce 3.6 terawatt hours of net electrical power annually—equivalent to five million barrels of oil. To replace all the oil burned in the world today would require 4300 marine colonies; replacing all coal would require 3000; replacing natural gas would require 2100 more. With 9400 sea colonies we could completely replace fossil fuels. Ten thousand colonies would produce the direct energy equivalent of 50 billion barrels of oil annually. This level of energy production is well within the world's ultimate projected ocean thermal energy capacity of 65 billion barrels.[15] There is ample room on the tropical seas for ten thousand marine colonies. If we built that many, each would be surrounded by seven thousand square miles of open ocean.

If the sea colonies are to replace coal and oil, we must convert electricity into some form of fuel. Hydrogen is the perfect fuel; abundant as sea water and clean as sunlight. It can be extracted from water, and when burned exhausts only steam. When liquefied, hydrogen can be transported long distances economically. Every day, each colony could produce 67 million $ft^3$ of liquid hydrogen.[16]

Numerous collateral benefits would include acid rain reduction, fewer oil spills, lower Middle Eastern induced world tensions, and reduced pollutants like ozone, methane, and carbon monoxide. With enough marine colonies, we can tip the ecological balance from catastrophe to sustainability. At little or no cost to the planet's base metabolism, the marine colonies can provide the critical margin of survival. By reducing pollution we can reverse the forces now pushing the planet over the brink. The marine colonies may be the straw (bale perhaps) which saves the camel's back. They will, at the very least, delay the planet's decline long enough for us to get a permanent toe-hold in space.

A thousand sea colonies will, of course, have some environmental impact. There is no way to do anything inside a closed ecosphere—even one as large as the Earth—without impacting its environment. The sea colonies will inevitably change their local environments. With thousands of them in operation they may change the global environment. Compared to burning coal or splitting atoms, however, the OTECs of Aquarius are benign—even beneficial.

### $CO_2$ Sponge

By the time they are constructed, the Aquarian colonies may be the only thing saving us from a runaway greenhouse effect. If accumulating $CO_2$ begins to trigger 'positive feedback' effects, we could enter a runaway greenhouse phase, with horrifying consequences.[17] As the temperature of the planet rises with increasing atmospheric carbon dioxide, water vapor is released from the oceans in small, linearly increasing amounts. However, at a critical sea surface temperature of 80.6° F., the release of water vapor jumps dramatically.[18] Water vapor is far and away the most important greenhouse gas.[19] It is up to a hundred times more abundant than carbon dioxide and absorbs infra-red energy over a broad spectrum.[20] Under older greenhouse models, a global temperature increase of 2 to 4° C. is projected. New models, which factor in the increased release of water vapor, project increases in global temperature of 8 to 24 degrees![21] A global temperature increase of 4° C., would be an absolute catastrophe; an increase of 24° C., might make the Earth uninhabitable.

The marine colonies attack this problem directly by cooling the equatorial waters. Marine colonies will be located where the temperature of the surface water is highest. Aquarian power plants work by pumping cold water to the surface (see Fig. 1.2). By simply discharging the cold water at the surface, instead of reinjecting it at depth, we could cool the warm tropical waters below the critical temperature.

If the world warms up faster than predicted, we may have to push the marine colonies hard enough to halt and then reverse the biosphere's collapse. This would require building 15,000 colonies. With enough colonies, we can actually sponge $CO_2$ out of the atmosphere. Each marine colony produces over 800,000 tons of algae per year (dry weight).[22] Since the algae are 20% carbon by weight, each colony removes 600,000 tons of $CO_2$ from the atmosphere each year. If these algae are then used as food, the carbon will return to the atmosphere in the exhalations of the people who eat it. If, instead of harvesting the algae crop, we allowed it to sink into the ocean's depths, it would take its load of carbon with it. Although a tremendous protein bounty would be lost in the process, carbon dioxide could thereby be cleansed from the atmosphere. With 10,000 marine colonies in operation, a billion tons of $CO_2$ could be removed from the atmosphere each year. This, together with the reductions resulting from replacing fossil fuels, could halt and even reverse the build up of carbon dioxide.[23]

Marine colonies may offer one of the only practical ways to absorb carbon dioxide. It is often suggested that excess carbon dioxide could be absorbed by planting more trees. Unfortunately, appealing as this idea is, it won't work. The problem is that trees, being terrestrial plants, must eventually decay, and when they do they release their carbon back to the atmosphere. To remove $CO_2$ from the atmosphere permanently, the carbon sink must be outside the active bio-cycle. Allowing the marine colony's algae crop to sink unharvested is really just a means to augment the process Gaia uses to maintain the atmosphere's carbon balance.

Our first great challenge in the New Millennium must be to find a means to tap into the energy and nutrient reserves of the ocean. To achieve this task will require nothing less than a quantum leap in the evolution of life.

**Cybergenesis**

Cybernetics: From the Greek *kybernetes*, meaning pilot. The comparative study of organic and computerized automatic control systems operating through feed-back loops.

Genesis: Creation—specifically of life.

Cybergenesis: The creation of cybernetic systems sufficiently complex to exhibit the fundamental properties of life—self-organization and replication.

Aquarius—the first sea colony, and all Millennial space colonies that follow her—will be nothing less than new macro-organic life forms. Today's city-as-organism is at the evolutionary level of a parasitic slime mold. Aquarius will catapult to the plane of a photosynthetic sea lotus.

In the process of vaulting between evolutionary planes, the Aquarian sea colonies will shatter the old limits of the zero-sum resource game. In a 'zero-sum' game there are limited quantities of some critical component. If one player has more, some other player must have less. Poker is a zero-sum game. If one player is getting richer, other players are getting poorer. The world today plays a sophisticated game of resource poker, in which some countries enrich themselves by impoverishing others. Aquarius breaks the rules of this game by dumping more chips on the table. It's like playing table stakes poker with a vacuum hose connected to Scrooge McDuck's vault.

Fracturing zero-sum barriers is nothing new for Life on Earth. New life-forms have always taken quantum evolutionary leaps—creating abundance out of nothing. When plants first colonized the land, vast carboniferous forests sprang up where nothing had grown before. In the same way, Aquarius will colonize a presently barren ecology—the open reaches of the mid-oceans. In the process, Aquarius, like the early forests, will create a wealth of new resources to be shared by all life.

***Fig. 1.1 - Cybergenic Organisms.*** *The modern megalopolis is a primitive macro-organism. Aquarius will elevate the city to a new evolutionary plane, as far removed from urban blight as the lotus is from the slime mold.* Slime mold photo by R. Carlton.

At this stage in its evolutionary history, the world needs a new macro-organic life form. A life form is, by definition, anti-entropic, and so can add to the total order of its system. Any viable solution to the global resource crisis must be anti-entropic. Any entropic approach to increasing food and energy, like Herculean irrigation projects or nuclear power plants, will inevitably place new demands on the pre-existing zero-sum resource base. Borrowing from the common resource pool can redistribute global problems, but it can't solve them; it always creates new problems to replace the old ones. To break out of this animal-based predatory cycle takes an entirely new approach—cybergenesis.

Through cybergenesis, Life on Earth will take a quantum leap up the evolutionary ladder; implementing the type of order expanding solution that Life has always used to broaden and deepen its hold on existence. Following the cybergenic pathway, Life can evolve itself out of its present resource trap.

**Power Plant**

The pulsing heart of Aquarius is an OTEC (Ocean Thermal Energy Converter). The OTEC produces electrical power by exploiting the temperature differential between warm surface waters and cold deep waters. Aquarius has a long tap root that penetrates to the cold deep waters of the sea. By taking in warm water from the surface and sucking up cold water from the depths, OTECs generate electrical power.

Most power generating facilities conform to the zero-sum rules. They consume more energy than they produce. A typical nuclear power plant consumes 3000 calories of energy for every 1000 it produces. This is not unlike the thermodynamics of a cow who consumes three pounds of grain for every pound of milk she produces. Unlike conventional power plants, OTECs are net energy producers. An OTEC consumes only 700 calories of energy for every 1000 it produces.[24]

This is a characteristic that OTECs share with most solar powered devices, including green plants. The OTEC consumes no fuel, so the only energy the system requires is that needed to construct and operate it. By virtue of its ability to absorb solar energy, and to use that energy to impose higher states of order on the materials in its environment, the OTEC, like a living plant, is able to operate in defiance of the second law of thermodynamics. Of course, the law is not violated in the broader universe, since the sun is providing the energy, and it is running down, just as the law demands. But it will be a long time before we have to include the fusion engine of the sun in our calculations of local entropy. For the time being, we can consider sunlight as a free good, outside the limits of our earth-bound system of energy accounting.

The anti-entropic nature of OTECs is what distinguishes Aquarius as a new cybergenic life form. Like a plant, Aquarius will grow organically from a seed by feeding directly on sunlight; it will create itself out of the raw amorphous materials in its environment; and, it will produce an abundance of precious resources out of little more than sunlight and sea water. In a broad sense, therefore, Aquarius can be considered a life form—a macro-organism, a super-plant.

Guided by on-board computer intelligence, Aquarius will grow its own structure out of the sea. This will be accomplished by a process akin to that used by shellfish. Building materials will be amassed by accreting minerals dissolved in sea water. Sheltered and nurtured like the cells in a silicate sponge, thousands of people can live comfortably in this organic structure.

In the process of producing power, the OTECs pump vast quantities of cold water up from the depths. This deep water is saturated with nitrogen and other nutrients. When this nutrient-rich water hits the warm sunlit surface, algae populations explode. The algae are cultivated in broad shallow containment ponds that spread out around the central island of Aquarius like the leaves of a water lily. The algae soak in the tropical sun, absorbing the rich nutrient broth from the depths and producing millions of tons of protein.

Aquarius will be the first of the new cybergenic life forms, but by no means the last. Once we have grown ten thousand of these colonial super-organisms, we will culture and harvest enough protein-rich algae to feed every hungry human on Earth. We will generate enough electrical power—converted into clean-burning hydrogen—to completely replace all fossil fuels. We will build enough living space to house hundreds of millions of people in self-sufficient, pollution-free, comfort. We will learn the harsh lessons of space colonization in the mellow school of a tropical paradise. And, we will unleash a torrential cash flow—large enough to underwrite any adventure in space we care to imagine.

### OTEC

All heat engines function on the simple proposition that energy will flow from a warmer to a cooler body. In conventional power plants, the temperature difference is hundreds of degrees. An OTEC operates on a temperature difference of only 40 degrees.

In the tropical seas, surface waters, bathed in the intense light of the equatorial sun, are heated to 80°+ F. (26.6° C.); deep waters, condemned to centuries in utter darkness, are cooled to 40°F (4.44° C.). This difference in temperature is enough to run a thermal engine, albeit at low efficiency. (The greater the difference in temperature, the more efficient the engine.)

A typical fossil fuel plant will convert 40% of the energy available in the fuel to electricity.[27] An OTEC, will convert only 2.5% of the available energy to electricity.[28] Usually, this would seem a ridiculously low level of efficiency not warranting any consideration as a realistic source of energy—but there is nothing usual about the sea. At sea, even very low levels of thermal efficiency are rendered practical by the sheer size of the available resource.

Expressed in electrical terms, the energy resource of the oceans represents a renewable power base of over 200 million megawatts.[29] By comparison, the global installed electrical capacity in 1978 was only one million megawatts.[30] In other words, the total electrical output of mankind represents only a half of one percent of the power latent in the world's oceans. Even at very low levels of net efficiency, OTECs could produce ten times as much electrical energy as every other current power source combined.[31]

Extracting this energy from the sea requires only a simple process: The warm surface waters boil a volatile 'working' fluid, thereby producing vapors which expand, driving a turbine; the spinning turbine powers a dynamo, producing an electric current. The process is akin to steam power, but it operates at low temperatures. Cold water is pumped up from the depths and condenses the vapor.

Warm surface water enters the OTEC at about 80° F. and is discharged at 70° F. (21° C.). Cold water enters at 40° F. and exits at 45° F. (7° C.)[32] The transfer of heat, from warm water to cold, generates net energy with no consumption of fuel.

The critical factor in OTEC technology is the temperature difference, 'Delta t ' (Δt) between surface water and deep water, 3000 feet below. The 40 degree temperature difference prevailing in equatorial waters provides the same potential energy as a 400 foot waterfall.[33]

It is theoretically possible to produce net power when the Δt is as low as 23° F. Present technology, however, limits potential OTEC sites to areas of the world's oceans where the Δt is at least 35 - 40° F., year round. There is a belt of warm water around the Earth where the surface water is at least 40 degrees warmer than the deep water. Twenty million square miles of ocean in the equatorial regions are optimal for OTEC platforms.[34] In some areas, this year-round temperature difference is as high as 44°. The higher the temperature difference the greater the net output of energy for an OTEC of any given size.

**Table 1.3**
**Δt. as a Function of Depth off Sri Lanka**

| Depth in Feet | Δt in °F. |
|---|---|
| 164 | 5 |
| 328 | 14 |
| 2300 | 38 |
| 2950 | 40 |
| 3300 | 41 |
| 3940 | 43 |

**Table 1.4**[35]
**Year Round Sea Surface Temperature at the Equator Near the Maldive Islands, Indian Ocean**

| Month | Temp. | Month | Temp. |
|---|---|---|---|
| Jan. | 80.6° | July | 80.6° |
| Feb. | 80.6° | Aug. | 80.6° |
| Mar. | 82.4° | Sept. | 80.6° |
| Apr. | 84.2° | Oct. | 80.6° |
| May | 84.2° | Nov. | 80.6° |
| June | 82.4° | Dec. | 80.6° |

*Monthly Avg. Temp. = 81.5° F.*

OTEC technology is very sensitive to water temperature because the gross output of the power plant varies as the square of the Δt; the net power output is even more sensitive.[36] A Δt of 42° F. is optimal for an OTEC in the 25 - 100 megawatt range. Raising or lowering the Δt by as little as 4° F. will change the net power output by 25%.[37] At the equator in the Indian Ocean, the mean annual Δt between water at the surface and water 3300 feet deep is 42.5° F.[38] Table 1.3 shows the Δt as a function of depth off the coast of Sri Lanka, 5° north of the equator.

The temperature at the sea's surface is fairly uniform throughout the equatorial region. Within about 15° of the equator, there persists a truly 'endless summer'. The effect of the Earth's tilt on its axis, 23.5°, is relatively minor at the equator. During the vernal and autumnal equinoxes, the sun is directly overhead; at the summer and winter solstices, when the sun is at its highest in the northern and southern hemispheres respectively, the sun is displaced less than 24° to the north or south at the equator. So, the equatorial belt of the planet, which is covered mostly by water, is bathed in a never-ending glow of direct summer sunshine.

**Open Cycle Process**

The OTECs of Aquarius will operate on an open cycle, using warm sea water as the working fluid. The secret to making the open cycle work is a low pressure chamber. Since the boiling point of any liquid is dependent on the ambient air pressure, you can lower a liquid's boiling point by lowering the pressure around it. In the open cycle, the warm sea water is admitted into a vacuum chamber where the pressure has been reduced; at this lower pressure, .43 psi, the boiling temperature of sea water is only 80°F. In the vacuum chamber, some of the warm sea water boils, producing low temperature steam. (The low temperature and low pressure mean that the 'steam' is essentially water vapor, so the low pressure chamber is called an 'evaporator' rather than a 'boiler'.) The expanding water vapor turns a turbine which spins a generator.

After leaving the turbine, the water vapor passes around the tubes of a condenser. Flowing through the condenser tubes is cold sea water pumped up from the ocean depths. The warm water vapor condenses on the cold tubes, producing distilled water. This change of phase, from vapor to liquid, causes the pressure to drop throughout the system; this in turn causes the warm sea water in the evaporator to boil, thus powering the OTEC. Only 1% of the warm surface water flashes to vapor, the balance passes out of the OTEC and returns to the sea. Despite the fact that only a small fraction of the water entering the OTEC is converted to vapor, enough low temperature 'steam' is produced to power the huge turbines. This is possible due to the tremendous quantities of surface water entering the OTEC.

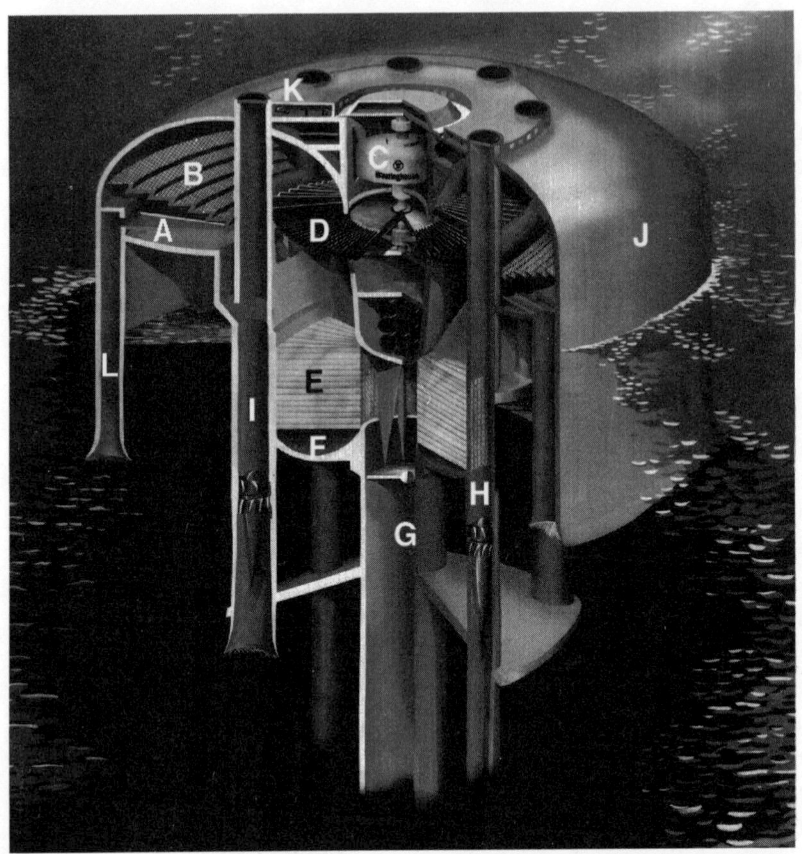


***Fig. 1.2 - Open Cycle OTEC.*** **A**- *Evaporator.* **B**- *Demister.* **C**- *Generator.* **D**- *Turbine (125 ft. dia.).* **E**- *Condenser Tubing.* **F**- *Fresh water collection.* **G**- *Cold water intake.* **H**- *Cold water discharge.* **I**- *Warm water discharge.* **J**- *OTEC Shell.* **K**- *People (to scale)* **L**- *Warm water intake.* D.O.E.

Cold water is brought up from a depth of 3300 feet through a pipe 40 feet in diameter, and is pumped through the condenser at a rate of 9400 cubic feet per second.[39] Since water has nearly neutral buoyancy, the cold water pump must overcome only the density difference between the cold deep water and the warm surface water, plus the friction losses in the pipe. Even though water is being raised over 3300 feet, it requires only the energy needed to pump an equivalent volume up 18 feet on the surface.

Nevertheless, an enormous propeller-type pump, mounted inside the cold water pipe, is needed to supply this motive force.

**Table 1.5**
**Cold Water Pump Specs**

| | |
|---|---|
| Inner Rotor Diameter | 20 feet |
| Outer Prop. Diameter | 39 feet |
| Prop. Speed | 38 rpm |
| Pump Volume | 4.2 million gal/min. |
| Power Consumption | 12.5 megawatts |

There are other power users on an OTEC as well. A 100 megawatt OTEC absorbs 41 megawatts for its own operation.

**Table 1.6**
**Parasitic Power Losses in Megawatts**

| | |
|---|---|
| Warm Water Pumps | 27.61 |
| Cold Water Pump | 12.55 |
| Deaerator Compressor | .61 |
| Fresh Water Pump | .12 |
| Total Losses | 40.89 |
| **NET POWER** | **59.11** |

OTEC components are large, highly specialized, and sophisticated pieces of machinery, which, at least initially, must be purchased for cash from outside suppliers. The costs of a 100 megawatt OTEC break down as follows:

**Table 1.7**
**Capital Costs for a 100 MW Open Cycle OTEC[40]**

| Item | Cost $/Kw |
|---|---|
| Flash Evaporator | $53.16 |
| Surface Condenser | 354.11 |
| Retubing | 113.41 |
| Air Removal Vent Condenser | 27.01 |
| Turbine | 220.90 |
| Generator Exciter | 40.42 |
| Sea Water Pumps | 132.73 |
| Condensate Pump | 2.21 |
| Air Removal Equipment | 181.43 |
| Fouling Control System | 57.71 |
| Auxiliaries, Power Conditioning | 50.89 |
| Hydrogen Production[41] | 340.00 |
| **Total Capital Cost Per Kilowatt** | **$1,573.98** |

At this price, the components for a 100 megawatt OTEC module will cost $157 million—installed. Aquarius will house seven 100 megawatt OTEC modules, at a cost of $1.1 billion.

Each module will produce 59 megawatts of net electrical power, for a total output of 413 megawatts. Of this power, 113 megawatts—most of the production from two modules—will be dedicated to internal uses. The 100,000 colonists living on Aquarius will require one kilowatt per capita, 100 megawatts; the remaining 13 megawatts will be used in various industrial operations in Aquarius.[42] This will leave a balance of 300 megawatts for export, at a production cost of 1/2¢/kwh.[43]

**A New River Nile**

An efficient OTEC can produce about .0017 kwh of net electrical power for each cubic foot of cold deep water it pumps up (.6 kwh/$M^3$). The OTECs of Aquarius will bring 67,000 cubic feet of cold deep water to the surface every second.[44] This is a volume of water equivalent to two-thirds the average flow of the River Nile.[45]

When the nutrient-rich water from the depths reaches the sunlit waters of the surface, it has much the same effect on the barren ecology of the open ocean, as the Nile itself has on the sterile sands of the Sahara. Just as the Nile created an oasis which nurtured one of mankind's oldest and most splendid civilizations, this new river of life will nourish mankind's newest and most dynamic civilization.

The deep oceans are virtual biological deserts—Saharas of the sea. Conditions in the mid-oceans are ideal for life, there is continuous warmth, and plenty of water and light; nevertheless, life there is very sparse.

Life teems around the margins of the continents, but most of the global ocean is deserted. Open ocean areas comprise 90% of the world's sea surface—two-thirds of the entire planet—yet these marine deserts account for less than one percent of the global fish catch.[46]

In true deserts like the Sahara, life is scarce because the environment is harsh; ironically, in the tropical seas, life is scarce because the environment is benign. Living conditions are so ideal in the warm sunlit waters around the equator that life grows explosively. As a result, all available nutrients are quickly absorbed. The surface waters of the tropical seas are devoid of nutrients in much the same way as the soils of tropical rain forests. In the rain forests, nutrients are continuously recycled, and so a huge concentration of life can be supported. The same is not true for much of the open ocean. (See Plate 1).

At sea, nutrients are not recycled the same way as on land. If they were, life in the seas would be far more plentiful than it is. At sea, the nutrient cycle is very indirect. A plant or animal will scavenge nutrients from its environment and bind them up in its body during the course of its life; then, when the organism dies, it sinks into the depths, taking its nutrients with it. These nutrients remain locked away in the depths until brought to the surface again—usually after a lapse of a thousand years or more.[47] Consequently, a millennium's worth of nutrients are concentrated in the deep waters of the world's oceans.[48]

**Upwelling**

Nutrient-rich deep water occasionally comes to the surface in places known as upwelling zones. In these rare locales, life is enormously abundant. The availability of nutrients causes single-celled plants to burst into riotous growth. Dramatic expansion at the base leads to corresponding increases in the rest of the food pyramid: myriad tiny crustaceans and other grazers of planktonic sea grass grow in proportion; shrimp and other organisms feed off these little plant eaters; these in turn are eaten by small fish that swarm in schools of millions; these sardines then serve as food for tuna, which finally end up as lunch meat for humans; here—excepting occasional human *hors d'oeuvres* for Jaws—the food-chain ends. The upwelling zones of the oceans are tiny in area, comprising only one tenth of one percent of the ocean surface, but they produce an enormous disproportion of the world's fish harvest—44%[49].

Some of these upwelling zones are familiar. Monterey Bay was one of the most productive sardine fisheries on the planet, until ruined by over fishing. The Bay overlies a huge upwelling zone created by an anomaly of currents and geology. The Monterey Canyon is a submarine slash in the earth twice as deep as the Grand Canyon. Monterey Canyon acts as a conduit, bringing cold nutrient-rich water from the depths to the surface.

This same upwelling of cold water gives San Francisco its unique foggy climate. The warm air of the Pacific passes over the cold upwelling water, and just like warm breath passing over a cold soda bottle, it condenses into the characteristic fog of the Bay area. The same upwelling of cold water that fueled the productive fisheries of Cannery Row was the cause of Mark Twain's famous remark that the coldest winter he ever spent was one summer in San Francisco.

Another prominent upwelling zone occurs in the waters around Antarctica. The cold waters around the Southern Continent do not 'upwell' so much as 'outcrop'. The oceans are stratified in layers by temperature, just like layers of sedimentary rock. These layers of water are most numerous and thickest at the equator, where there is a wide variation between temperatures at the surface and at depth. Near the poles, however, there is no such difference. The surface waters are almost as cold and almost as rich in nutrients as the deep waters. This happy circumstance leads to one of the richest concentrations of life on the planet: vast shoals of Antarctic krill, and the flocks of penguins and pods of whales who feed on them.

The OTECs of Aquarius will create an artificial upwelling zone by bringing a river of nutrient-rich cold water to the surface. When the nutrient-rich broth of deep sea water (see Appendix 1.7) is exposed to sunlight, there will be an explosion of plant growth comparable to that obtained when fertilizers are sprayed on land crops. Since the growth of phytoplankton is almost always limited by the availability of one or more vital nutrients, bringing deep water to the surface will provide the raw material for algal growth. The addition of nitrogen to the surface waters will enhance primary productivity by 160 times.[50]

**Nitrogen**

The key nutrient for all plants, both marine and terrestrial, is nitrogen. Plants form the base of the pyramid of life, both at sea and on land. Plants capture the radiant energy of the sun, and through the magic of photosynthesis, convert it into proteins, sugars, and fats, providing food for all the myriad creatures in the pyramid above. An increase in the nitrogen supply broadens the plant base and expands the biomass of the whole living pyramid.

Nitrogen is vital to plant metabolism. It is the basic building block of all amino acids—the backbone of proteins. Nitrogen availability is usually the limiting factor in plant growth, both on land and at sea. Available nitrogen in the surface waters of the tropical seas is very quickly absorbed by phytoplankton—single celled plants like diatoms and other algas. If there is an abundance of nitrogen available, algae reproduce exponentially and the nitrogen is quickly absorbed.

Food production is essentially a nitrogen supply problem. If there is enough nitrogen available, then ample food can be produced. Most plants can't use elemental nitrogen, $N_2$, directly from the air or water because it is chemically unreactive and therefore difficult to metabolize. Plants require 'fixed' forms of nitrogen: nitrite ($NO_2$), nitrate ($NO_3$), or ammonia ($NH_3$). On land, agriculture depends on producing nitrogen in fixed forms so plants can use them for growth. This is accomplished by recycling nutrients to the soil in the form of humus or manure, by planting nitrogen fixing crops like legumes, or by using man-made fertilizers.

To free himself from the limitations of the natural nitrogen cycle, man uses fossil fuels to artificially fix nitrogen from the atmosphere and spread it on the soil. Plants respond to the greater availability of nitrogen with rapid growth and increased crop yields. Much of the credit for the Green Revolution must go to the increased world-wide use of fertilizers, and to the development of new plant strains which can utilize the extra nitrogen.

Using fertilizer has allowed man to double his food supply, but only at the cost of an enormous increase in the consumption of fossil fuels. This is obviously a dead-end street. If we are to redouble our food supply—an imperative if we intend to feed everyone in the future—then we must find a way to increase the nitrogen supply without burning fossil fuels.

Fortunately, the oceans contain a virtually limitless pool of nitrogen in fixed nutrient forms. Because plants and animals take their nutrients into the depths with them when they die, the concentration of nitrates increases rapidly with depth, reaching a peak at around 3300 feet.[51] Below 3000 feet, nitrate concentrations remain relatively constant throughout the world's oceans—around .4 grams per cubic meter, or .00005 ounces per gallon.[52]

Less than half a gram of nitrate per cubic meter, is an extremely dilute nutrient medium by any standard. To comprehend how little nutrient we are talking about, imagine that you are preparing a recipe of deep ocean water. To one gallon of fresh water you would add about nine tablespoons of sea-salt.[53] To this mixture you would add two ten-thousandths of a teaspoon of nitrate. This would be a speck no larger than a single grain of salt.

How can such a vanishingly tiny trace of nutrients fuel the explosive growth of phytoplankton? The answer lies in the tiny size of these plants. Compared to their land equivalent—grass—phytoplankton have an enormously greater capacity for absorbing dilute nutrients. This is not due to a superior biochemistry, but is rather a function of geometry. A single plant, a blade of grass, for example, might have a volume of 10 cubic millimeters. An equivalent volume of algae would require 10,000 plants, each with a volume of .001 $mm^3$. The algae—despite having the same volume as the blade of grass—will have a hundred times the surface area of the grass, and 1000 times the grasses' ability to absorb nutrients. The effective concentration of nitrate in sea water is, therefore, more like .05 ounces per gallon; enough to create a lightly concentrated plant food. To the micro-organisms it nourishes, deep sea water is like fertilizer.

The deep waters of the ocean are an eternally renewable resource. The circulation is in equilibrium now, replacing about .1% of the deep water every year through down-welling in the polar seas. The oceans of the world hold some 300 million cubic miles of sea water.[54] About three-quarters of this water lie at depths greater than 3000 feet, so there are 225 million cubic miles of nutrient rich deep water in the oceans. Of this volume, 225,000 cubic miles, are replaced every year. In the cold waters of the North Atlantic alone, a cubic mile of water sinks into the depths every twenty minutes.[55]

Each marine colony will exhaust the energy and nutrient potential of 13.5 cubic miles of deep sea water every year. Ten thousand marine colonies will remove 135,000 cubic miles of deep water from the oceanic storehouse each year—three-fifths of the replacement supply. More than 15,000 marine colonies could pull from the deep water vaults indefinitely without ever reducing this enormous resource. By tapping this vast renewable reserve of nitrogen, we can feed the world.

### Food Glorious Food

The world is approaching the Malthusian wall. We are running out of arable land, cheap energy, and time. In the immediate future we face crucial shortages of everything but people. Our resource base is shrinking. The planetary ecosystem is already collapsing under the weight of five billion people. How then are we ever going to feed twice that number?

Aquarius is the answer. The OTECs of Aquarius are essentially gigantic nitrogen pumps. With enough of them we can feed the world.

In the past, when man has looked to the sea to feed his teeming multitudes, he has thought in terms of netting the fish. This amounts to strip mining the ocean, and it has no future. If we are to tap the sea's food potential, we must abandon our hunter-gatherer mentality and adopt the same principles of agriculture and animal husbandry which have succeeded in feeding us on land. If we are to feed the future's hungry billions, we must learn to farm the sea.

A marine colony pumps up 43 billion gallons of nitrogen laden water every day. Though the nutrient concentration is low, the huge volume brings 70 tons of organic nitrogen to the surface.[56] This bounty is a free byproduct of the OTEC's energy production process. Phytoplankton, one of the most efficient life forms in existence, can convert 78% of the available nitrogen into protein.[57] At this rate, about 55 tons of nitrogen will be converted into protein every day. Of the dry weight of algal protein, 13.4% is nitrogen; so 55 tons of nitrogen makes 410 tons of algal protein. Aquarius will produce 195,000 tons of dried algae per year, of which 118,000 tons will be high-grade protein.[58]

The standard for protein consumption set by the U.N.'s Food and Agriculture Organization (FAO) is a minimum of 1.2 ounces per person per day.[59] To provide a human with this much protein requires just under two ounces of dried algae powder. The protein production from one marine colony could supply the protein needs of ten million people! If algal protein were used only as a dietary supplement, a single marine colony could eliminate protein malnutrition in 20 to 40 million people. A thousand marine colonies could save billions of people from protein malnutrition. At a global population of 10 billion, the Millennial marine colonies could virtually eliminate protein deficiency in the world.

In 1975, the world required 36 million tons of animal protein for human consumption. But this was in a world where 60% of the human population were getting less protein than they needed.[60] If the world population tops out at 10 billion, the total human demand for animal protein will be 90 million tons per year—assuming the same per capita consumption levels that prevailed in 1975. We cannot, however, presume that the impoverished peoples of the world will be content to remain that way. Eventually, every one of those ten billion people will want to enjoy a diet as rich in protein as that now taken for granted by Americans. Americans consume 3.4 ounces of protein a day, 2.3 of which are of animal origin.[61] To provide 10 billion people with this much protein would require 390 million tons per year—over ten times what was required in 1975. Producing such a mountain of protein from our already over stressed planet might seem impossible; yet this entire demand could be satisfied by the algae production from just 2,600 colonies.

**Blue-Green Algae**

The marine colonies will be able to produce food so abundantly because they cultivate the simplest, hardiest, and most prolific plant in the known universe—blue-green algae. The blue-green algas are perhaps the most remarkable forms of life on the planet, being both the simplest and the most successful. The blue-green alga is such a primitive life form that its cell doesn't even have a nucleus; its genetic material simply floats freely within its body. These primitive plants have existed, essentially unchanged, for practically the entire history of Life on Earth.[62] Fossils of blue-green algae, nearly identical to their modern descendants, have been found in rock formations of Western Australia, dating to 3.5 billion years.[63] The blue-green algae had already lived out 94% of their present history when the dinosaurs were just beginning.

During their long history, blue-green algae have adapted to virtually every ecological niche where there is light and moisture. These astonishing plants can be found in hot springs, where the water temperature exceeds 160° F.; in brine lakes, where the salinity is so great that no other life can exist; in perpetually frozen Antarctic lakes, under ice 18 feet thick; even inside lichens, where they live in symbiotic harmony with their fungal hosts.[64] Even in the harshest reaches of the Sahara desert, 48 varieties of algae have been discovered living under a single rock, subsisting on nothing but the morning dew.[65]

Algae's vitality and antiquity are surpassed only by their abundance and importance. Just as the 'Earth' is covered mostly by water, so too is 'Life' mostly made of algae. Algae form not only the base of the pyramid of Life, but comprise the bulk of the pyramid as well. Seventy percent of the world's biomass is algae.[66] Together, with the other algas, the blue-greens account for as much as 90% of the photosynthetic activity on earth.[67] The algae are both the main source of oxygen in the atmosphere, and the primary regulators of carbon dioxide. In large measure, the algas, the blue-greens in particular, are the very cause of Life on Earth. Without them, Gaia would be a lifeless hell-hole not much different than Venus. The blue-green algae have been responsible for the survival of Life on Earth in the past, now they will be integral to its future salvation.

Algae's simplicity makes them ideal as crop plants. As single-celled organisms they do not waste any of their growth on stems, leaves, roots, flowers, or any other potentially inedible part. One hundred percent of edible algae mass can be used as food. Each cell is also a seed, so it is very easy to cultivate. The life-cycle of these primitive plants is extremely simple, with no complicated reproductive cycle or dormant period as in terrestrial plants. Algae reproduce like bacteria through asexual mitosis—one parent cell dividing to become two daughter cells.[68] Given the right conditions and nutrients for growth, algae will continue to populate at an exponential rate until the nutrients are exhausted. If there is a continuous supply of nutrients, as in the case of the deep water brought to the surface by Aquarius, this population explosion can be sustained indefinitely.

In Aquarius we will culture a particular species of algae—*Spirulina platensis*. Spirulina is a blue-green alga that has a venerable history as a human food. The Aztecs used to harvest an alga, believed to be Spirulina, from Lake Texcoco in Mexico. Natives around Lake Chad, in Nigeria, still eat it. Spirulina is already being produced in Mexico, Thailand, Kuwait, and Hawaii, for distribution to the health-food industry

Spirulina is the best choice for algae cultivation on a Millennial scale for several reasons: Though spirulina's preferred habitat is highly alkaline warm lake-water, it will grow in a variety of media, including sea water.[69] Spirulina grows in a coil-shaped group of six to twelve cells, making it relatively large, and consequently easier to harvest than other species which grow as independent cells. Spirulina is an ideal human food—uncannily so. Human dietary needs and the nutrients found in Spirulina could hardly have been more perfectly matched if they had been designed that way on purpose—it makes one wonder.

It seems a peculiar coincidence that land-dwelling mammals like us should have evolved the capacity to thrive on a diet of one-celled aquatic algae. Spirulina is highly compatible with human digestion because it contains no cellulose (the indigestible part of plants that makes straw and paper). This strain of algae is so primitive that its cell wall is still composed of a 'mucopolysaccharide,' which is 85% digestible. This gives spirulina a tremendous advantage over other varieties of phytoplankton since it can be easily assimilated by the body without any necessary preparations. By contrast, many other algae must first be broken down to strip the nutrients out of their indigestible shells. Diatoms, for example, are encased in a glassy pill-box of silica. Krill and other zooplankton thrive on diatoms, but they, unlike humans, can break the tiny glass cases open and extract the nutrients.

Spirulina will make an ideal food supplement, particularly in the Third-World where it will cure many of the diseases brought on by malnutrition. Protein is the most important nutrient in the human diet, and spirulina is the heavy-weight champion of protein—65% by weight. Compared to soybeans at 35%, beans at 22%, or alfalfa at 18%, spirulina is something quite remarkable. Virtually all vegetable proteins are deficient in one or more of the eight essential amino acids. For example, wheat, corn, and rice are low in lysine, while soybeans, which are high in lysine, are low in methionine. It is difficult to supply the body with enough protein on a purely vegetable diet. Vegetable sources are also deficient in other nutrients. Vitamin $B_{12}$ is particularly scarce in vegetables, since it is synthesized only by molds, bacteria, and other very low orders of life. At 1.5 micro-grams per gram, Spirulina is the most concentrated source of $B_{12}$ available, providing twice the $B_{12}$ found in animal liver.[70] Spirulina contains all eight essential amino acids in quantities equivalent to meat, milk, or eggs.[71]

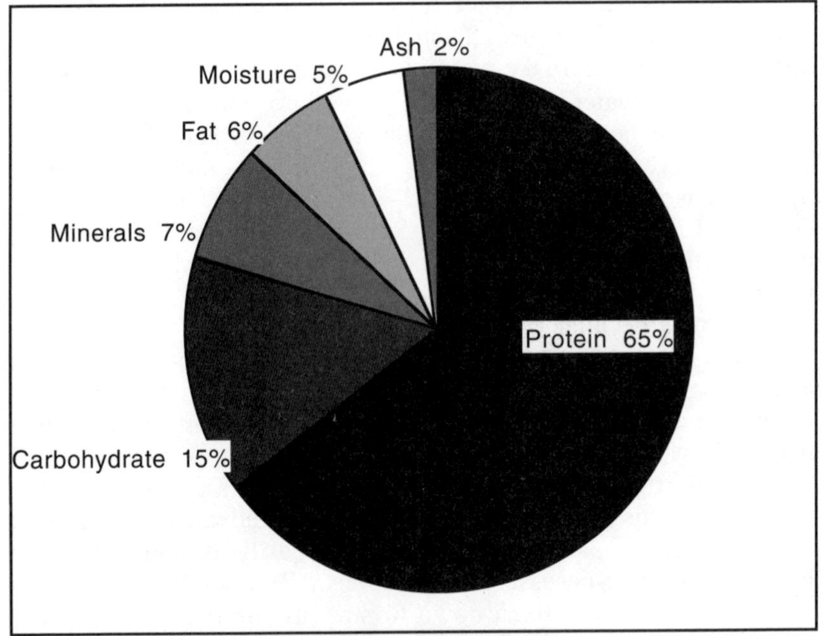


***Fig. 1.3 - Nutrient Composition of Spirulina.***

One tablespoon of spirulina powder will provide 500% of the USRDA for Vitamin A in the form of Beta Carotene.[72] Beta Carotene is not only an ideal source of Vitamin A, but has also been shown to be a powerful anti-carcinogen. A tablespoon of Spirulina will also provide two grams of Vitamin E—80% more than raw wheat germ. Vitamin E is an important anti-oxidant which protects cells from molecular damage by scavenging deadly free radicals. Recent research indicates that most diets, even in industrialized countries, provide less than optimal quantities of these versatile anti-oxidants. As environmental insults, like increased ultra-violet radiation multiply, the need for these vitamins as supplements to our diets will grow dramatically. Future demand for Beta Carotene particularly is apt to increase substantially.

Spirulina has already been shown to be an effective treatment for kwashiorkor, a protein deficiency disease plaguing Third World children.[73] (It is this disease which causes the distended belly, characteristic of starving children.) Spirulina should also prove an effective treatment for beri-beri caused by $B_1$ (Thiamine) deficiency, and pellagra caused by Niacin deficiency. Spirulina could also eliminate Vitamin A deficiency related blindness in Third World infants. Babies who receive supplemental Vitamin A in their first year are almost 90% less likely to die than those who are deficient in this vital nutrient.[74] Spirulina could save the lives of millions of undernourished children every year.

The lack of complete proteins in the diets of Third World mothers, infants, and children is especially tragic, because it is essential to the proper formation of the brain. A diet deficient in complete proteins can lead to impaired brain development, even if the diet provides sufficient calories and other nutrients. A lack of adequate protein is condemning generation after generation of Third World children not only to the ravages of disease, but also to a permanent mental handicap. Spirulina from Aquarius can put an end to this monumental human tragedy.

Spirulina is also low in fat, 6%, and virtually devoid of cholesterol-0.013%. Each tablespoon of Spirulina contains just 36 calories and only 1.3 mg of cholesterol, but it yields 6.5 grams of protein; that's about equal to the protein in a large egg. But an egg has 80 calories and packs 300 mg of artery-hardening cholesterol. So, while Spirulina is the ideal food for supplementing the diet of the protein starved Third World, it wouldn't be a bad supplement for the fat-gorged First World.

The animal protein equivalence of spirulina means that no intermediate steps in the food chain are necessary. Spirulina, which grows from a simple combination of sunlight, water, and nitrogen, can be harvested and used directly as a source of complete 'animal' protein. None of the losses normally attendant to feeding forage to livestock need be suffered. In cattle feeding operations, for example, five pounds of vegetable protein must be fed to produce a single pound of beef. This means that spirulina is effectively five times more productive than plants, which, to produce protein of equivalent value, must first be fed to animals.

**Table 1.8**
**Protein Percentage by Weight of Selected Foods**

| Food | % Protein |
|---|---|
| Spirulina (dry) | 65 |
| Powdered Eggs | 47 |
| Powdered Milk | 36 |
| Soy Flour | 37 |
| Wheat Germ | 27 |
| Peanuts | 26 |
| Beef (wet) | 22 |
| Wheat Flour | 14 |
| Tofu | 8 |
| Brown Rice | 8 |

**Table 1.9**[75]
**Vitamin, Mineral, and Nutrient Content of Spirulina**

| Nutrient | mg/100g |
|---|---|
| Vitamin A | 170.00 |
| Vitamin $B_1$ | 5.50 |
| Vitamin $B_2$ | 4.00 |
| Vitamin $B_3$ | 12.00 |
| Vitamin $B_6$ | 0.30 |
| Vitamin $B_{12}$ | 0.20 |
| Vitamin E | 19.00 |
| Pantothenic Acid | 1.10 |
| Vitamin F (Folic Acid) | 0.05 |
| Potassium (K) | 360.00 |
| Calcium (Ca) | 100.00 |
| Phosphorous (P) | 960.00 |
| Iron (Fe) | 43.00 |
| Zinc (An) | 2.50 |
| Manganese | 2.00 |
| Chlorophyll (a green photosynthetic pigment) | 680.00 |
| Caratenoids (primarily Beta-Carotene) | 340.00 |
| Phycocyanin (a blue photosynthetic pigment) | 3 - 10 % |

Using algae as a food source increases by two orders of magnitude—100 times—the availability of nutrients from a given biosystem. In normal food chains, there is a characteristic ratio of ten-to-one between the mass of one link in the chain and the next.[76] For example, the copepods, which feed on algae, weigh about 10% of the mass of the algae itself. A population of carnivorous predators feeding off the copepods, sardines for example, will in turn weigh only 10% of the mass of the copepods—1% of the weight of the algae. A population of predators higher in the chain, tuna for instance, will weigh 10% of the mass of the sardines—.1% of the algae's biomass. Looking at it this way, a can of tuna represents 250 pounds of spirulina. So the next time you casually eat a tuna sandwich, bear in mind that you are consuming enough production from the biosphere to feed 1000 people. By using a primary producer like spirulina—an 'autotroph,' which transforms solar energy directly into food—we can magnify our harvest from the sea a thousand fold over the primitive rape and pillage methods used today.

**Green Eggs and Ham**

Spirulina will be harvested from the containment ponds around Aquarius by the kiloton (see Appendix 1.8). Once harvested and dried, the spirulina is almost ready for consumption. Raw spirulina powder is a nourishing food which could easily be added to staples like flour; unfortunately, it has one serious drawback as a universal food additive—it is green. Spirulina is not just green, it is <u>really</u> green. People eat a lot of green foods, but bread is usually not one of them.

Many well-intentioned efforts to introduce cheap nutritious foods to the Third World have broken on the rock of arrogance, when would-be Samaritans wrongly assumed that hungry people will eat anything. People are almost universally finicky about what they eat; starving people are only slightly less finicky than anyone else. It will do us little good to produce the ideal food if the people who most need it won't eat it. Since we can't assume that hungry peoples of the world are going to be willing to eat green bread, we must further process our spirulina powder.

In a two-stage, alcohol-based process, virtually all color and flavor can be removed from the algae. The end result is a flavorless, odorless white powder which is 85% protein. This will reduce the weight of the final product by almost a quarter. The final output will be 320 tons of protein concentrate per day. The color of raw spirulina is in the photosynthetic pigments phycocyanin and chlorophyll, which also contain most of the beta carotene. These pigments are removed intact, and form a valuable by-product. It is ironic that these pigments will have to be marketed separately to the same people who will only accept the spirulina in their bread in its reduced and colorless form. But humans are not Vulcans, and they can seldom be expected to do something simply because it is logical.

Removing the chlorophyll and other pigments from spirulina is probably necessary for marketing reasons and is certainly desirable for economic reasons. Powdered spirulina protein concentrate will be worth $300 million a year. Ironically the beta carotene, removed to make that product acceptable, could be worth ten times as much as the protein powder. Together, chlorophyll, Beta carotene, and phycocyanin represent 7.5% of the mass of dried spirulina. Aquarius will be producing 40 tons of these pigments a day. Beta carotene is presently worth $136 a pound, and chlorophyll and phycocyanin are worth about the same.[77] At such prices, these 'by-products' would be worth $4 billion annually.[78]

These are not the only by-products Aquarian algae will produce. Algas which lack nuclei are highly receptive to genetic engineering. By modifying the algae's DNA, we can program various new strains to produce virtually any biological compound in tremendous quantities.[79]

The algal protein concentrate stores extremely well. A test batch of the material retained 95% of its protein content after seven years in storage.[80] This property will make it possible for countries to stockpile the protein concentrate in good years, against the inevitable bad years. Governments could buy the powder to be added to flour in central milling plants. It will probably be most economical to use the supplement in large urban areas where commercial mills provide bakeries with their flour.

The protein powder could be added to flour at the rate of three tablespoons, 1.5 oz., per pound of flour.[81] At this rate, a person will obtain 100% of his daily protein needs from the supplement in his bread. Added to the other sources of protein in his diet, this will enable the average consumer to enjoy a relatively high protein diet at a small cost. The powder can also be added to water as a cheap supplement to infant formula.

The finished protein powder could be wholesaled to governments and central distributors at a price of $1/lb. At this price, the cost to supplement a pound of flour will be about 10¢. A pound of flour, or its carbohydrate equivalent, is the daily per capita consumption in the Third World. People will therefore be able to fulfill their daily protein needs for a dime.

Eventually, as we bring more marine colonies on line, and the supply of protein powder increases, we will begin to distribute it inland. Much of the Third World population is rural and isolated.

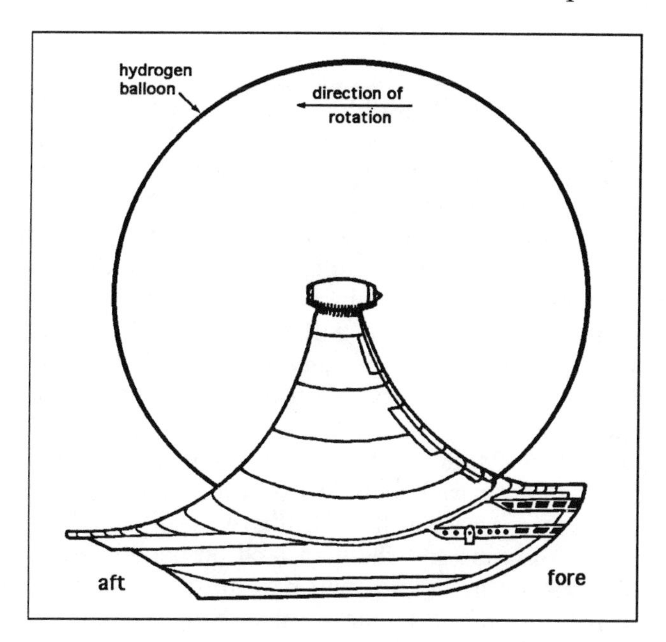


***Fig. 1.4 - Aquarian airships*** *will incorporate a radical new design concept. The large spherical balloon is filled with hydrogen and provides enough buoyancy to counter the weight of the ship. Extra lift is generated by spinning the balloon along its horizontal axis in a direction opposite to the direction of travel. A fleet of these airships will ferry protein powder to hungry people in remote parts of the world.*[82]

Most undeveloped countries lack adequate roads and rails. To overcome distribution bottlenecks, we will eventually deploy our own fleet of airships to carry protein concentrate into the remotest regions of Africa, Asia, and South America. Using our own hydrogen-powered dirigibles, we will be able to fly into remote areas and reach isolated villages. There, protein concentrate, packaged in bags of edible chitin, can be dropped directly to the people below.

Protein powder is one of the few commodities it will actually be practical to distribute in this way. Although it is a bulk commodity that can be easily stored and roughly handled, protein powder is nevertheless light enough and valuable enough to warrant air-mail delivery. In practice we will probably deliver protein concentrate to the world's needy as a free good, in exchange for use of the oceans—"the common heritage of all mankind".

**Shellfish**

One of the potential problems associated with the algal containment structure provides one of its great opportunities as well. It is the problem of biofouling. While the spirulina is a free-floating form of algae, and will be removed as rapidly as it is produced, the same cannot be said for algae which will adhere to the bottom and sides of the containment ponds. On these surfaces, long filamentous 'grass algae' will be encouraged to grow. These benthic (bottom-dwelling) algae will respond to the availability of nutrients in the deep water in the same way as the spirulina. These algae will be difficult or impossible to harvest directly in any easy or economical way.

There is an attractive solution to this potential problem—sea sheep. Like their terrestrial counterparts, sea grazers will provide an excellent way to use a crop that is otherwise difficult to harvest. Certain species of crabs and shellfish feed on benthic algae. Chief among these grazing herbivores will be *Mithrax spinosissimus*, the "Caribbean King Crab". Like the Alaskan King Crab, it grows to a large size and produces a delicate snowy white meat. This crab has a short simple life-cycle that is very adaptable to mariculture. Populations will be seeded in the containment ponds where they will graze on the abundant crops of algal grass. Other algae grazers like abalone and queen conch will be cultured in smaller quantities.

Total production of crab and other shellfish from benthic algae will amount to nearly 60,000 tons per year.[83] Obviously, crab meat, an expensive delicacy, retailing for $15 a pound, is of little use to starving people. However, the production of this and other high value commodities will provide additional revenue streams of no small dimensions to the colony. With these revenues, we can accelerate the expansion of the marine colonies, a result which <u>is</u> of use to starving people.

We will also engage in a variety of other mariculture operations. Species of filter feeders will be cage cultured in the algae ponds. Dense populations of clams, oysters, scallops, and mussels, can be nourished on the rich broth of spirulina. If 15% of the nitrogen in the deep water were fixed by filter feeders, every thousand gallons of sea water would produce .186 oz. of shellfish meat. Total meat production would amount to 155 tons a day. Of this amount, about 30% will be scrap. The shells of these maricultured species will be used for the production of lime needed for finishing plaster and to process magnesium. Crab, shrimp, and lobster shells will be processed for their chitin. Chitin is a versatile material that can be used as a substitute for cellulose in the production of paper and other products. The colony will be self-sufficient in chitin based paper goods.[84]

**Pearls**

One containment segment will be devoted to the culturing of pearl oysters. With a modest allocation of our mariculture resources, we can produce thousands of pearls a day.

***Fig. 1.5 - Gems of the ocean.*** *Pearl culture will allow Aquarians to transform algae into jewels—a powerful means of gleaning wealth from the sea.*
© 1992 Fred Ward.

Quality cultured pearls retail for anywhere from $25 to $4000 apiece—wholesale to pearl brokers is about a quarter of that.[85] We will also harvest oyster shells and process them for their valuable lining of mother-of-pearl. In addition to the fine white pearls of oysters, we will also culture the lustrous colored pearls of abalone and conch. Pearl farming is hardly an activity that will aid the suffering masses of the world, however, like the production of crab, abalone, and conch meat, the production of pearls will provide a robust cash flow for the colony. Pearl oysters will absorb less than 1% of total algae production, while providing over $400 million in revenues.

**Sea Silk**

In the outer mariculture containments, we will cultivate seaweeds rich in the substance algin. Seaweed will clean up the deep water effluent, and provide feed stock for the production of sea-silk. The seaweeds fix both the nitrogen left unused by the algae and the ammonia excreted by filter feeders. When the deep water is discharged from the outer containments, the seaweeds will have thoroughly stripped it of excess nitrogen and cleansed it of any metabolites from mariculture operations.

Twenty-two percent of the original load of nitrogen in the deep water remains unused by the spirulina. In addition, the secondary producers—crabs and shellfish—fix only a third of the nitrogen they absorb. The tertiary producers—lobsters, shrimp, and fish—fix only 10.5% of the nitrogen they take in, excreting the rest as ammonia. Spirulina can not use nitrogen in the form of ammonia, but seaweed can. Between that portion unused by micro-algae and that excreted by higher level producers, the deep water will still retain 24% of its original nitrogen load when it enters the outer containment ponds. Assuming that seaweeds—which are also species of algae—can utilize as much of the available nitrogen for growth as spirulina, then the seaweeds will fix about a quarter of the total nitrogen. Total seaweed production will be around 1,264 tons per day.[86]

The primary reason for culturing seaweed is to produce algin.[87] Algin is useful for producing a wide variety of products: films, gels, rubber, linoleum, cosmetics, polishes, and paints. Algin is also useful as a creamy food additive. The Aquarians will use algin mainly for the production of textiles. Algin's molecular structure, a long chain of interlinking rings, makes it easy to spin into long fibers, suitable for the production of textile threads. Spun algin fibers have the fineness and texture of silk.[88] When chemically bonded with trace amounts of certain metals, algin fibers gain in tensile strength and become insoluble in both acids and bases. Some of the algin silks are highly colored by their constituent metals: copper and nickel alginates are emerald green, cobalt alginate is ruby red, and chromium alginate is sapphire blue. These colors are inherent in the chemistry of the fiber, so these fabrics need not be dyed and are absolutely colorfast. Other alginates, although colorless, have an affinity for basic dyes, and can take any color.

In addition to their silky texture and bright long-lasting colors, alginate fabrics are also non-flammable.[89] You have probably seen alginate textiles without realizing it. Strict fire codes in the United States demand that the curtains used in movie theaters be non-flammable. Some of these curtains are therefore made of algin. In Aquarius, we will use algin based fabrics to meet practically all of our textile needs: curtains, awnings, carpets, upholstery, draperies, and clothing. We will be self-sufficient in textiles, and Aquarius will be virtually fireproof.

We will mostly cultivate *Laminaria digitata*, a variety of kelp that is almost half algin by weight.[90] Total algin production will be nearly 600 tons a day. Internal demand for textiles for all uses in Aquarius will only be 1800 tons a year, so the balance of 215,000 tons will be exported.[91] The value of this product will be over $2 billion a year.

**Fish Farming**

There will be a lot of scrap animal protein produced by the mariculture operations: dissected pearl oysters, shell-fish culls, wastes from packaging and processing crabs, conchs, etc. This residue could create a waste disposal problem, but like other problems in Aquarius, it also presents an opportunity. These scrap meats will be processed and pressed into dried, pelletized fish feed. Suspended in cages in the seaweed containments will be populations of lobsters, shrimp, tuna, salmon and other species of marine carnivores, fed on the prepared fish pellets. When these high-value fish are in turn harvested and processed for shipment, any waste products from their own processing will be recycled into pellets. Assuming 30% of all shell-fish production ends up as scrap, there will be about 100 tons available for processing as fish food every day. Fifty percent of the pellet's weight will be composed of spirulina, scraps from the colony kitchens, seaweed, etc., so there will be 200 tons of fish pellets to feed each day. Total fish production will exceed 60 tons per day.[92] About a quarter of this will be consumed by the colony, with the balance being exported.

**New Age Diet**

Before we colonize outer space, we must develop the art of creating artificial foods to a high degree of perfection. We must learn how to create a variety of satisfying foods, using the simple organic building blocks algae provides us. It will be difficult, perhaps impossible, to grow extensive gardens and orchards in our early space colonies. It will also be impractical to supply ourselves with foods imported from the Earth. Accordingly, we are going to have to learn to feed ourselves on the foods we will have available. Not surprisingly, the foundation of our diet in space will be our familiar friend spirulina. A diet of algae would certainly get monotonous after a while—probably a short while—so we must learn to create a variety of tastes and textures from simple ingredients.

This won't be overwhelmingly difficult. Most flavors are produced by aromatic proteins in the plants and animals we eat. Through genetic engineering, we can program simple bacteria, like *Escherichia coli*, to produce these 'flavanoids' on command. For example, we can easily isolate that group of special aromatics that give banana its distinctive flavor. This flavor can then be produced in quantity by engineered *E. coli*. Once mixed with textured carbohydrate from algae, it can create a highly believable banana substitute. The advantage of this system is that it avoids mimicking flavors with artificial chemicals.

Seaweed and spirulina will provide base materials for the creation of synthesized foods. Seaweed is typically 38% ash and fiber and 9% protein. Using carbohydrates, fats, and proteins extracted from spirulina, in combination with fiber from seaweed, and flavors from engineered *E. coli,* we will be able to synthesize a wide variety of foods. With practice and research, we can eventually reproduce the entire range of food tastes and textures, from apricots to zucchini.

A diet of synthesized foods is an alien concept to most of us; nevertheless, adapting to such a diet will be one of the most fundamental life-style changes essential to space colonists. Just as it will be out of the question to have cows in orbital colonies, it is out of the question to have them in Aquarius. Happily, one need not give up the luxury of eating steak, since this can be duplicated synthetically.

In Aquarius, we will slowly adapt our diets, consuming a higher and higher proportion of synthetic foods, until they comprise about 50% of our consumption of fruits and vegetables and 100% of our consumption of meats—other than seafood. Fruits and vegetables will be grown hydroponically in extensive greenhouses, orchards, and gardens on a breakwater barrier surrounding the colony. Enough fresh produce will be available to enliven the Aquarian diet while techniques for producing palatable foods from algae are perfected.

In addition to fresh and synthetic foods, the Aquarians will necessarily include an abundance of fresh sea-food in their diet. Marine colonists will have to endure banquets of snowy crab legs, sautéed abalone steaks, lobster bisques, and fresh oysters on the half shell—space colonization is a tough business, but somebody's got to do it.

**Dollar Power**

Aquarius is the first step on our stairway to the stars. It is the foundation underpinning all else. Aquarius, and the other marine colonies, will provide the essential ingredient needed to attain the stars—money.

The economic demands of space colonization are awesome. Though we will do things differently than NASA, it will nonetheless cost billions and billions of dollars to get into space. Aquarius and the other marine colonies explode the previous limits on productivity that have constrained our land-based economies. In so doing, they produce the food and energy the world needs to survive, even at tremendously increased numbers and standards of living. This sort of production will not go unrewarded. In an entropic universe like this one, it is a surety that you can best do well by doing good. By unleashing the limitless power of macro-evolution, and tapping into the oceanic reservoirs of nutrients and energy stock piled at sea, we will liberate a flood-tide of food, power, and money.

The slate of mariculture products produced by Aquarius will be diverse and enormously valuable—worth over $6 billion a year. Total Aquarian income will be around eight billion a year, from eight principal sources: mariculture, hydrogen, magnesium, distilled water, tourism; services in the medical, engineering, consulting, and data processing industries; and the sale and development of new technologies.

**Table 1.10**
**Mariculture Products of a Marine Colony**

| Product | Tons/Yr. | Value-MM$ |
|---|---|---|
| Algal Protein Concentrate | 118,000 | $236 |
| Beta Carotene et al. | 14,625 | 2,000[93] |
| King Crab | 36,000[94] | 540[95] |
| Abalone | 18,000 | 270 |
| Queen Conch | 4,200 | 45 |
| Green Turtle | 4,200 | 45 |
| Oysters | 9,430[96] | - |
| Scallops | 9,430 | 47[97] |
| Clams | 9,430 | 47 |
| Mussels | 9,430 | 47 |
| Anchovy | 9,430 | 47 |
| Lobster | 3650[98] | 55[99] |
| Shrimp | 3650 | 55 |
| Mackerel | 3650 | 15[100] |
| Herring | 3650 | 15 |
| Cod | 3650 | 15 |
| Salmon | 3650 | 36[101] |
| Sea Silk - Copper Green | 35,800 | 358[102] |
| Sea Silk - Cobalt Red | 35,800 | 358 |
| Sea Silk - Chromium Blue | 35,800 | 358 |
| Sea Silk - White | 107,400 | 1074 |
| Sea Weed Paper | 263,000 | 131[103] |
| Pearls - White Oyster | | 334 |
| Pearls - Green Abalone | | 103 |
| **TOTAL** | | **$6,231** |

**The Floating Island — Laputa Sails the Waves**

One of the fundamental ideas of the cybergenic process is to build in conformance with the underlying symmetry of the universe. We must use chaos to our advantage rather than trying to overcome nature by brute force. Cybergenics is a sort of design *Ju-jitsu*, in which the natural tendencies of materials and systems are followed rather than opposed.

*Never hesitate to discard your complex solutions for simple ones, over and over again until it all appears so obvious people will say "anybody could design that." And they will never know what you went through, how much God went through before evolving his hydrogen atoms and blades of grass and eggs--*

**R. Buckminster Fuller**

Buckminster Fuller's elegant words are a guiding light for any designer. In nature there is an utter simplicity of design which springs directly from the geometry of space and time. As Fuller prescribes, the process of design should be one of discovery. Once the discovery is made, the design will create itself in a perfect and irreducible form, like a snowflake or a soap bubble.

There is an order and harmony to the universe which underlies even the most apparently chaotic systems. These 'strange attractors' have their own internal beauty; as structurally sound as any artifice designed by man.[104] This internal harmony is beautifully expressed in the sublime structure of a soap bubble. If you look at a soap bubble in oblique light, you will see an amorphous swirl of colors as the soap and water molecules move with random abandon over the surface. The movement of all these molecules is as chaotic as that in any cloud. Yet, somehow, in accord with a deeper symmetry—a symmetry imposed by the fabric of space itself—these swirling motes, together, form a sphere of absolute geometric perfection. This is the mysterious marriage of chaos and order which permeates our universe.

*It turns out that an eerie type of chaos can lurk just behind a facade of order—and yet, deep inside the chaos lurks an even eerier type of order.*

**D. Hofstadter**

The highest levels of order, like drops of water running to the sea, derive from apparently chaotic systems. Gaining access to the universe's chaotic bedrock requires clearing away the overburden of complexity. As Thoreau advises: "*Simplify, simplify.*"

To approach an ideal engineering solution, cybergenic design looks to natural harmony. Accordingly, the design of Aquarius must harmonize with nature, minimizing material requirements, while maximizing volume, usable surface area, and dynamic stability. Looking to cybergenic design, we find an optimal solution in the simple bubble float.

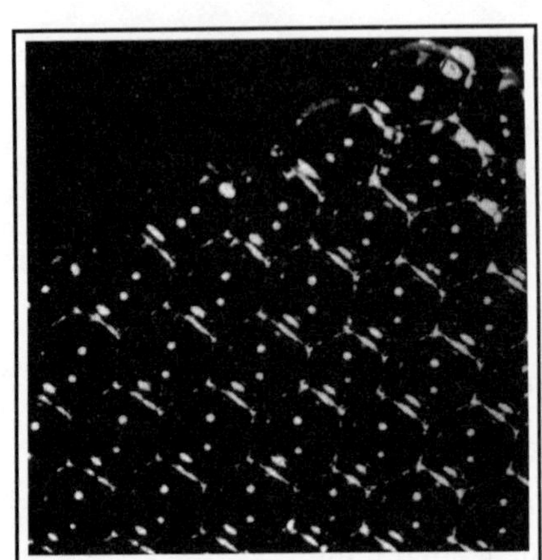

***Fig. 1.6 - Bubble Float***
©1967, Tensile Structures, Frei Otto, ed., MIT Press.

Certain aquatic snails use this design to build buoyant platforms from which they suspend themselves. The design is simple and efficient. When bubbles are packed together, they naturally form a hexagonal grid. This then is the inevitable template for a floating city like Aquarius.

The surface structure of Aquarius is composed of a cluster of nested six-sided columns rising from hexagonal floats. The center column rises the highest. Surrounding it are eight rings of columns, each ring lower than the preceding one by its natural Fibonacci number.[105] This design produces a buoyant structure with inherent dynamic stability.

**Modularity**

All Millennial colonies will follow the same development philosophy: start small and scale up by replicating similar modules. For the sake of clarity and brevity, this book describes colonies at fully mature stages of development. It is important to remember, however, that these structures are not carved out fully formed—like Ymir being licked from the ice.[106] Each Millennial colony grows from an original seed and develops to maturity. The process is not one of spontaneous generation, but rather of developmental evolution. Cybergenesis, like organic growth, is a flexible, adaptable, and plastic process, unthreatened by change or unexpected contingencies.

The starting seed for Aquarius will be a single OTEC module. The energies involved in Ocean Thermal systems are very dilute—warm water instead of high-pressure steam. Therefore, huge volumes are required to extract energy efficiently. OTEC components are correspondingly large. A single OTEC platform, will be 284 feet (86.56 m.) across. The first platform will be grown up to a height of 210 feet before the next ring of modules is started. After the first ring has grow to a height of 130 feet, the next ring of modules is started. Aquarius grows, one module at a time, until it achieves its full size—not unlike a plant.

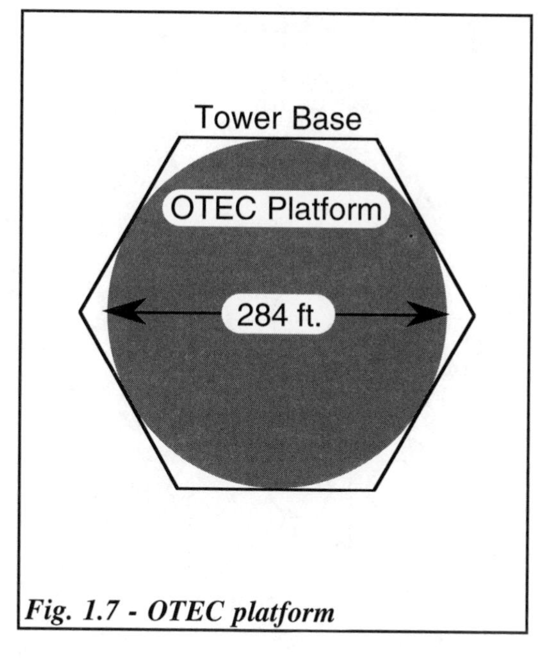

*Fig. 1.7 - OTEC platform*

The central tower, and the six surrounding towers, are supported by OTEC platforms. Each of the seven OTECs can produce 100 megawatts of electrical power, with the dimensions of the OTEC machinery determining the size of the towers.[107] Each OTEC platform will provide a net buoyant force capable of supporting 360,000 tons.[108] This excess buoyancy is an asset of tremendous value. The seven OTECs form the core of the island from which the rest of Aquarius grows.

The OTECs are the first elements of the structure to be constructed. Each power plant is brought on line before the next in the series is started. When all seven OTECs are in place, the rest of the island is grown around them.

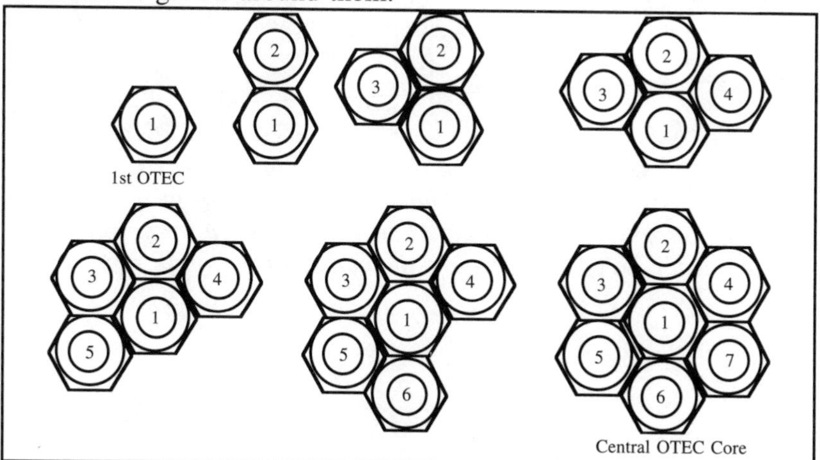

***Fig. 1.8 - Modular OTEC Growth***

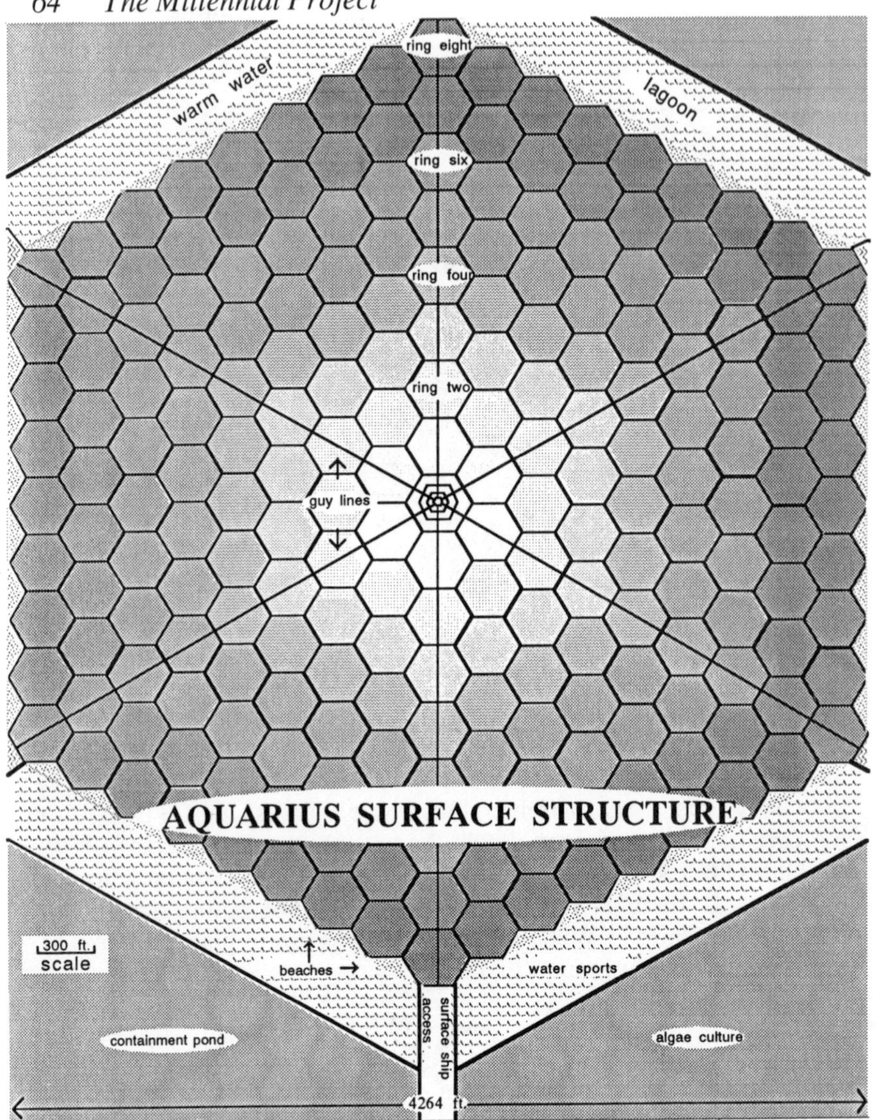


***Figure 1.9 - Aquarius Surface Structure Plan View.*** *The floating island consists of a cluster of buoyant cells, each of which supports a hexagonal tower. The island is surrounded by a lagoon of warm surface water, beyond which are the mariculture ponds.*

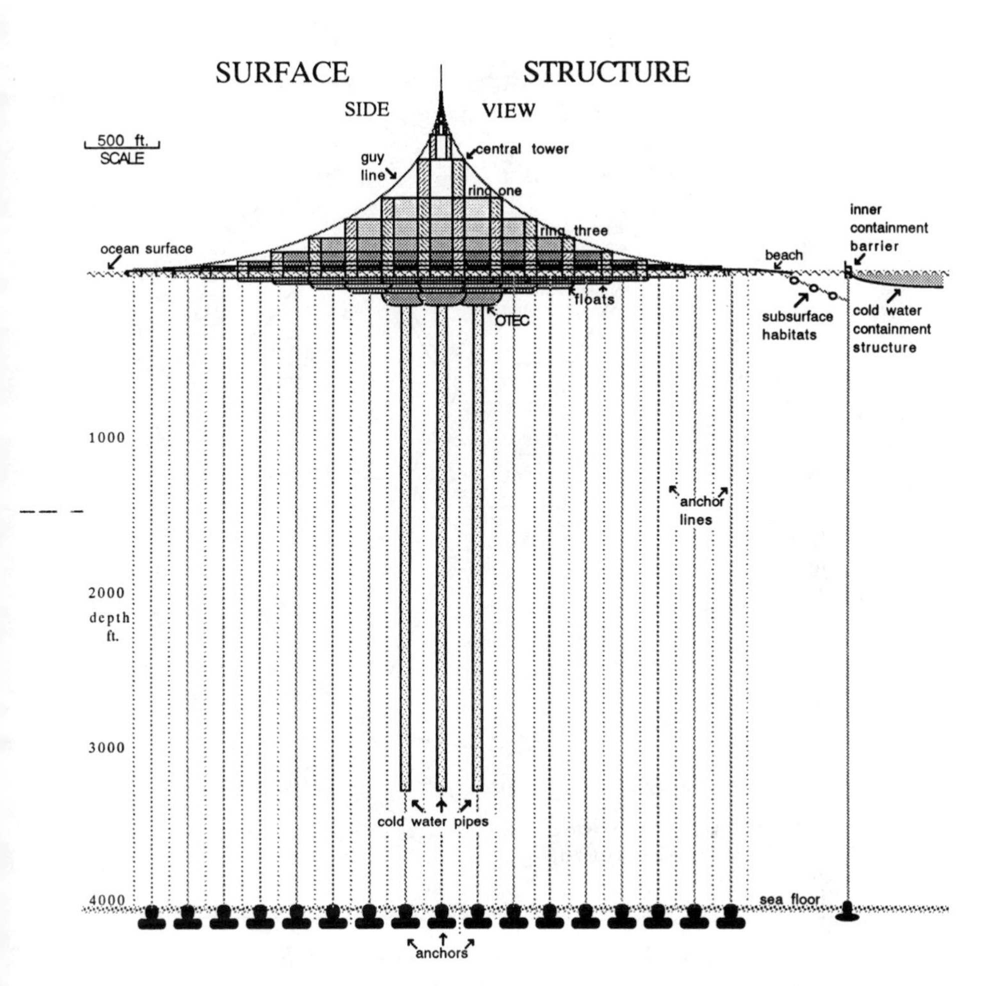


***Figure 1.10 — Aquarius Side View.*** *The structure floats, buoyed by the OTECs and submerged platforms. Beneath the warm clear waters around the margin of the island are numerous subsurface habitats. Here, Aquarians learn to live in outer space by living under water. Dead-weight anchors embedded in the sea floor hold the island in place. Gigantic pipes bring cold water up from 3300 foot depths to power the OTECs and nourish marine life.*

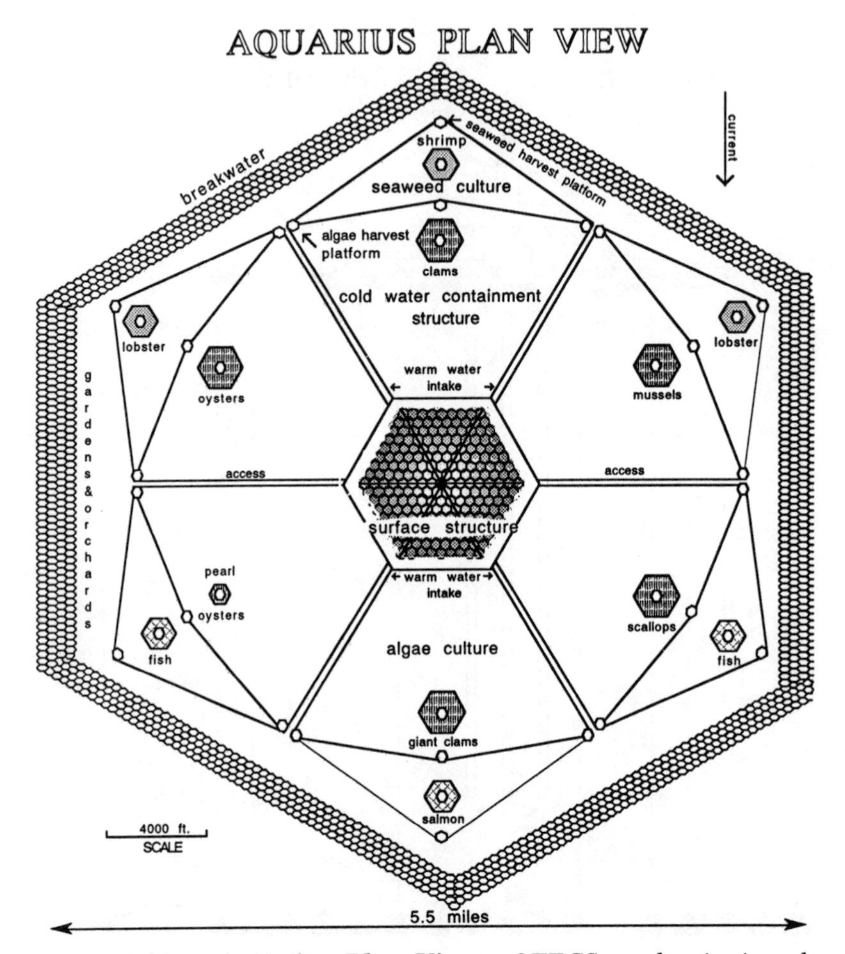


***Figure 1.11 — Aquarius Plan View.*** *OTECS, underpinning the central surface structure, bring up cold nutrient-rich water which is retained in the surrounding containment ponds. There it supports intensive algae growth and mariculture. A protective breakwater surrounds the complex.*

**Living Room**

The central tower, rising 550 feet above the sea, is a high-rise building, resting, not on concrete piles, as it would on land, but atop the buoyant OTEC platform. (See Plate No. 2)

Every story of each tower, will provide just over an acre and a half of floor space.[109] The central tower will be 55 stories high, with total floor space of nearly four million square feet.

Altogether, Aquarius will provide over 216 million square feet of floor space.[110] There is enough working, living, utility, and recreational space in the surface structure to accommodate up to 100,000 colonists.

At this population density—a maximum—the per capita allotment of floor space would be 2,167 square feet. Homes in the U.S. average 592 square feet for each resident. In Europe, the average is 323 $ft^2$; in Russia it is 215 $ft^2$.[111] The Chinese experience true overcrowding: average residential living space in Beijing is only 35.5 square feet per person.[112] The allocation of personal living space in Aquarius will be a roomy 600 square feet per capita—larger than the typical American home.

**Table 1.11**
**Average Square Footage of Commercial Buildings Per Capita—U.S.[113]**

| Type | $Ft.^2$/Capita | No. Bldgs. | Total* |
|---|---|---|---|
| Factories | 23.5 | 457,000 | 5.5 |
| Schools | 25.8 | 177,000 | 6 |
| Restaurants** | 8.8 | 380,000 | 2 |
| Hospitals | 9.7 | 61,000 | 2.3 |
| Hotels | 9.6 | 106,000 | 2.2 |
| Stores | 44.6 | 1,071,000 | 10.4 |
| Offices | 36.2 | 575,000 | 8.5 |
| Warehouses | 30 | 425,000 | 6.8 |
| Other | 36.6† | 696,000 | 8.6 |
| TOTALS | 224.8 | 3,948,000 | 52.3 |

**Billion $Ft.^2$ **Includes food stores. †Includes commercial buildings used for residential purposes: 10.5; vacant buildings: 14.3; and other types of commercial buildings: 11.8.*

People need more than just living space, however. All of life's multifarious activities require elbow room. In addition to residential space, each marine colonist will require 250 square feet of office, industrial, and support space—about the same as in the United States.[114] The balance of 1,317 square feet per person is available for open space, gardens, and hydroponic greenhouses.

Exactly how much room is necessary or desirable can only be determined by the residents. The less space required, the more cost effective the structure; but only the people who eventually live there can decide what is appropriate. A population of 100,000 is only a rough guess. A structure this size could accommodate far more people, or far fewer—it will be up to the Aquarians to decide.

The design is flexible. Residents will live either in apartments in the surface structure or in pavilions on top of the towers, or out on the breakwater. If people prefer life among the gardens to life in the tower apartments, the surface structure can be built so that it is lower and wider, providing more open terrain and less internal floor space. In a similar vein, the breakwater area can be extended indefinitely to provide additional open space for gardens or pavilions. Much of the space inside the surface structures will be available for purely aesthetic and recreational purposes. These communal spaces are dedicated to enhancement of the quality of life.

In conventional cities, the building space is subdivided into a myriad of tiny cubicles. The compartmentalization of modern societies demands that each little realm of activity be hermetically sealed away from all others. The very intention of such designs is to separate individuals. The net result is an atmosphere of intense isolation and alienation. By contrast, Aquarius' structure will encourage floor plans characterized by openness, with large areas partitioned by soft hangings, if at all. The exception to this will, of course, be individual living space, where privacy and solitude are desirable. Working areas, entertainment complexes, gathering places, service sectors; all these can be housed in wide-open, unrestricted areas, similar in concept, if not in execution, to suburban malls. Vaulted regions with high over-head space will give the interior a feeling of openness and spaciousness, that is completely lacking in most urban settings.

The many lakes, ponds, steams, and fountains on the tops of the towers, will be underlain by skylights of thick plexiglass. These bodies of water will do double duty, enhancing the living environment, and providing clear windows into the atrium spaces in the structures below. The generous allotment of internal space and the use of water skylights will flood the interior with rippling light.

In Aquarius there will be substantially less feeling of congestion than is typical of modern urban centers. In a modern city, dominated by skyscrapers, each structure stands alone and demands its own zone of space around it. Huge volumes are thus created that are entirely useless to everyone but the pigeons. Even worse, the little ground space remaining is ruined. In a modern city, the ground dweller gets the acute impression that he is standing at the bottom of a well. Looking up, he can see only a patch of sky peeking through the maze of buildings. Skyscraper cities are unlivable places for a simple if not obvious reason: they are built upside down. The open space is at the bottom and the buildings are on top. Consequently, deep concrete canyons squeeze air, space, and light out of existence. Aquarius, by contrast, is built right-side up: the buildings are at the bottom and the open space is on top.

Aquarians will also be free from the tyranny of the automobile. A modern city is designed more to accommodate the mechanisms that inhabit it than the people. The car, and its mass surrogates the bus and truck, are incredibly demanding of space. Not only do they require space to move around, but even more space is demanded to park them. None of this space will be necessary in Aquarius. The maximum distance from point to point in the surface structure is just under a mile. Under your own power, you are at most a 20 minute walk from anywhere. By riding the integrated system of transport pods, moving walkways, and elevators, you can reach any point in the city within five or ten minutes. If you are in a real hurry, or need to move around a lot, you can travel by electric go-cart or on powered roller blades.

In addition to providing access, the transportation corridors of the colony will present a recreational resource of no small dimensions. From the top of the central tower to beach level is a drop of 55 stories. Pathways will wind through and across the structure with grades of varying steepness. These paths of smooth seament will provide irresistible expanses of "white ice" for high speed roller bladers and skate boarders.

The entire outer surface of Aquarius is devoted to park-land, recreational, and residential space. Altogether, the tower tops provide 346 acres of open space. Atop some towers will be gardens with walking paths, flower-beds, groves of trees, intimate secret coves, ponds, fountains, and water-falls; others will have playing fields; some will have dance pavilions, outdoor cafes, and open-air amphitheaters. All of this outdoor space will be easily accessible to anyone in the colony. No matter where you are in Aquarius, you will only be an elevator ride away from the peaceful sanctuary of nature. (See Plate No. 3)

The space at the top of the towers is the most accessible, but not the only open area. The perimeter of the colony is lined by 2.75 miles of white-sand beaches. There is a warm-water zone between the colony structure and the mariculture containments, for swimming, sailing, wind surfing, and water sports. Surrounding the mariculture containments is the breakwater barrier which provides abundant open space.

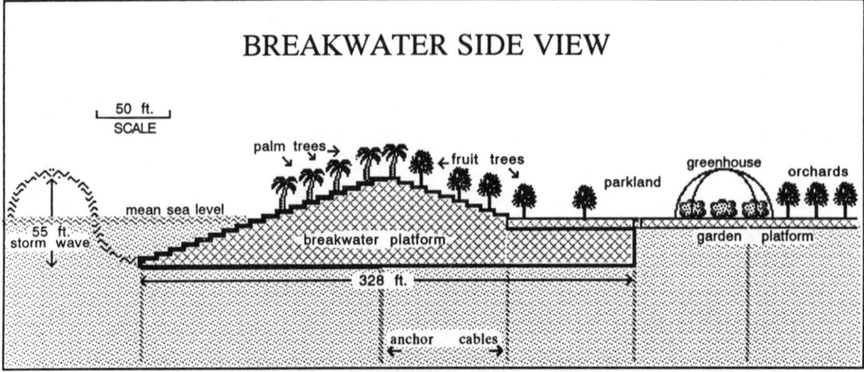


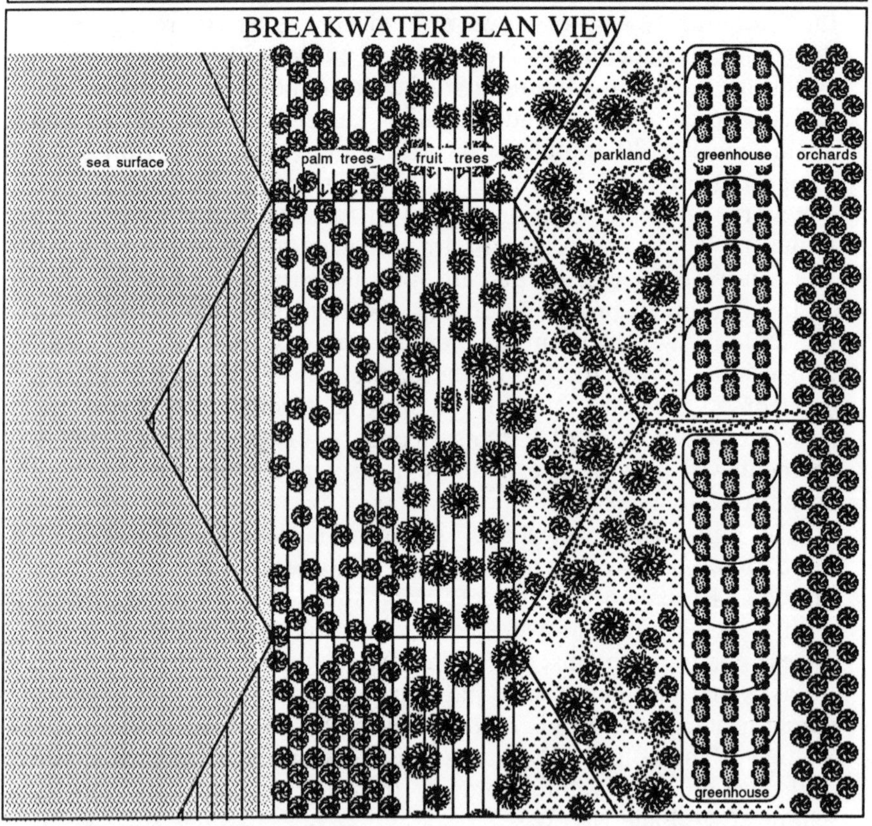


***Figure 1.12 - Breakwater.*** *View of a section of the breakwater, seen from the side and from above. Note that each hexagonal section of the breakwater is of the same planar dimensions as the hexagonal cells that make up the surface structure (Fig. 1.9). The outer row of platforms is very robust, and can absorb 55 foot storm waves. The thin platforms behind the outer row are primarily for garden and residential space.*

The breakwater (see Appendix 1.3) also provides space for the cultivation of fruits and vegetables. 767 square feet are required to provide each colonist with fresh produce.[115] This assumes that two-thirds of his needs will be met by mariculture and by synthesizing algae into substitute foods, with the balance provided by growing fresh plants in hydroponic greenhouses, gardens, and orchards on the breakwater. These gardens and orchards will be designed in a free-form style, with pathways winding through them and many secluded private spots. Residential pavilions set among the trees and gardens of the breakwater will provide homes for those who prefer life on the beach.

The surface structure grows from the top down, making it infinitely expandable. New floors can be added to the bottom of the structure at any time. Each new bottom floor will add 15 million square feet of space to the colony. (It may not be desirable to push the colony up more than five stories at a time without adding another ring of modules along the outside of the structure for stability.) If four new floors were added to the bottom of the colony as configured, it would increase the space available in the structure by 60 million square feet.[116] If a ninth ring of modules were added, with five additional stories beneath it as well, the total area of the structure would be expanded by over 200 million square feet—more than doubling its effective size.[117] With this expansion, the colony could accommodate 300,000 inhabitants. Additional expansion is always an option.

**Under The Sea**

In Aquarius we can adapt to the space colonist's way of life. Though Aquarius is on Earth, it is a space colony nonetheless. Psychologically, socially, and economically Aquarius will not be much different from colonies on the Moon. Aquarians will experience a sense of isolation and removal from the terrestrial world that will be common to all space colonists. Of all the novel sensations in space, this may be the one that takes the most getting used to. The inhabitants of Aquarius will be living in a comparatively self-contained society, dependent on the outside world for few basic needs. The social and personal aspects of life in a space colony will be the most important and difficult to adapt to. In Aquarius, we can evolve the unique cultural systems such a life will demand. Out at sea, we can meet these social and individual challenges without having to face the rigors of life in a vacuum.

In Aquarius we will also develop many of the physical systems needed to deal with life in outer space. An expanse of warm surface water—in essence a lagoon—surrounds the perimeter of the surface structure (see Fig. 1.9). Separating the surface structure from the algae containment ponds, this lagoon provides a large playground for water sports, and serves as a staging and training area for submarine activities and research. No marine culture can survive and thrive on the ocean's surface without highly developed submarine skills. Aquarian construction activities, and many industrial and maricultural pursuits will require a substantial human presence beneath the waves. Aquarians will necessarily be adept submariners. The warm-water lagoon is our portal to the submarine world.

Rows of underwater bubble habitats, simple hemispheric membranes holding air, will be anchored to a cable lattice at the bottom of the warm water lagoon. (See Plate No. 4) The habitats do not have to be very strong. The bubbles are filled with air at the same pressure as the surrounding water, so stresses on the membranes are minimal. These underwater habitats, serve as rest areas, workshops, and even temporary dwellings for the sub-marine Aquarians. Transparent bubbles will be clustered atop the newest ring of hexagonal modules, at a depth of just ten feet. Below these, other bubbles will be anchored at various depths. Down to 40 or fifty feet, the water will be crystal clear, warm, and brightly sunlit. Here, the colonists will work and play in a submarine environment, learning the skills and developing the tools they will later put to use in space.

Living and working beneath the sea is actually more dangerous than living and working in outer space. The most dangerous thing about space is not the vacuum; it is the pressure differential between the inside of a space habitat and the vacuum outside that is deadly. In space, that differential will never be more than 15 psi (pounds per square inch)—atmospheric pressure at sea level. A sub-mariner working under 34 feet of water has the same pressure relationship with the surface that a person at atmospheric sea level has with outer space. In practice, air pressure in space habitats and vehicles will seldom be more than 5 psi. To simulate the 5 psi pressure differential between the inside of a space colony and the outside vacuum, requires submerging to a depth of only eleven feet.

Aquarian sub-mariners will have to learn to deal with the dangerous potential inherent in pressure differentials. Before any space cadet heads for orbit to earn his wings, he will first come to Aquarius and earn his fins. By coming and going through airlocks, by decompressing in pressure chambers, by living for long hours in diving suits, by swimming freely in the zero-gravity environment under water, by all these exercises and a thousand more, Aquarians will learn the skills needed for life in outer space. In fact, life in a vacuum will seem safe and easy compared to life under water.

**Germination and Growth**

When completed, Aquarius will be one of the largest structures of any kind, and certainly the largest floating object ever built by man. Construction of such an enormous megalith by means of conventional techniques is appalling to contemplate. Building the structure in sections on land and hauling the pieces out to sea, or building the structure in place, both involve logistic and engineering problems of nightmarish proportions.

As in all cybergenic processes, however, Aquarius begins as a seed. In this case the seed is a 100 megawatt ship-mounted OTEC. Any large vessel can be converted to serve as the seed ship. (The superpowers are in a race to dismantle their armies and navies; maybe we can get a good deal on a used aircraft carrier.) The seed ship will motor to the site chosen for Aquarius, a spot on the equator off the east coast of Africa. After anchoring on station, the ship will lower its cold water pipe into the depths. With power from its main engine, the ship will pump up cold water and jump-start the OTEC plant carried on board. Once the OTEC is up and running, it will crank out 60 megawatts of net power. Ten megawatts of this power are consumed in surface related activities: providing power to fabrication workshops on board the ship, electrolyzing hydrogen to power surface vessels and aircraft, charging batteries of submarine vessels, and other uses, leaving 50 megawatts of power available for the accretion of Aquarius' shell.

**Sea Shell**

Aquarius will not be built so much as she will be grown. Like many man-made structures, Aquarius will be made of reinforced concrete; but instead of being poured or assembled, her cement will be accreted out of sea water—like a seashell.

The accretion of 'sea-ment' is accomplished by applying an electric current to a metal grid; calcium carbonate and other mineral ions dissolved in sea water bond electrochemically to the charged metal, forming a cement-hard coating.

Most surface waters, particularly in tropical seas, are saturated with calcium carbonate, ($CaCO_3$). This mineral is very familiar in its common forms: calcite, marble, limestone, seashells, and

Portland cement. In sea water, calcium carbonate occurs as a positively charged ion. If a positively charged anode, and a negatively charged cathode are suspended in sea water, with an electric current flowing between them, ions of calcium will combine with carbonate ions and accrete on the cathode. As long as the electric current flows, calcium carbonate and other minerals will continue to build up.[118] One kilowatt hour of electric power will result in the accretion of 4.2 pounds of "sea-ment".[119]

A sheet of metal mesh, supported by conductive reinforcing bars, forms the cathode. As current flows, the wires of the mesh become progressively more and more heavily encrusted with sea-ment. The gaps are gradually filled, forming a solid slab of reinforced concrete. If sheets of wire mesh, separated from each other by around 1/2 inch, are built up in layers, slabs of sea-ment can be grown to any desired thickness.

Sea-ment is stronger and lighter than conventional concrete, and does not lose any of its strength when it dries.[120] Typical concrete, of the type used in sidewalks and foundation walls, has a breaking strength of 3500 psi (pounds per square inch). Sea-ment, accreted on 1/2 inch hardware cloth, has an average strength of 4267 psi—20% stronger than conventional concrete.[121] Samples of sea-ment vary in strength, depending on the time taken to accrete them, the current density used, and other factors. Samples range in strength from 3720 to 5350 psi.[122] With very slow deposition—up to a year or more—strengths of 8000 psi have been attained.[123]

The current density needed to cause sea-cement to form is very low. Thirty amps at 6 volts forming a current density of 1.3 milliamps per square inch is sufficient.[124] Higher current densities can be used, up to 50,000 milliamps (mA) per square foot, but rapidly accreted sea-cement is weaker.[125]

At a current density of 189 mA/ft$^2$, a tenth of an inch of sea cement will accrete on 1/2 inch wire mesh in 170 hours.[126] At this rate of accretion, .0005 in./hr., the spaces in 1/2 inch mesh will be completely filled up in 500 hours—three weeks.[127] At 12 volts, a kilowatt of power can generate a current density of 189 milliamps per square foot over an area of 441 square feet.[128] A 100 megawatt OTEC, producing 59 megawatts of net power, can thus electrify 26 million square feet, accreting 124 tons of sea-ment per hour.

At peak production, the OTECs could produce 8.5 million tons of sea cement over the course of a year—approximately equal to the total mass of Aquarius[129] In practice, only 70 MW of the available power will be devoted to seament production. With this power budget, the entire structure can be grown in 6.5 years.

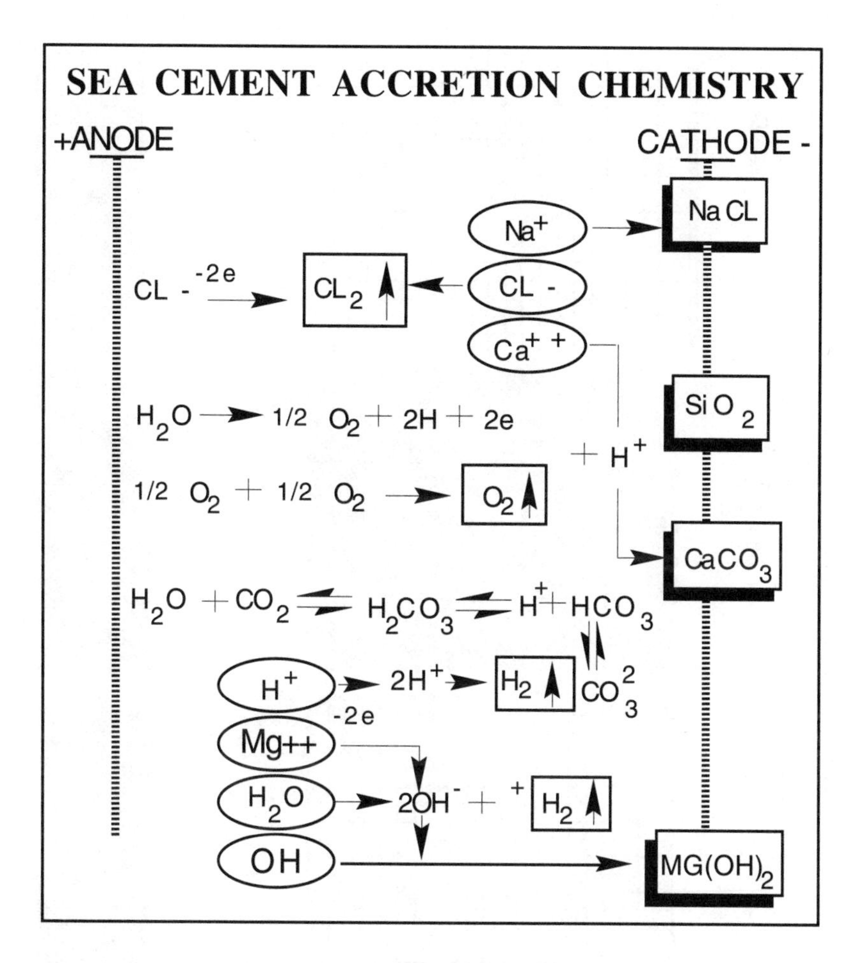


***Figure 1.13 - Sea Cement Schematic.***[130] *Sea water forms an electrolytic conductor connecting anode and cathode. As current flows through this circuit, the reactions detailed in the schematic take place. The result is the accretion of a mineral matrix on the cathode, together with the production of hydrogen, oxygen, and chlorine gas. These gases can be captured as valuable by-products of the process.*

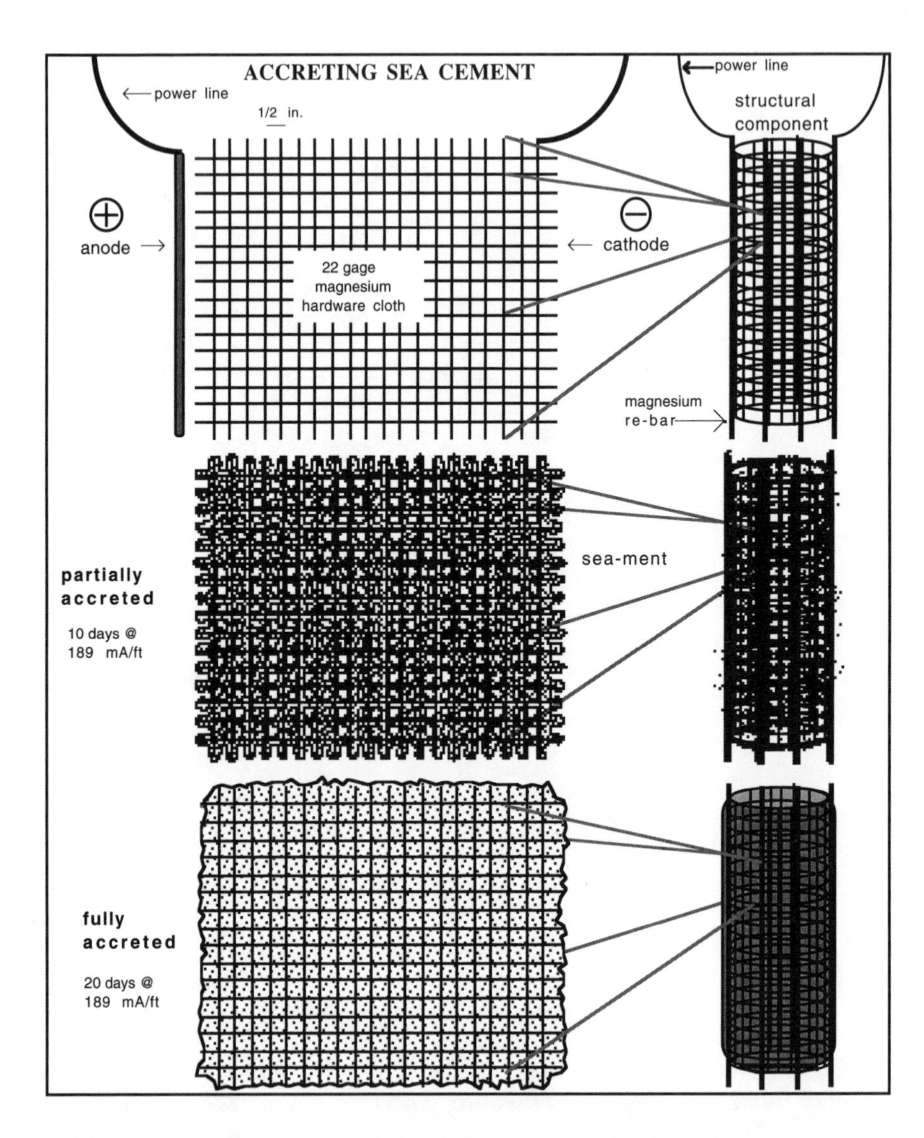


***Figure 1.14 - Sea Cement Formation.*** *Magnesium wire mesh with half-inch spacing is electrified at 189 milliamps per square foot. Over 21 days, spaces in the mesh are filled in with sea-ment. The mesh can be formed around magnesium reinforcing rods, and accreted in place to form structural building components.*

**Table 1.12**
**Mass of Aquarius' Shell Structures**

| Component | Mass* |
|---|---|
| Breakwater | 2.80 |
| Surface Structure | 2.71 |
| OTEC Platforms (plus machinery) | 1.15 |
| Foundation Floats | 1.06 |
| Anchors | .30 |
| Breakwater Floats | .15 |
| **TOTAL** | **8.17** |

***millions of tons***

**Skeletal Structure**

Sea-cement requires three components: electricity, which the OTECs provide; calcium carbonate, which is abundant in sea water; and metal, to form the conductive mesh and provide reinforcement. The metal mesh is a crucial component: it is the electrical cathode on which the seament agglomerates, it forms the framework which defines the shape of the finished piece, and it provides the reinforcement critical to the strength of any concrete. The metal framework is the skeleton of Aquarius.

For Aquarius to be truly cybergenic, she must be able to grow her skeleton from the materials in her environment. Aquarius, like all space colonies, must be a study in self-sufficiency. Once the core OTECs are operating, we should be able to create an entire sea-borne civilization with minimal dependence on outside inputs. If we rely on bringing in ship-loads of steel from foundries in Korea, or Europe, we will be hobbled by at best a philosophical, and at worst a physical Achilles' Heel. To avoid this, we must find an on-site source of suitable metal. One would think this unlikely in the middle of the ocean; but, as we shall see over and over again, ideal solutions are at hand; we need only discover them.

The skeleton of Aquarius will be formed of magnesium extracted directly from sea water. Sea water is, in fact, the principal mineral ore for most magnesium produced in the United States. The process of extraction is very simple: sea water, which contains .13% magnesium, about 2.6 lbs./ton, is mixed with calcium oxide and the solution is electrolyzed.[131] The only inputs to this system are sea water, calcium oxide, and electricity; the only outputs are magnesium and water.

Extensive mariculture operations will produce a mass of sea shells, 87,000 tons per year, as a byproduct. This might constitute a waste disposal problem, except for the happy circumstance that sea shells are mostly calcium carbonate—the feed stock needed to form calcium oxide. Calcium oxide, known more prosaically as lime, can be manufactured by heating sea shells in the presence of oxygen.[132]

Magnesium is a wonderful metal. It's the lightest of all structural metals—a third lighter than aluminum, and only a quarter the weight of steel. Yet, despite its light weight, it is surprisingly strong. When alloyed with aluminum and other metals it has a tensile strength, of over 50,000 pounds per square inch.[133] This compares favorably with good structural steel, which typically has a tensile strength of 64,000 psi. To substitute magnesium for steel will require using 30% more by volume, but 30% less by weight.

The reinforcing bars (rebar) in sea cement must be both strong and electrically conductive. To form sea-cement, voltage must flow freely through the reinforcing mesh. Magnesium is a good conductor; it is about 60% more resistive than aluminum, but only half as resistive as steel. Voltage will readily flow through magnesium, making it suitable as the reinforcing metal for sea-cement.

Magnesium does have shortcomings. In its pure form, for example, it is soluble in sea water. This could be a big problem. If Aquarius dissolves, we will all get very wet. Fortunately, magnesium can be rendered highly corrosion resistant when alloyed with manganese, and can be made virtually corrosion proof when anodized.[134] A manganese alloy will be used for the reinforcing mesh which is also protected from corrosion by its thick coat of seament. The small amounts of manganese needed can be obtained by dredging manganese nodules from the sea bottom.[135]

The other problem with magnesium—the one that first flashes to mind—is that in certain forms this metal is highly explosive. An exploding island is, if anything, worse than a dissolving one. There is little danger of igniting Aquarius' magnesium skeleton, however. Magnesium is only highly flammable in its pure form and then, only when powdered or ground into fine shavings. In this form, magnesium is better known as the 'flash powder', used by early photographers. While magnesium in powdered form is highly explosive—it is an important component in many types of bombs—in solid form it is no more explosive than aluminum. Most of the magnesium on Aquarius will be sealed inside a thick coat of sea-cement, and all of it will be alloyed with other metals and cast in bulk form as bars or wire mesh. The skeleton of Aquarius will be in no more danger of spontaneous combustion than the ribs of a modern airliner.

Magnesium, which is a highly valued metal in the world economy, will cost us only 4.25¢ a pound to produce, since it requires no outside inputs. Its high strength-to-weight ratio, resistance to corrosion, and ductility make it a material of choice in aerospace applications. Magnesium is really just a form of concentrated electrical energy—8.5 kwh per pound.[136] We will use some of our surplus power to produce magnesium for export. This would be highly advantageous since magnesium is worth about $1.50 a pound, $3000 a ton, and thus relatively insensitive to transport costs.

Aquarius' skeleton will require 565,000 tons of magnesium.[137] By devoting 170 MW its production, 87,000 tons of the metal can be produced each year.[138] After completion of Aquarius, magnesium production for export could continue. Aquarius could supply 25% of the world's present demand for this extraordinary metal.[139] At $1.50 a pound, magnesium exports could be worth $260 million annually.

**Raw Materials**

There is no question of exhausting the materials needed to grow Aquarius. Every 17 days, enough dissolved solids pass through Aquarius' growth zone to supply all the materials needed.[140] If all the solids in the world's oceans were precipitated out of solution they would form a crust over the land 450 feet thick.[141] Calcium carbonate and the other solids used to grow Aquarius would form a layer 22 feet deep. The eight million tons of solids needed to grow Aquarius amount to only a fraction of the 150 million tons carried into the world's oceans by rivers each year.[142] Of the 250 trillion tons of these solids dissolved in the world's oceans, Aquarius will use only .0000004%.[143] There is enough dissolved calcium carbonate and other building material available in the world's oceans to grow over two million marine colonies the size of Aquarius. There is little danger that the growth of marine colonies could ever deplete or even seriously impact the reserves of these solids in sea water.

Electro-deposition of sea-cement provides an optimal technique for building structures at sea. All work can be done on-site, avoiding the logistical problems of hauling huge prefabricated pieces hundreds of miles out into the ocean. There are no transportation or other costs for the raw materials which are already in place—delivered free to the job-site, by a bountiful nature. Raw materials are extracted, processed, and positioned in a single step, requiring relatively little labor. Assembly will be rapid by terrestrial standards, since all that needs to be constructed is the wire mesh framework—electrical fields actually do the heavy construction.

The process is inherently cybergenic since it is as much an organic growth process as it is an engineering construction process. It can take full advantage of Millennial technologies: a central coordinating computer controls the operation, and most of the grunt work of forming the metal grid can be done by robots. This process takes advantage of our strengths while avoiding our weaknesses. While sea-cement uses massive quantities of raw materials we have in abundance, it needs only small inputs of those perennially scarce commodities: labor and money.

As the structure grows, essential infrastructure can be prepositioned. For example, plumbing, ventilation, electrical, pneumatic, and telecommunication conduits needed to service the interior of the structure can be wired directly to the form mesh and run along the shortest possible routes. When the walls are formed, the conduits are grown in, becoming an organic part of the whole. Access ports are preformed in the walls to allow maintenance and expansion of the conduits. If there is trouble in an inaccessible place, a surgical incision can be made in the wall, the conduit serviced, and the cut location 'healed'.

All of these advantages work synergistically to render construction of an epic monolith like Aquarius not only possible, but fast, simple, and cheap.

**Morphology**

The mesh and rebar form will allow us to create internal space with almost complete freedom. Since the technique is entirely divorced from the stick-building methods used on land, and since the actual application of the massive building material is performed electronically, we can build free from the rigid dictates of cubic geometry. The interior spaces of Aquarius, can be laid out in a free-form manner corresponding more to biology than to architecture. The interior of Aquarius will resemble the inside of a sea shell; lines will be smooth and curved with a flowing organic look.

***Fig. 1.15 - Aquarian interior.*** *The organic nature of Aquarius will be apparent in its interior architecture. Natural free-flowing lines will be evocative of her shared heritage with sea shells, corals, and other crustaceans. This artist's sketch illustrates only one of an infinite variety of potential treatments. The cybergenic growth process will free architects to create designs tailored to the physical and psychological needs of human beings, unconstrained by the rigid dictates of geometry. The end result will be a living environment that is intimately connected to the people it shelters.*

Organic design will influence all aspects of Aquarius' interior appearance. The lighting will be recessed and indirect, suffusing the walls with a shadowless glow. The colors will come directly from the palette of mother nature—the ultimate interior decorator. We will follow her recommendations to the letter, using the oyster whites, nautilus tans, coral pinks and other subdued hues from her exquisite collection of shell tones; all punctuated by startling splashes of neon blues, yellows and reds, from her book of reef colors. Much of the decor will actually be alive. A multitude of aquaria will be set into walls, ceilings, and floors. Brilliant corals, anemones, and exotic fish, swimming behind thick panes and bubbles of sea-glass will enliven Aquarius' interior.[144] All the aquariums will be interconnected to a central circulatory system which sustains them with abundant streams of fresh surface water and algae.

This free-form approach to interior design will also allow us to make much more efficient use of space than in cubicle architecture, where huge volumes of space go unused. If you set up a camera and photographed the human usage of a typical room, you would find that half the space is entirely wasted. What for example are the upper corners of rooms ever used for? Yet you pay for all of that unused space. In space colonies, the efficient use of space will be one of the inescapable necessities we must all live with. As in all things, we will begin to conform to this necessity in Aquarius.

The operative principal in modern architecture is to enclose a cube of space and then fill it with paraphernalia we think necessary for a comfortable existence. In Aquarius, which is not so much a structure as an organism, the guiding philosophy will be to engineer the space itself to insure the comfort of its inhabitants.

Furniture will be a largely obsolete concept. Take for example the dresser my mom bought me when I was a kid. I still have it, and by the standards of its era, it's an admirable household fixture. It is a massive construction of maple wood, expertly joined with cunningly cut pieces, fitted and glued with the strength of iron. It is set with massive brass fixtures, and looks today—discounting the dust—as new as the day it was purchased, a quarter century ago. So far, so good; a fine piece of furniture you might say. But let's look at it objectively, as a machine, as an object with a purpose. Here sit a hundred pounds of hardwood with a compressive strength of 1500 psi, jointed by an expert craftsman into a rigid box that would easily support a bull elephant. And what is the sole purpose of this massive crate, this monument to a dead tree?—it holds my socks.

Not only is it blind engineering overkill of epic proportions, it is also an environmental disaster. The home to generations of squirrels, a sentinel post for falcons, an autumnal banner of golden glory, a living creature, was chopped down to enshrine some underwear. This, my friends, is no way to run a planet.

Most of the artifacts that surround us are obsolete, as are the preconceptions about living which they reflect. These things and ideas are not only obsolete here on Earth, but are even more so in the new frontier of space.

Imagine for a moment that you're preparing for departure to a new life in the Martian colonies. You are deciding what few precious personal possessions you can afford to take along: a video-disc of family photos, a couple favorite T-shirts, a ring, a pressed leaf. You decide that you can't live without your Rolling Stones LP, *Sticky Fingers*, the one Mick Jagger personally autographed with a lewd remark under the zipper—priceless treasure. Unfortunately, including it in your baggage puts you 26 grams over MGWA (max gross weight allowance). So you decide to leave the vinyl record behind—obsolete anyway—and take just the autographed album cover with you. But, you are still 3 grams over weight. You are busily engaged in cutting three inches off the handle of your tooth-brush when your loving mate comes in wanting to know where you're going to pack the dinette set. Dinette set?!

We can't haul this garbage with us, either physically or conceptually. Almost everything we have and everything we do is obsolete. We must leave it all behind, first on the land and eventually on the planet. We cannot take our maple-wood furniture into space with us, and we must leave even the conception of it behind when we take our first giant step, out to Aquarius.

This does not mean that we have to leave behind the function of these things. We do not have to throw our socks on the floor, or do without a comfortable place to sit. We simply have to find more sensible ways to achieve the same ends.

Virtually all furnishings in Aquarius will be built into the structure as an organic part of the whole. A couch, for example, will simply be a platform area which flows naturally from the curvature of the wall, set with removable, washable cushions. Desks, tables, cupboards, sinks, bathtubs, shelves, even light fixtures, will be grown in place as permanent parts of the internal structure.

**Circulatory System**

Prewired, the inside of Aquarius will serve the needs of its inhabitants, in much the same way the body provides for the needs of its cells. As a cybergenic organism, Aquarius will have an intelligence all its own. Rooms will be equipped with sensors so that temperature and humidity can be controlled to maintain a comfortable environment. Lights will switch on and off automatically as people leave and enter a room. In living quarters, provisions for the necessities of life are built into the structure. Laundry is disposed of in a pneumatic tube which whisks it away to a central facility where it is automatically washed, dried, pressed and returned by tube to its owner. A microchip sewn into the clothing, no bigger than a fleck of dandruff, tells the laundry's computer whom it belongs to and how much starch to put in the shorts. You can order food by video menu from a central kitchen. "Three slices of deep-pan pizza, mushrooms, pepperoni, extra cheese, hold the anchovies," is keyed or spoken into the computer. Minutes later, your pizza arrives from the central kitchen via pneumatic tube, still piping hot.

You can shop the same way; peruse a video catalog for the things you need and order them on the spot. They too, will, depending on their size, be automatically pulled from storage and sent to you, usually within a matter of minutes, by pneumatic tube. Your credit account is charged automatically.

Most of the junk we own only gets used occasionally. It will make sense, economically and environmentally, to share stuff. The cybergenic advantages of a space colony like Aquarius, make this possible. The pneumatic delivery system makes getting things about as quick and easy as turning on a tap. In Aquarius, goods are more accessible from a central warehouse than they would be in the back of your own closet. Therefore, you might as well return things when you're not using them, and so share them with others. This not only saves cash and minimizes environmental degradation, but also conserves one of the world's scarcest and most threatened resources—closet space.

**Central Nervous System**

The cybergenic city is an "intelligent environment," continuously interacting with its inhabitants to maximize safety and convenience while minimizing consumption and costs. In all cybergenic systems, the key components are the central and peripheral nervous systems. Aquarius will exemplify this model. In Aquarius, the central nervous system will be a massively parallel computer. By the year 2015, when Aquarius is fully developed, computers will have increased their power 10,000 fold over their present capacities.[145] Today, computer chips are at the VLSI (very large scale integration) stage of development, and pack a million components on their surfaces. By the year 2000, chips will have advanced to GSI (giga-scale integration), with a billion components. By 2010 they will pack 10 billion components on a chip.[146]

The central computer on Aquarius will be built as a neural network. Instead of a single massive central processor, Aquarius' brain will be built of a network of many linked processors. This network will consist of up to a million co-processors all joined as one enormous super-computer. Neural networks have many advantages over conventional computers, especially in the realm of artificial intelligence.[147] The need for a central nervous system in a cybergenic living city like Aquarius will demand that we break new ground in the construction of a massively parallel neural network.

To construct such a computer we will use a chip architecture that will be ten or fifteen years old in 2015. Older chips, having already paid out their R&D costs and profits, can generally be purchased for their manufacturing costs, plus a modest profit margin. For example, the Intel i486™ chip hit the market in the early 1990s. This chip has 1.2 million transistors and can execute 20 million instructions per second.[148] When it was first introduced, it cost $1500 dollars. But most of that price was for the research and development it took to create the chip in the first place; the chip's actual manufacturing cost is a few pennies. Ten years from now, when the cutting edge market is off exploiting another chip, the 486 will be available for its production cost plus profit. Accordingly, in 2015 when we are building Aquarius' brain, we will be using chips that were state of the art in the year 2005.

The central computer of Aquarius will be built of one million GSI chips linked in parallel. It will have a computing capacity of $10^{16}$, ten-thousand trillion, machine instructions per second (mips). Such a machine will be equivalent to ten million of today's super-computers.

It is just as well that the data utility has such a staggering capacity for computation, since it will probably incorporate seventh or eighth generation parallel architecture. VLSI single processor computers of today are the fourth computer generation. Like their predecessors, fourth generation machines are capable of handling only exactly quantified data. In the fifth generation, parallel processors will begin to analyze 'information'. Information, unlike data, is usually imprecise. In the language of Artificial Intelligence programmers, it is "fuzzy". For example: "What is the distance between Rome and Paris?" is a precise question that can be answered with an exact piece of data—540 mi. By contrast, something like: "How will the United Nations react to colonies outside the jurisdiction of any national government?" is a fuzzy question for which there is no definite answer. Yet a powerful computer with access to enormous amounts of information on the U.N., the *Law of the Sea*, the tendencies of governments to resist change, etc., would be able to give an informed response to the question.

Fuzzy logic requires an altogether different sort of machine than the glorified toasters we now call computers. In the next generation, with the advent of neural networks and true parallel operating systems, computers will begin to achieve real AI (Artificial Intelligence).

AI is enormously demanding of computer power. The Japanese have established the awesome goal for their fifth generation machine of handling 100 million lips (logical inferences per second). Since a single lip is equivalent to 100 standard machine instructions, the Japanese computer will have to run at 10 billion mips.[149] Higher generations of AI computers will certainly be even more demanding of raw power. The central computer in Aquarius, with a million times this capacity, will be able to handle even these demands.

The capabilities of such a computer will be mind boggling. It will handle all the internal metabolic functions of the colony: environmental control, operating automated processes, tracking the medical, nutritional, and psychological, needs of individual colonists, monitoring the power production facility, maintaining platform stability, and a myriad other tasks. In addition, it will serve as a work-station for colonists involved in data-processing and telecommunications.

In addition to these computational powers, Aquarius' brain will require a memory to match. The central computer will be equipped with a vast bank of optical storage devices. On this optical medium will be recorded much of the available knowledge of mankind. By 2015, most of the world's written information will have been converted to a digital format. A large percentage of this information will be down-loaded into Aquarius' central data storage facility. This data base will include most published works: books, periodicals, patents, programs, films, videos, and audio recordings.

This will provide Aquarians with instant access to virtually any piece of information in existence.

As of the late 1980s, typical magnetic disks could hold 800,000 bytes, equivalent to 400 typewritten pages of information in a square centimeter.[150] In state of the art optical systems, lasers can now focus on data bits 1/2 a micrometer in diameter, and can follow a track with an accuracy of .1 micrometer.[151] This precision allows for a data density ten times that achievable in magnetic media. Optical disks can store 8 million bytes, 4000 pages in a square centimeter. At this data density, a 13 inch optical disk can store 2 billion bytes, 1 million pages of information, equivalent to 2000 books. Assembled in stacks of platters, spaced every half inch, an assembly of such disks in an 8 foot module would hold 384 billion bytes—192 million pages.[152] At this density, we can store 200,000 books in a single square foot of floor space.

The U.S. Library of Congress contains 20 million volumes, 80 million items including periodicals, photographs, and maps, amounting to 9.8 X $10^{15}$ bytes of information.[153] To contain such a volume of information would require 25,500 optical disk modules, occupying just over 50,000 square feet. In Aquarius we will install about twice this amount of data storage capacity. This would just about fill two floors of the central tower. At a data storage cost of 3¢ per megabyte, we could install the entire databank for around $600 million—not counting the cost of the data itself.[154] As storage technologies mature, ever greater amounts of data will be stored in ever smaller packages.[155] At atomic densities, the contents of the Library of Congress could be recorded on a single disk.[156]

The massive concentration of data and computing power in Aquarius will enable the colony to function as a sort of global computer utility.[157] Operating via satellite links, at first commercially leased from private satellite operators, and then eventually through our own satellite network, we can communicate with and send data to anyone on Earth, providing engineering, data processing, and consulting services. Aquarius will be linked intimately with the global economy, our isolation in the middle of the Indian Ocean notwithstanding.

Communications will become one of the major occupations of the colonists. Data, a massless commodity, will flow through Aquarius in a prolific stream. Many of those who inhabit Aquarius, will be engineers, consultants, and data specialists, able to enjoy the benefits of living in a marine colony while still performing the high-paying jobs that today require them to live in a terrestrial city.

The main living area in a typical Aquarian home will be equipped with large high-definition video panels. A video screen is the colonist's personal terminal for interfacing with the central computer, able to retrieve any information the computer has at its disposal. A colonist can call up any programming in the central library, from old re-runs of *Mr. Ed* to instructional tapes on neuro-surgery. The screen can contact any other member of the colony, or can allow the user to interface with working groups or attend electronic meetings of the colony's general assembly.

Much of the work done in Aquarius will involve data synthesis. Many colonists, will work in the comfort of their own apartments. They will pull information from the central data bank through their own computer screen, process that information, and send it back. The terminal will even enable people to enjoy some of the companionability of the work-place. The large format, and digital split-screen output will enable individuals to see and talk to a number of their colleagues simultaneously. For the added flexibility and sociability of the office environment, people might prefer to get together with co-workers who are actually in Aquarius. But if your colleagues are located in another colony or are elsewhere, the electronic interface will enable you to come face to face at a moment's notice. In short, the computer panel is your looking glass to the world.

**Living in Aquarius**

A variety of lifestyles and types of accommodation will be available to Aquarians. Apartments on the outer surface of the colony will necessarily be snug. Many people will prefer living in quarters where they can walk out onto a balcony, smell the tropical sea breezes, and watch dawn break over the ocean. The trade-off will be between snug little apartments with an outside view and more spacious accommodations in the interior. There will be combinations of both amenities available in greater or lesser proportions. In addition to apartments in the surface structure, pavilions of various sizes and types will be available in the tower gardens and on the outer breakwater.

For each person, Aquarius will be comfortable, convenient, economical, and interesting. As residents of a super-organism, every colonist's physical needs are supplied by the colony. Food, shelter, clothing, medical care, education, recreation, privacy, and meaningful work are all available. Personal needs, on social, spiritual, emotional, and sexual planes may be fulfilled by individuals, as them deem appropriate.

The colony does, however, provide a social atmosphere in which these needs can easily be met. In modern societies, the individual is, except perhaps for the family unit, completely on his own. In a Millennial colony, by contrast, the individual is a part of an integrated whole. His individuality remains intact, and may even be amplified by his identity within the colony. The difference being, that within the colony the individual is not condemned to the lone struggle for survival.

In the colony environment, it is easy to make friends. It is especially easy for people to aggregate themselves together into sub-groups bound by ties of special interest. Similarly, it is easy to find companionship on all levels, intellectual, spiritual, social, or sexual. The guarded suspicion that characterizes interpersonal attitudes in societies contaminated by crime and brutality will, in Millennial colonies, be as absent as the behaviors which engender it. Colonialism allows the individual to retain his essential identity intact without having to encase himself in an impenetrable shell of defensive social armor.

**Garden Spot**

Aquarius will be located off the East coast of Africa, near the equator for three main reasons: First, the equatorial zone provides warm water to power our OTECs, and has a benign climate, free of strong winds, waves, and storms. Second, proximity to Africa puts us within easy reach of primary markets for food, water and energy. And finally, this location will place Aquarius about mid-way between our colony of New Eden in the Seychelles (see Chapter 8) and our orbital launch facility on Mt. Kilimanjaro (see Chapter 2).

The area of ocean 5° on either side of the equator, is known as the 'doldrums'. As the name implies, nothing much ever happens there. The average temperature varies little, seldom climbing over 80° or falling below 75°. Rainfall is plentiful, about 40 inches a year, but so is sunshine, over 2000 hours a year. The whole region is meteorologically a dead bore—not such a bad thing if one big storm might sink an entire city.

When one contemplates life on a floating city, one might justifiably worry about what would happen to such a structure in a hurricane. Aquarius will be designed to withstand a force 12 storm (Beaufort Scale). The moorings will be strong enough to hold in 75+ mph winds and currents of 2.5 knots. The breakwater platforms will be built to take waves 50 feet high. In such a storm, the cold-water containment membrane would probably be destroyed. The outer breakwater, however, would certainly remain intact, dissipating most wave energy well away from the surface structure. If extreme waves did reach the inner structure, they would be unable to penetrate beyond the outermost ring of modules. These modules stand 10 feet out of the water and are each protected by a beach. The beach will force any wave to break before hitting the outer wall. At worst the broken waves, might cover the entire outer module to depth. But the waves couldn't propagate any further, and would be entirely dissipated by the time they reached the second ring, a hundred yards away. The second ring of modules is 20 feet high, and presents an absolute limit, beyond which no waves could go. So in an absolute worst case disaster, the colony would only sustain peripheral damage. Though it will be built to withstand them, Aquarius will never face such storm conditions.[158]

Winds along the equator are characteristically light, waves are correspondingly small, and surface currents are tame. In this region, winds remain under 'force four' up to 40% of the time. A force four wind on the Beaufort Scale is a 'moderate breeze,' generating waves with an average height of five feet. Near the Seychelles islands, the calculated storm frequency for a force eight blow is from 0 to .1.[159] This means that in a typical decade there will be only one such storm. A force eight storm on the Beaufort scale is not something the Ancient Mariner would bother including in his Rimes. A storm of force eight magnitude, characterized as a 'fresh gale,' will generate winds of 39-46 mph and waves up to 25 feet high. The probability of a force 12 storm along the equator, with winds over 75 mph, and waves over 50 feet high, is essentially zero.

Weather conditions along the equator are mild for a couple of reasons: Global circulation patterns tend to cancel each other out at the equator. The great currents of air that form the trade winds circulate in opposite directions in the northern and southern hemispheres. Because of this, there persists a 'zone of convergence' where the northern and southern trades push air into the same region. Another reason is that air, warmed by the relatively hot equatorial water, tends to rise, thereby creating a low pressure zone along the equator. This zone presents an atmospheric energy valley that is barometrically broad and flat. This lack of pressure topography creates a region of great atmospheric stability.[160]

The best thing about the equatorial zone is that the chance of encountering hurricane conditions within 5° of the equator is completely nil. Cyclonic storms, like hurricanes, are formed and propagated by the 'Coriolis effect', which has no force near the equator.

Named for the French physicist who first explained it, the Coriolis effect is the product of the apparent motion of winds as the earth turns under them. The effect can be demonstrated by standing on a moving merry-go-round and throwing a ball across it. To someone standing on the ground, the ball will appear to go straight, but to the person on the turning platform, the ball will appear to follow a curved path. In the same way, winds around low pressure zones appear to twist to the right, clockwise, in the northern hemisphere and to the left, counter-clockwise, in the southern. If you don't believe me, go drain the tub. The water will swirl clockwise if you are in Winnipeg, counter-clockwise if you are in Brisbane. At the equator, the water will just flush down the pipe, and won't swirl either way. The Coriolis effect is most pronounced near the Earth's poles and disappears at the equator.

Hurricanes and typhoons are formed when a low pressure area develops due to warm air rising over warm water. As air rushes into the resultant zone of low pressure it curves in accordance with the Coriolis effect. If the swirl of air around the low-pressure 'eye' catches up with itself, it can form a self-propagating cell that grows more powerful. As the winds approach the eye, they move at the same 'real' velocity. But the fact that they have to move around a smaller and smaller circle, causes them to move at a faster and faster relative velocity—just like a spinning skater pulling in her arms. Within 5° of the equator, the Coriolis effect is too slight to enable hurricanes or other cyclonic storms to form.[161] This means that the doldrums are not only free of normally high winds, but are free also of hurricanes, typhoons, cyclones, waterspouts, and even maelstroms.

**Autonomy**

An absolute requisite of all Millennial colonies is total political autonomy. If we are to colonize the stars, we must evolve a new society, adapted to the peculiar demands of the ultimate frontier. It will be impossible to develop such a social structure within the context of any pre-existing political system. If we are to fulfill our ultimate destiny, we can not be fettered by the unwitting blindness and stubborn intransigence of an entrenched bureaucracy.

True democracy is crucial to the concept of Colonialism. It is essential that the super-organism be free to express its communal will through pure democracy. If this is thwarted in any way, then cybergenesis cannot follow its natural path to a higher evolutionary plane. If we submit to any existing political jurisdiction, we will inevitably be subjugated by that authority. Evolving a new society under such constraints is like trying to grow a palm tree in a paper cup. It simply won't work. Cybergenesis must have free reign to evolve in harmony with the cosmic forces guiding Life's evolution. Aquarius must therefore be free from political coercion.

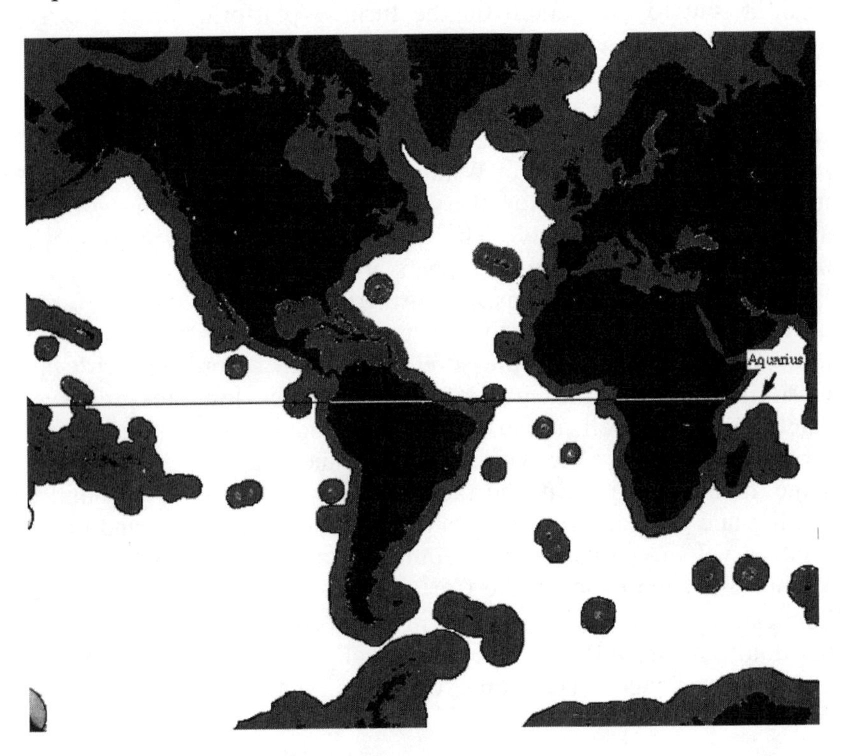


***Fig. 1.16 - Map of international waters.*** *Land masses are black, territorial waters are gray; the white areas are virgin territories, unclaimed by any sovereign power. These vast regions of the planet's surface are still available for colonization.* Reproduced from The Times Atlas of the Oceans, by permission of Von Nostrand Reinhold.

Apart from the frozen barrens of Antarctica, the sea remains the only open frontier on our crowded planet. On land the world is a closed shop. Every valley, every island, every place, is already under some political dominion. There is not an acre of ground anywhere on earth, over which some country does not assert its sovereignty. If we are to find free space before leaving Earth, then we must forsake the land and go to sea.

Article II of the Geneva Convention on the High Seas reads:

> "*No state may validly purport to subject any part of them (the high seas) to its sovereignty.*"[162]

This international statute bars any state from asserting its sovereignty over international waters. This same Convention guarantees powerful maritime nations the right to pillage the ocean, so it is one international law certain to be upheld. The 1958 Geneva Convention on the High Seas specifically includes, as one of the "Freedoms of the High Seas", the right to construct artificial islands.[163] Accordingly, Aquarius will be positioned in international waters beyond the clutches of any existing government.

The location of Aquarius is determined in part by the political territories of Somalia, and Kenya. These countries claim a territorial limit of 3 to 55 miles, and an exclusive fishing zone out to 100 miles.[164] In addition to these political perimeters, they also enjoy an "exclusive economic zone" out to 200 miles. These countries would certainly be within their rights under the international law of the sea to demand sovereignty over any floating island that came within their 200 mile zone. For example, the United States, and other maritime nations, consider themselves free to board and search any vessel within 200 miles of their shores. Essentially, the 200 mile zone marks the outer limits of a nation's sovereign and police powers. Obviously, we are not going to evolve a new society, dedicated to the freedom of the individual, within jurisdiction of any nation's police. Therefore, the colony must be located at least 200 miles from the African coast.

The other factor determining Aquarius' exact location is the necessity to anchor her to the bottom. Eventually, as we perfect the technology of growing sea colonies, we will learn to anchor in very deep water. Ultimately we may even build unanchored colonies; these Laputian cities of the sea could move freely from place to place propelled by the currents and the power of their OTECs.

Aquarius, and her early sisters, however, will be firmly anchored to the sea bed.

Therefore, it is desirable to find the easiest anchorage that also meets all the other siting criteria. As in most things Millennial, we find that we have been munificently provided with an ideal location. A few hundred miles due north of the Seychelles, sitting astride the equator, are the Coco-de-Mer seamounts. These undersea mountains rise to within 3500 feet of the surface. There are seven or eight submarine peaks in the Coco-de-Mer range, all within a couple of hundred miles of each other. This cluster of underwater mountains will provide bases for the first group of Aquarius' sister cities.

**Arkquarius**

Every space colony is an ark. Aquarius will serve not only as the jumping off point for the colonization of space, but also as a reserve ark for the potential recolonization of Earth.

Life must be preserved; we, the caretakers of Life, must survive to preserve it. Our planet is threatened by a variety of calamities. Some are developing as slowly as an eroding atmosphere; others could come as suddenly as a nuclear blast. In any case, if catastrophe does overtake us, there must be some seed of surviving human culture from which to rebuild civilization. Without man, or some other intelligent tool user, Life will be condemned to remain bound to this single tiny planet—perhaps forever. If Life is to survive then we too must survive.

Aquarius can fulfill the role of planetary ark admirably. Sheltered in the warm equatorial waters, our floating marine colony can survive almost any conceivable disaster, including nuclear war. The warm waters of the tropical oceans will give up only a fraction of their heat even if the rest of the planet is plunged into nuclear winter for months. Similarly, the catastrophic climatic effects of nuclear winter that will ravage the mid-latitudes with colossal hurricanes, typhoons, and tornadoes, will leave the stable equatorial belt more or less unaffected. Aquarius is self-sufficient in energy, food, and most other things. While the rest of humanity attempts to cope with a shattered world: unable to raise crops, perhaps for years; ravaged by social and political anarchy; decimated by plagues, pestilence, floods and droughts; and slowly poisoned by radiation in the food chain; the people of Aquarius will be relatively unharmed.

The social dissolution following nuclear war is likely to be as destructive to life and civilization as the war itself. No trace of social order is likely to survive in the countries actually blasted by the bombs. Massive upheavals will shred the already tattered social fabric even in countries left unmolested. By the time radiation, starvation, riots, poverty and disease have reaped their grim harvest, there will be only isolated pockets of survivors clinging to life here and there around the world. At that point, it will be up to the scientists, technicians, artists, poets, and philosophers of Aquarius—just as in the legends of Atlantis—to recolonize this planet.

**Cost of Living**

One of the great advantages of all space colonies is that, once established, they can provide life support to their inhabitants at a minuscule cost. All space colonies must be largely self-sufficient. In outer space, they must also be completely closed, recycling everything but energy. Aquarius will not be a closed-loop system, but it will, nonetheless, provide its inhabitants with many of the economic advantages of other space colonies.

Assume that an average Aquarian space colonist has the earning power of an American in the second highest income quintile for 1985—$31,000.[165] Of this amount, the American spends over 90% on basic life support.

**Table 1.13**
**Annual Expenditures for American Making $30,967 a Year**

| Item | % | Amount |
|---|---|---|
| **Shelter** (Including $1762 for supplies, & furnishings)[166] | 20 | $6,078 |
| **Transportation** (Including $1,314 for gas and oil) | 18 | 5,632 |
| **Food** (Including $549 for alcohol & tobacco) | 14 | 4,365 |
| **Life Insurance** (Including Social Security & retirement) | 11 | 2,909 |
| **Personal Taxes** | 9 | 2,789 |
| **Utilities** (Including fuel and public services) | 6 | 1,836 |
| **Health Care** (Including $455 personal care products) | 5 | 1,536 |
| **Entertainment** (Including $476 for TV, radios & sound equip.) | 5 | 1,560 |
| **Clothing and Services** | 4 | 1,300 |
| TOTAL | **92** | **$28,005** |

In Aquarius, all of the things in the table above, or substitutes, are provided for the colonist. The structure provides his shelter, and furnishings. Transportation costs within the colony are virtually zero. Food comes from the sea, as do power, light, and even clothing. Entertainment is provided mainly by the data utility and the other colonists. Health care is provided in the medical center. Taxes are, of course, irrelevant. Therefore, the resident of Aquarius needs to make none of these expenditures.

Colonialism, as used here, is a term which must not be confused with 19th Century 'imperialism'. In terms of the Millennial Project, 'colonialism' is meant in its macro-organic sense. Just as cells in the primordial sea congregated together to form 'colony' organisms, so shall we individual humans congregate to form the colonial macro-organisms of Aquarius and the other space colonies.

Understanding the colonial system requires a conceptual leap. Aquarius' social/economic structure must be every bit as novel as its physical structure. Just as Aquarius demands new approaches to its growth and sustenance, she demands new approaches to the organization of the community inhabiting her. Aquarius is a super-organism, and as such she provides for the needs of her people, just as we provide for the needs of our own cells. The social organization of the colony will reflect this organic reality.

In economic terms, life in a super-organism means never having to pay the rent. It is hard to comprehend how 100,000 people can live in style and comfort with all of their basic needs fully met, and yet not be called upon to spend a penny in the process. It is impossible to fully understand 'colonialism' in terms of any of the old economic paradigms, but they can help to illustrate it.

Colonialism is not capitalism, but it can be at least partly understood in those terms. Imagine that Aquarius is a corporation. Now say this corporation has a monopoly on providing all of the basic goods and services to 100,000 people. (Instinctive capitalists will salivate at the prospect.) If we assume that the resident in Table 1.13 is average, and that 60% of the population works, then the corporation can anticipate a gross income of close to 1.7 billion dollars a year.[167] The corporation would be in the position of selling every meal, every item of clothing, renting every apartment, providing all transportation and all the multifarious other services of a city. The corporation is without competition. The costs of providing these goods and services in Aquarius amount only to the maintenance costs of the OTECs and other infrastructure, and some labor. The potential for profits would seem mind boggling—

except for one thing. Again this is only by way of illustration, so it is an incomplete picture, but one can at least get a feel for colonialism if one makes the final assumption that the corporation of Aquarius is wholly and exclusively owned by the people who live there. Once this final assumption is made, the scenario will have folded back on itself and colonialism can be seen in its organic context. Obviously, if the residents of Aquarius own the corporation, their main interest will be in minimizing the costs of goods and services to themselves. So the 'profit' really takes the form of reducing each colonists 'cost of living' to an absolute minimum.

Here the capitalist paradigm fails us, even as a means of illustration. If we followed that paradigm, the next step in our thinking would be to conclude that by minimizing consumption, we could also minimize the necessity for production. This would enable the inhabitants of Aquarius to support themselves luxuriously while working only a couple of hours a day. This results in the vision of an idyllic floating paradise, where the inhabitants have little to do but loll by the limpid waters of the lagoon, sipping spirulina coladas.

Such an outcome is entirely possible, and some marine colonies may eventually adopt it as a way of life—it has a certain appeal. It is not, however, the vision of the Millennial Project. Rather than using the economic slack Aquarius provides us to kick back and coast, we will instead use the bounty thus created to fuel our ascent to the stars.

### The Bottom Line

Table 1.14 below, details the income streams flowing into Aquarius:

**Table 1.14**
**Aquarian Revenue Streams**

| Source of Revenue | Income (mm$) |
|---|---|
| Mariculture Products | $6,231 |
| Tourism[168] | 1,000 |
| Services[169] | 450 |
| Magnesium | 260 |
| Hydrogen | 37 |
| Distilled Water | 64 |
| **TOTAL REVENUES** | **$8,042** |

Outside cash expenses for the colony will be only a fraction of the gross revenue stream. The colony is completely self-sufficient in energy, food, clothing, shelter, transportation, medical care, and virtually all other internal services. The only requisite outside expenditures will be for goods and machinery that the colony cannot or does not produce itself. These should amount to no more than 10-15% of the total income of the colony, leaving a surplus balance of from $4 to 5 billion a year.[170] This tremendous surplus revenue will allow us to undertake the next great stride toward fulfilling our Cosmic destiny—the leap to space.

# CHAPTER 2

# BIFROST

***Fire all of your guns at once,***
***and explode into space.***
**Steppenwolf**

**Bifrost Bridge** - Norsemen called the rainbow "Bifrost"—the bridge linking Earth and Heaven. The Valkyries bore the souls of worthy warriors on their winged steeds, and carried them up Bifrost to paradise in Asgard.

The ultimate purpose of this Project is to spark a human migration into space. Aquarius will buy us enough time and resources to make this possible. To actually get mankind off the planet, we will need to build a highway to heaven that is simple, reliable, and cheap.

We will build our own version of the mythic Bifrost Bridge. The Bridge's design is extremely simple: A launch capsule is accelerated to high speed inside a vacuum tube by means of electromagnetic levitation. When the capsule emerges from the end of the launch tube, atop a high equatorial mountain, it is propelled on into orbit by an array of lasers. The Bridge will be built with revenues and powered with electricity from Aquarius and her sister colonies. Once constructed, the Bifrost Bridge will provide a path into space broad and smooth enough to make large-scale space colonization feasible.

**Retro Rockets**

Thus far in the 'conquest of space', mankind has always ridden skyward on chemical rockets. These chariots of fire will not, however, suffice in the New Millennium. Rockets are too complicated, too expensive and too apt to blow up. A 'rocket' might loosely be defined as a bomb with a hole poked in one end. Rockets are about as safe to straddle as any other bomb. Rockets are almost completely reliable; they can usually be counted on to explode—whether the explosion turns out as planned, however, is more problematic. In the case of the unfortunate Space Shuttle, the probability of an uncontrolled explosion turned out to be about the same as losing a game of Russian Roulette played with four or five revolvers. As for being cheap, going into space aboard a modern rocket is like throwing away a gold Rolex™ every time you check your watch.

Rockets are not a wholly bad idea. They do provide a way to achieve high velocities and generate thrust in a vacuum. Eventually, we will use fusion rockets to open up the solar system, and ultimately anti-matter rockets will enable us to colonize the galaxy. But primitive chemical rockets, like those in use today, have too many short-comings to provide us with an open highway to space.

The trouble with a rocket is that it has to carry its fuel with it, and the energy available in chemical fuels is strictly limited. A rocket needs fuel to lift its payload, of course, but it also needs fuel to lift that fuel. This necessitates carrying fuel to lift the fuel to lift the fuel, etc. The result is a vehicle consisting mostly of fuel and the tanks to hold it.

The mighty Saturn V of NASA's glory days is a good example. This enormous rocket—36 stories high—weighed 3,212 tons. It was capable of placing 120 tons in Earth orbit or sending 45 tons to the Moon. Thus it took 3,092 tons of rocket to put 120 tons of payload in orbit—a ratio of 25 to one. Putting it another way, 96% of the mass of the vehicle is the fuel and fuel tank. Before it had even traveled its own length, the Saturn V burned a greater weight of fuel than the total weight of the payload it would deliver to the Moon.[171]

Profligacy on this scale is expensive. To put a ton of payload into orbit with a rocket requires about 20 tons of fuel. In addition to the fuel, huge tanks, built to aeronautical standards are required. Even in reusable systems like the Space Shuttle, the fuel tank is discarded after a single use.

The Space Shuttle was supposed to provide a cheap, reliable, and safe way to space, but it has become an orphan of obsolete parent technologies. The Shuttle has all the limitations of a rocket with none of a rocket's versatility; it has all the complexity of an airplane, with none of a plane's autonomy. The going rate for orbital payload on the Space Shuttle is $8800 per kilogram and rising.[172] A ticket to ride on the Shuttle would cost you a cool half million dollars. Since a tube of toothpaste will run you an extra $800, you should plan to travel light.

Traveling at all, presumes you could book a seat in the first place—don't hold your breath. Since the *Challenger* disaster, NASA officials talk about civilians in space "within 20 years." After they blow up the next one, they won't talk about it at all.

NASA's proposed alternatives are not much better. There are designs floating around for BDRs (Big Dumb Rockets). Optimists project BDRs could get the cost of an orbital flight down to $1000 a pound.[173] At that price, a ticket to space will only set you back a couple hundred thousand dollars. Even this is wishful thinking. The BDRs will not be designed for human cargo. There is talk now of an orbital space plane, which may offer some real hope for the future. It is a limited solution at best, however. Space-planes will not have the heavy lift capability needed to establish and support space colonies.[174]

Ever since Goddard and his German disciple Wernher Von Braun came to dominate human thinking about space travel, we have been locked into a rocket-based consciousness. We have ridden to the Moon on the shoulders of these giants, but to claim space as our own, we must transcend even these colossi. We must change our approach. To do that, we must first change how we think about space and how to get there.

**I Feel the Need for Speed**

Getting into orbit is not a question of altitude; it is a question of speed.

There are people around who still believe that the Earth is flat. They argue that it couldn't possibly be round; if it was we'd fall off the edge. In a way, these people are right. Stand out in the middle of a prairie and look off at the horizon, eight kilometers away.[175] You are, in effect, looking downhill. Because the Earth is curved, the ground at the horizon is five meters lower than the ground where you are standing.[176] If you could transport yourself out to the horizon instantly, you would find yourself five meters off the ground. The reason you don't fall off the curved Earth, is that it takes you only 1 second to free fall five meters. If it takes you over an hour to hike out to the horizon, you will have worked your way downhill so gradually that the ground seems flat. To fall off the Earth, you must get to the horizon in less than a second.

Imagine being fired out of a cannon five meters above the ground. If the cannon propelled you fast enough to reach the horizon in one second you would still be at a height of five meters. In the second you were hurtling toward the horizon, gravity would have pulled you down in free-fall five meters, but the Earth's surface would have curved away five meters beneath you.[177] After another second, if you continued flying at the same speed, and the ground remained perfectly flat, you would be another eight kilometers down range, but still five meters off the ground. If you kept up your speed, and didn't run into any obstacles, you could continue to fall around the Earth this way. You would, quite simply, have achieved orbit—albeit at an altitude of only five meters. As long as you move eight kilometers horizontally every second, you will maintain the same altitude with respect to the Earth's curved surface. Eight kilometers per second (kps) is—not coincidentally—orbital velocity. Getting into orbit, therefore, is simply a matter of moving fast enough to fall off the edge of the world.

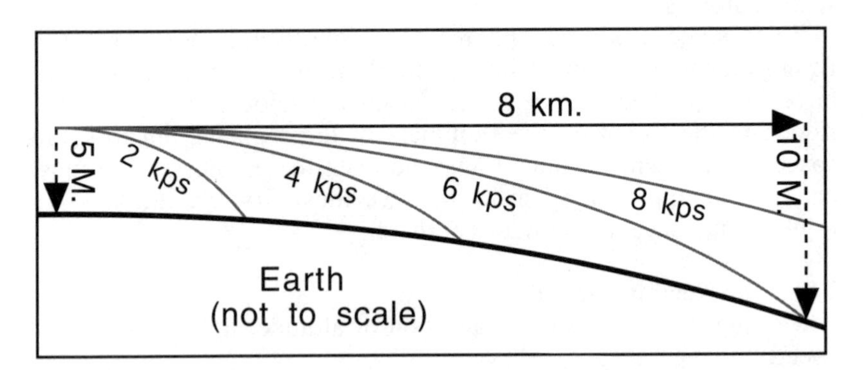


***Fig. 2.1- Curved horizon.*** *You can indeed fall off the edge of the world, but to do so, it is necessary to get over the edge faster than you would fall the same distance.*

Once one realizes that orbit is more a question of speed than of altitude, one is free to look at alternatives to rockets. What we need is a device that can achieve the same speed as a rocket, but without the rocket's disadvantages. The ideal solution would be a system in which the fuel stayed on the ground and only the payload moved. A system like this would enable us to pump all the energy into the payload, rather than wasting 95% of the energy on the useless acceleration of rocket fuel.

**Cannon-Ball Express**

What kind of vehicle doesn't carry any fuel? An electric locomotive is one answer that whistles through the mind. Instead of dragging around a car-load of coal, the electric engine gets its power through the rails.

We can use the basic principal of the electric locomotive for space travel, but we need to modify the vehicle's design a bit. Wheels and tracks would disintegrate at the required velocities. Our locomotive requires a frictionless means of suspending the vehicle above the rails. Superconducting electro-magnets can do the job. The image of a superconductor hovering magically over a smoking dish of liquid nitrogen is an icon of technological promise.

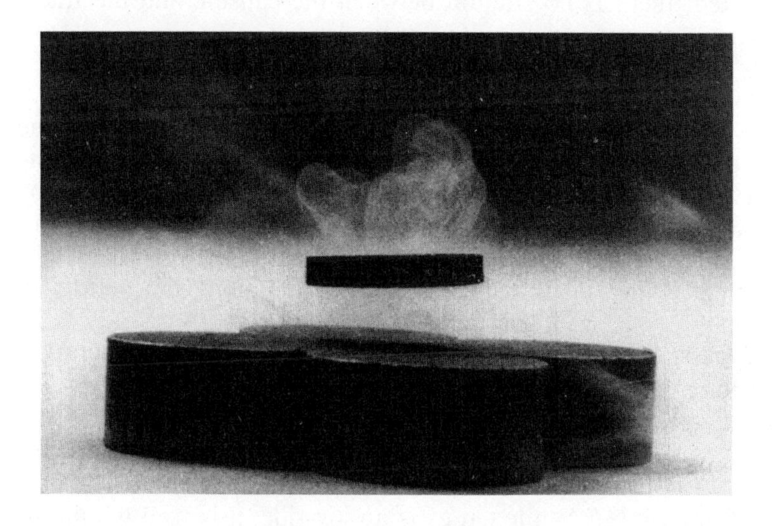

***Fig. 2.2 - Levitating superconductor.*** *A pellet of superconductive material, cooled to -196° C. (77° K.) by liquid nitrogen, levitates above an array of magnets.* Photo by Argonne National Laboratory.

Super-conductors levitate because they repel lines of magnetic force. This property will be harnessed to create a magnetic cushion on which a launch vehicle can ride at high speed without friction.

The use of magnetic levitation allows for the elimination of the motor, as well as the fuel. The same magnetic force fields which support the vehicle can also propel it. The payload capsule rides a wave of magnetic force along the superconducting track. Electromagnets embedded in the rails ahead of the capsule attract it, while magnets behind repel it. The two forces, acting in synchrony, will accelerate the payload to high velocity.

For raw speed, few machines can compete with an electromagnetic accelerator. Similar devices are used to accelerate atomic particles, and have achieved velocities close to the speed of light. Bifrost's electromagnetic launcher will be accelerating much larger masses to much smaller velocities, but the basic principal is the same. When used to motivate things larger than protons, these devices are called 'mass drivers'.

To eliminate air resistance, the electromagnetic track will be enclosed in an evacuated tube. All these features will combine to make the mass driver incredibly efficient. No fuel need be accelerated; there is no friction between the capsule and the track; and there is no atmospheric drag in the launch tube.

**Ridin' the *g* Train**

The launch tube will run horizontally inside an underground tunnel for most of its length, turning up at the end and terminating atop a tall mountain. The tube will be constructed to withstand external pressures in excess of 15 pounds per square inch. This, and high-speed airlocks, will allow a permanent vacuum to be maintained. The superconducting track, runs along the bottom of the tube.

The passenger launch tube will initially be 125 kilometers (78 miles) long, and will later be extended to 250 km.[178] At first, all those going into orbit will be specially trained personnel with a high degree of physical fitness. These space cadets will ride the launch capsule at an acceleration of just over ten *g*s—ten times the force of gravity.[179] While ten *g*s is no joy-ride, it is well within the envelope of human tolerance. Untrained individuals, without *g*-suits, are easily able to tolerate 10 *g*s for more than two minutes.[180] During this time, the subjects are able to see clearly, respond to commands, remain mentally alert, and retain unimpaired mobility of their hands and feet. Astronauts in acceleration couches have taken 30 *g*s without any damage. Football players routinely experience deceleration impacts of two-thousand *g*s for fractions of a second. A trained space cadet should be able to tolerate ten or more *g*s for up to seven minutes in a conventional acceleration couch. In a *g*-tank, in which the traveler would be immersed in water, accelerations of nine *g*s could be tolerated for almost an hour.[181] Those riding the Bifrost mass-driver will be subjected to just over ten *g*s for 50 seconds, plus a single bone-jarring jolt of 255 *g*s, lasting for two seconds.[182] It is recommended that passengers remove their dentures prior to boarding the space craft.

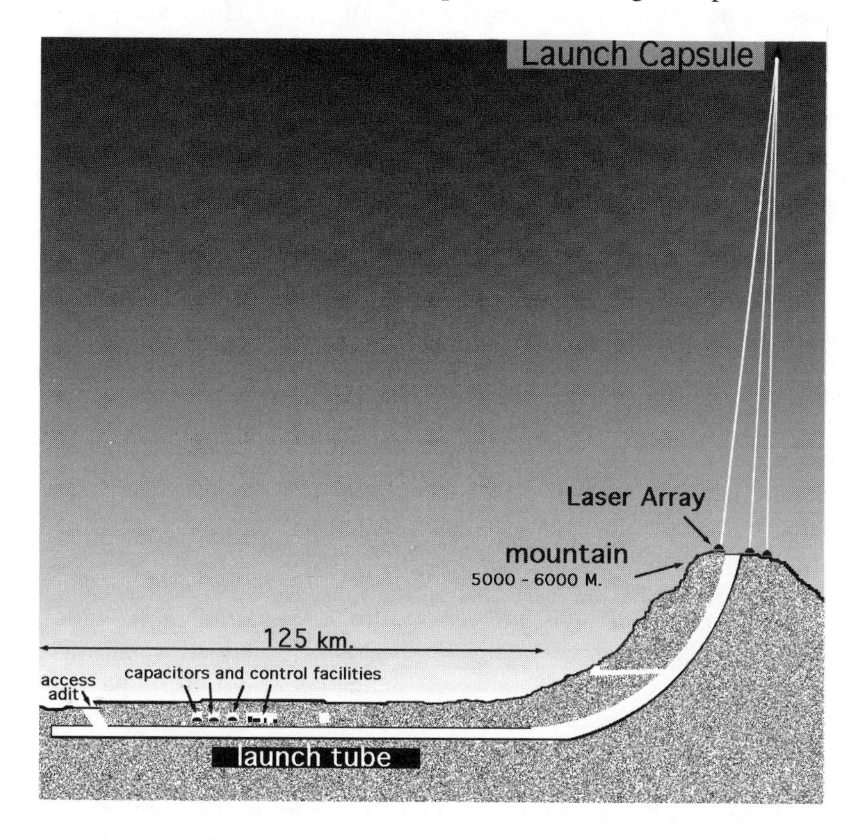


***Fig. 2.3 - Bifrost Bridge Profile*** (not to scale). *The launch tube tunnel is bored through the rock at sufficient depth to remove it from the ecozone. Energy storage and support facilities are also built underground. The tunnel runs longitudinally for most of its length and then rises in a gradual curve through the interior of a mountain. A launch capsule is accelerated through the tube to a velocity of five kps. When the capsule emerges from the tube at an altitude of 5000 - 6000 meters, it is already above half the atmosphere. Beams from a laser array then propel the launch vehicle up out of the atmosphere, accelerating it to orbital velocity.*

Later, the launch tube will be extended, and the radius of curvature widened to provide a more comfortable ride. In a hydrodynamic *g*-couch, wearing an over-pressure suit, even a relatively frail person will be able to ride the longer mass launcher.

We can begin to use the launch tube long before it is complete. At the maximum practical acceleration for human beings—10+*g*s—the launch tube must be 125 km. long. For inert cargo, however, the tube can be much shorter.

Colonizing space requires huge tonnages of mundane materials like fiberglass, aluminum, tools, water, etc. These commodities are virtually immune to the forces of acceleration, and can be blasted into space at very high *g*-loads. Therefore, the upper 45 km. of track will be constructed first. This section of the tube will run up the inside of a mountain to the muzzle. The linear induction motors in this section of the track will be able to accommodate more voltage and so handle higher magnetic thrust than the balance of the launcher. Launch capsules will be accelerated up this 45 km. track at accelerations of up to 30 *g*s. This will send the capsules blasting out of the launch tube muzzle at 5000 meters per second. Even before the lasers are installed, cargo can be sent into orbit at a tremendous savings over rocket launches. After the lasers are operational, this section of the track can still be used for cargo launches.

**Launch Energy**

It requires surprisingly little energy to accelerate mass to orbital velocity. If the launch system were 100% efficient, it would take only nine kwh of electric power to launch a kilogram of payload into orbit. Of course, no such system can be 100% efficient; power is lost to friction with the air, deceleration by gravity, transmission losses, and other inefficiencies. Even with these losses, a kilogram of payload can be sent up the Bifrost Bridge to orbit for less than a dollar's worth of electricity.

One of the great advantages of the magnetic launch tube is its immunity to 'gravity drag'. While a rocket is in vertical flight, no matter what its means of propulsion, it is under the influence of the Earth's gravity, and is therefore under a constant downward acceleration. For every minute a rocket spends in vertical flight, it will lose 588 M/sec. of velocity. Without the launch tube, the capsule and its payload would lose 1000 M/sec to gravity drag. While it is inside the launch tube, the capsule accelerates in a mostly horizontal direction. Consequently, the energy saved during this phase of the launch amounts to 20%.

At a launch weight of 14,000 kg. and an acceleration of 10.2*g*, the launch tube will require a four gigawatt power supply. To accelerate the launch capsule to 5000 meters per second will require 60,000 kilowatt hours of electricity.[183] This is a lot of power, but the OTECs on Aquarius have power to spare. Aquarius has a net power output of 300 megawatts. Enough power can be supplied from this inexhaustible source for a launch every twelve minutes.[184]

An electric power line, run directly from Aquarius, will supply energy to the launch facility.[185] Generating four gigawatts of power necessitates storing electricity in large underground capacitors. Designs for massive superconducting capacitors already exist in connection with the United States' Star Wars system. A full-sized storage ring, able to power missile defense lasers, would be 2.4 km. in diameter, and could store 5 million kilowatt hours.[186] Happily, the energies required for launching spacecraft are more modest than those required to shoot them down. Bifrost's mass launcher will need only one or two relatively small ring capacitors, each able to store 50,000 kwh. The laser array will require the storage of an additional 230,000 kwh.

The storage density of capacitors has been advancing rapidly. Since 1985, storage density has increased by a scale factor (ten times). At present, one kwh can be stored in a capacitor with a volume of six cubic meters.[187] It is a reasonably safe assumption that capacitor technology will allow a further scale factor improvement in storage between now and the time we build the Bifrost Bridge. If so, ring capacitors with a volume of 182,000 cubic meters will be required to store the 300,000 kwh needed to launch a 14 ton capsule. A 50,000 kwh capacitor will require a superconducting cable just over six meters in diameter running around an underground containment ring a kilometer in circumference. Six such rings will be able to store all the power needed for a launch.

Mass driver payload is carried in a ballistic capsule which rides a sled equipped with superconducting magnets. It is the sled which is actually accelerated, taking the capsule along for the ride. After the sled leaves the launch tube, it separates from the capsule and parachutes back to earth, where it is recovered.

The Bridge's mass driver will launch a five ton payload of human passengers or cargo. The passenger capsule can carry a crew of six and 25 passengers with limited cargo space. The payload bay will be about 8 meters long. The capsule itself will weigh about five tons, or 625 kg. per meter.[188] Passengers and or cargo will weigh five tons. Gross weight of the fully loaded capsule will be 14 metric tons, including laser propellant.

**Laser Propulsion**

Capsules will be launched into orbit in two phases. In the first phase, the capsule is accelerated to five kilometers per second by the mass driver. In the second phase, the capsule is propelled out of the atmosphere and into orbit by an array of ground-based lasers. The lasers provide the energy needed to overcome atmospheric friction, and accelerate the capsule to eight kps—orbital velocity.

When the payload capsule emerges from the launch tube, it will be at an elevation of 6000 meters (20,000 feet), and moving at a speed of five kilometers per second. To punch through the atmosphere and attain orbital velocity, the capsule must continue to accelerate. Having left the magnetic impulse of the launch tube behind, the capsule must now be impelled by other means. On the second leg of its journey into space, the launch capsule will ride an array of laser beams.

Unlike a conventional rocket, which can be extremely complicated (see Appendix 2.1), a laser propelled rocket is the soul of simplicity: a beam of light and a block of ice. The ice, when heated by the laser, serves as the propellant. The laser strikes the ice block, and the water—super-heated to 10,000° C.—flashes to steam.[189] The superheated steam expands at 10,000 meters per second, propelling the space capsule with a specific impulse of 1000 seconds.[190] The beauty of this system, in addition to producing a high specific impulse, is that it has no moving parts.

Lasers, among their many other miraculous properties, enable us to beam energy across space. With a laser propelled rocket we can leave the fuel and engines on the ground where they belong.

This is not to say that a laser rocket doesn't need any propellant; it does, but only a fraction of what chemical rockets require. The thrust of a rocket is determined solely by the mass and velocity of its propellant. If the propellant is moving relatively slowly—as in the case of a chemical rocket—then it takes a huge mass of fuel to provide the desired thrust. If, on the other hand, the propellant is moving very rapidly, it takes a disproportionately smaller amount.

The speed of the propellant is a function of its temperature. In most chemical rockets, the temperatures available range from 1300 to 4000° C. In a rocket powered by the combustion of liquid hydrogen and oxygen (LHOX), temperatures reach 2500° C.[191] At that temperature, the exhaust gas—water vapor—exits the rocket at 3500 meters per second.

The rocket thrust, divided by the mass of fuel used each second, determines the rockets 'specific impulse' ($I_{sp}$). Specific impulse is a measure of a rocket's performance and is expressed in seconds of time. The specific impulse of a LHOX rocket is 357 seconds at sea level.[192] A single kilogram of LHOX fuel can exert one kg. of thrust for 357 seconds, or 100 kg. of thrust for 3.57 seconds. For example: if the specific impulse of a high performance LHOX rocket is 400 seconds, and we want to put 5 tons of space capsule and 5 tons of payload into orbit, it will require 60 tons of liquid hydrogen and oxygen.

Increasing the specific impulse dramatically lowers the fuel requirement. If our propellant generated a specific impulse of 1000 seconds, we would need to carry only four tons of propellant in order to orbit the same 10 tons of capsule and payload.

To achieve a specific impulse of 1000 seconds requires a temperature of more than 10,000° C. Such temperatures are not attainable through most chemical reactions, and are difficult to contain in nuclear reactions. It is, however, a simple matter to generate such temperatures with lasers. Laser propulsion can generate thrust three to seven times as efficiently as chemical rockets. The exhaust velocity of a laser driven propellant can be over 10,000 meters per second.[193] The specific impulse of a propellant at this velocity is 1000 seconds. Laser induced specific impulses as high as 2000 seconds are theoretically possible.[194]

Any material that will vaporize at temperatures over 10,000 degrees will serve as a propellant.[195] Since there aren't many materials that won't vaporize at such temperatures, our choice of propellants is virtually unlimited. An ingot of iron would make a suitable—though less than ideal—propellant. The ideal propellant will be cheap, easy to handle, readily available, environmentally safe, and of low molecular weight.[196] Water, in the form of ice, meets all these criteria.[197]

A four ton slab of ice, 40 centimeters (16 in.) thick, will be mounted on the after end of the launch capsule.[198] The ice will be completely vaporized during the two minute burn of the laser launch phase. Stresses on the ice amount to 900 atmospheres, or 930 kilograms per square centimeter.[199] The ice will have to be made about half as strong as vitrified brick to withstand such pressures. This can be done by super-cooling the ice to liquid nitrogen temperatures, and reinforcing the ice with a honeycomb of tough plastic.[200]

At 40% efficiency, a laser will produce 100 Newtons of thrust per megawatt of laser power.[201] (A Newton is the force required to accelerate one kg. at one meter per second per second.) When the capsule emerges from the launch tube, it weighs 14 tons. At burnout it will weigh ten tons. The rate of propellant burn is constant, so its average weight will be 12 tons. Accelerating 12 tons at 1.25*g* will require one and a half gigawatts of laser power.

Rather than having one gigantic laser, we will build six smaller lasers, each producing 250 megawatts of power. This will minimize scale-up problems and reduce certain effects like Ramman scattering and thermal blooming, which are associated with lasers in the gigawatt range.[202]

Generating enough laser induced thrust to launch capsules into orbit requires 160 megawatt hours of power—three times that required to accelerate the capsule through the launch tube. This is because all forms of rocketry are inherently inefficient. In the launch tube, virtually every kilowatt of power can be converted directly into motion, since there is no friction or electrical resistance. In rocketry, however, not all the available energy goes into the accelerated exhaust gas. Some of the energy is invariably wasted as heat, turbulence, and noise. In laser propelled rockets, efficiencies as high as 90 to 100% are theoretically possible. We will use the assumption here though that only 40% of the energy in the laser beam is converted into thrust.[203]

**Free Electron Lasers**

The Bifrost Bridge will require massive lasers, blazing awesome streams of power. Producing such lasers will be challenging, but the task is well within our technological reach. The most promising technology for high power lasers is the Free Electron Laser (FEL).

FELS are a nascent technology. Not even the theoretical basis for such lasers existed until the mid '60s.[204] A free electron laser works by harnessing the enormous powers available in modern cyclotrons. A cyclotron is a closed magnetic ring in which charged nuclear particles, like electrons, are accelerated to terrific velocities. The cyclotron works very much like Bifrost's launch tube, but at an atomic scale. Electrons are suspended in a frictionless magnetic field and propelled to high speeds by waves of magnetic force. Cyclotrons can pump huge amounts of energy into a stream of electrons. In the case of a one gigawatt laser, the electrons will be pumped up to an energy level of 50 million electron volts (50 MeV).[205]

Energized electrons leave the cyclotron as a beam of particles moving at nearly the speed of light. The electron beam is then fed through a device called a "wiggler". Inside the wiggler is a series of closely spaced electromagnets which 'wiggle' the electron beam as it passes through. As the beam is bent back and forth by the magnets, the electrons are forced to change directions and so give up some of their energy. When an electron gives up energy, it does so by releasing photons of electromagnetic radiation—light. The magnets impart the same wiggle to all the electrons. Therefore, the photons released are all of exactly the same wavelength; hence the light is coherent—lased.

There are many elegant features that make FELs the certain choice for a laser propulsion system: Because the free electron laser gets its energy from the cyclotron, FELs can be scaled-up to produce almost any conceivable amount of power. An electron beam can store a theoretically infinite amount of energy, dependent only on the cyclotron's capacity to contain it. While electrons can't exceed Einstein's cosmic speed limit, there is no limit to the amount of energy they can absorb. Electrons, like any material bodies, will increase in mass as they are accelerated. The one gigawatt (GW) FEL is already on the horizon; the 10 GW FEL cannot be far behind; and the 100 GW FEL is an eventual certainty. (If you wonder what on earth, or more aptly, what off earth, a 100 gigawatt laser is good for, just stay tuned.)

The second great advantage of free electron lasers is their efficiency. In typical chemical or gas lasers, only a few percent of the total power input is converted into laser light.[206] For example, it takes 33 gigawatts of power to run a 1 GW carbon dioxide laser. By contrast, FELs can theoretically be made up to 99% efficient.[207] FELs have already been produced with efficiencies as high as 40%.[208] FELs are so different from other lasers because much of the stored energy in the electron beam can be recovered. In chemical and gas lasers all the waste energy must be dumped. In a FEL, however, the electron beam is sent back into the cyclotron after passing through the wiggler. This way, the beam's residual energy can be recycled instead of wasted. The energy savings of the free electron laser will therefore be enormous. For example, producing light at 5000 angstroms—a visible wavelength—requires accelerating the electron beam to an energy of several hundred million electron volts. Assuming we can achieve 90% net efficiency, a 1.5 GW laser operating for 245 seconds will require

120,000 kwh of electrical power—ten times less than an equivalent chemical laser would require.

The third great advantage of free electron lasers is that they are 'tunable'. They can be adjusted to produce light at any desired wavelength. Chemical lasers cannot be tuned because they produce light at only one set wavelength, which depends entirely on the energy levels of their lasing materials. Electrons in a FEL can be made to wiggle at any frequency, and so produce light at any wavelength.[209] This is accomplished by adjusting the spacing of the electromagnets in the wiggler, or by increasing the energy of the electron beam, or both.

The FEL's tunability will allow us to choose the optimum wavelength for the laser. The choice of wavelength is critical for three main reasons: ① The laser beam must carry a tremendous load of energy; ②the beam must remain tightly focused at long ranges; and ③ the energy must pass easily through the atmosphere. All three of these critical requirements are dependent on wavelength.

① The amount of energy carried by a laser beam, or any other form of electromagnetic radiation, is dependent on its wavelength. The shorter the wavelength of the radiation, the greater its energy. FEL beams of 5000 å (ångstroms, ten-billionths of a meter) have already been produced in a superconducting accelerator by circulating the electron beam through the accelerator twice.[210]

② During the laser burn, the launch capsule will travel 1600 km. down range. The ability to focus the beam on the propellant at this distance depends on the wavelength of the lased light and the diameter of the focusing mirror. The longer the wavelength of the light and the greater the distance to the target, the larger must be the mirror. Correspondingly, the shorter the wavelength of the lased light, the further away a spot can be focused. At a wavelength of 5,000 angstroms, and with a mirror 10 meters in diameter, a laser can easily remain focused on the launch capsule at a distance of more than 30,000 km.[211]

③ Of the radiation produced by the sun, the earth's atmosphere is most transparent to wavelengths in the visible portion of the spectrum—between 3800 and 7800 angstroms. We will tune our free electron lasers to produce light at a variety of wavelengths within the visible range. We will be firing six lasers, each tuned to a slightly different wavelength. Each laser will accordingly produce a visible beam of light of a different color. The longest wavelength, at 7500 å will be red, while the shortest wavelength, at 4000 å will be violet.[212] Other spectral colors, orange, yellow, green, and blue, will be produced by lasers of the appropriate wavelengths. When the colored beams combine at the focal point on the propellant block, they will form the white light most easily absorbed by ice. Using several beams at different wavelengths also balances the changing focal requirements as the capsule moves rapidly away. Distributing laser power across the visible spectrum is one way to deal with the engineering problems peculiar to laser propulsion. It also produces a launch system of singular poetic beauty, which is, perhaps, as important as any technological consideration. We will, in effect, be riding a rainbow into space; hence—the "Bifrost Bridge". (See Plate No. 5)

**Kind of a Drag**

Penetrating the atmosphere is one of any launch system's biggest problems. The seemingly thin and tenuous cell membrane of the living planet presents a tough barrier to high speed projectiles. While protecting us from most incoming meteors, the atmosphere also impedes outgoing spacecraft. The capsules emerging from the mass driver's launch tube will hit the atmosphere at 15 times the speed of sound. At that velocity, thin air is like a brick wall. The capsule has to punch through this wall, while gaining speed.

Obviously, we will want to ram our heads through as thin a wall as possible. The Earth's atmosphere is deceptively thick. 'Space' proper doesn't really begin below an altitude of 700 kilometers. Everything below space is atmosphere, so that makes Earth's gaseous rind about as thick as the distance from Washington D.C. to St. Louis, MO.

Fortunately, the aerodynamic resistance impeding the projectiles is not uniform through the entire depth of the air mass. Half of the atmosphere is squashed into the bottom 5600 meters (18,000 ft.). Atmospheric density falls exponentially with height, cut in half every 5.6 kilometers. At an altitude of 100 km. (60 mi.), atmospheric pressure is only a millionth of that at sea level. Above 100 km., atmospheric drag ceases to have much effect on the launch capsule.

Most atmospheric friction in hypersonic flight is caused by formation of a sonic shock wave in front of the vehicle, creating 'pressure drag'.[213] Parasitic drag and skin drag, caused by friction with air flowing over the body of the capsule, are relatively minor detriments. Compression of the shock wave will cause the nose of the capsule to become superheated to around 30,000° C. (54,000° F.). Such temperatures seem beyond the tolerance of any conceivable material—except perhaps the mythical "unobtanium". The hellish conditions can be handled, however, by ablating materials. Such materials were used for the heat shields of the Gemini and Apollo space capsules. Even the material science of the mid-'60s was able to cope with reentry temperatures comparable to those the Bifrost capsules will encounter. Ablating materials deal with heat by sacrificing themselves in carefully controlled layers. Energy is absorbed in large amounts when the ablating material vaporizes and carries heat away in the slip stream. As long as some ablating material remains, the structure behind it will be protected from the inferno. The nose and leading edges of the Bifrost launch capsules will be sheathed in a protective shell. Even a simple material like metallic tungsten would work, but it is very heavy.[214] It is likely though that the material scientists of the New Millennium will come up with some innovative recipes for lightweight and effective materials. The body of the pod will be insulated from heat in the boundary layer by a skin of reinforced carbon-carbon.

Carbon-carbon, the material used in the nose and leading edges of the space shuttle, can stand temperatures of up to 1650° C. (3000°F.).[215] The skin temperature of the capsule will climb as high as 1730°C. The amount of heat absorbed by the insulating layer will be small because of the short transit time through the atmosphere.

The launch tube will terminate atop a tall mountain, ideally at an altitude above 5600 meters. At this altitude, the capsule will emerge from the vacuum tube above the denser half of the atmosphere. The capsule's average velocity on the way out of the atmosphere is a blistering 6.5 kps—20 times faster than a speeding bullet. At this speed, the trip from Washington to St. Louis would take less than two minutes. Atmospheric stress on the capsule falls by three orders of magnitude (1000 times) within the first ten seconds. In another ten seconds, after another thousand fold reduction, frictional heat and stress have practically vanished at a millionth part of their original values. The capsule will have effectively penetrated the planet's atmospheric membrane in 15 seconds. So, while the heat and pressures are extreme, they won't last long.

The optimum launch profile is a path 20° from vertical.[216] This angle maximizes gains from the earth's rotation while minimizing transit time through the atmosphere. Total drag losses for a vertical launch from sea level amount to around 6% of the total kinetic energy, while launches 50° from the vertical lose 10% of their energy.[217] A 20° launch from high altitude should lose only 2% of its energy to atmospheric drag.

After the capsule leaves the launch tube, at a speed of five kilometers per second, it must be accelerated to an orbital velocity of eight kilometers per second. One and a half gigawatts of laser power will impel the capsule upward with a sustained acceleration of 1.25 *g*. At this acceleration, 12.2 meters per second per second, the launch capsule will attain orbital velocity in just over four minutes.

Additional energy must be added to the launch capsule to make up for gravitational drag. The capsule will be flying more or less vertically for only about 20 seconds before angling over into horizontal flight. To make up for this drag loss requires an extra 14 seconds of burn time.[218]

The lasers will also have to supply enough extra energy to the capsule to make up for atmospheric drag losses. These losses could amount to as much as 20% of the capsule's initial velocity if the launch were made from sea level.[219] The Bifrost capsules, however, will emerge from the launch tube at an altitude of almost 6000 meters. The lower half of the atmosphere presents four times more resistance to the capsule's passage than the upper half. Of the 20% of velocity lost during a flight from sea level, only 5% is lost in the upper half of the atmosphere.[220] To overcome atmospheric drag, the lasers will provide an additional impulse of 1800 meters per second.

When the capsule collides with the atmosphere, upon exiting the launch tube, it will create an intense shock wave. The capsule is moving at Mach 15, so the 'sonic boom' will be dramatic. Within 40 meters of the launcher muzzle, the overpressures will be high enough to rupture eardrums. At distances over 700 meters, the shock wave will have dissipated to a level that is typical of supersonic aircraft.[221] At a launch angle of 70° (20° from vertical), the capsule will have left the atmosphere before it has traveled much more than 100 km. horizontally. Therefore, the sonic footprint will be mostly confined to the slopes of the mountain. (See Fig. 2.5) The mountain itself will act like a gigantic reflecting cone, having the opposite effect of a megaphone. The mountain sides will bounce the sonic boom away from lower altitudes, damping and dissipating the shock wave.

**Surfin' Bird**

NASA's press releases describe the space shuttle as a reusable system that flies into space like a rocket and then "glides" back to earth like a plane. In reality, the space shuttle does not 'glide' so much as it plummets. In its nominal glide phase, the space shuttle actually drops faster than a human body in free fall. If you fell out of the shuttle while it was 'gliding', it would reach the ground before you did.[222]

Rather than being configured as a 'glider' like the space shuttle, the Bifrost launch capsules will be built as 'wave-riders'.[223] Wave-riders are true gliders, though their operative principle is very different from that of most aircraft. Wave riders are designed to 'surf' on their own shock waves. At supersonic speeds, the wave rider's passage through the atmosphere creates a shock wave, which is trapped in an aerodynamic trough on the underside. The wave rider derives lift from the shock wave. Tapping into this powerful sonic force gives wave riders ten times the lift-to-drag ratio of conventional hypersonic designs.[224]

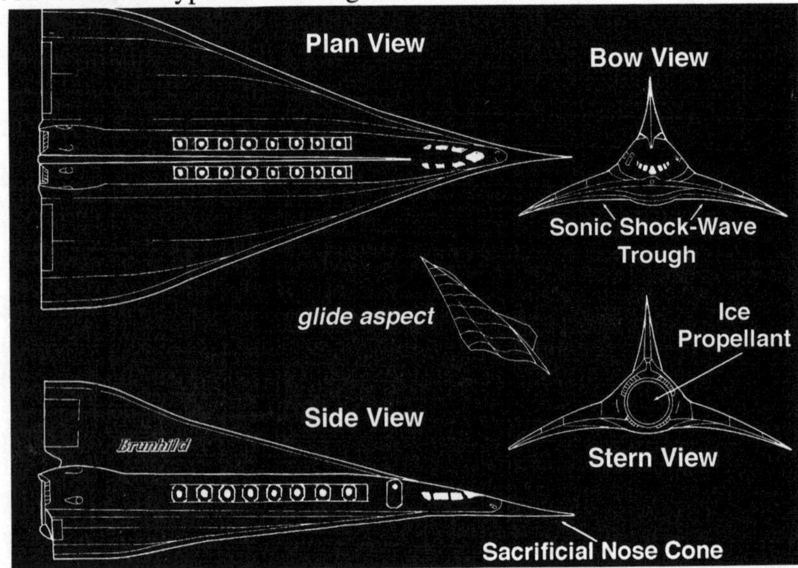


***Fig. 2.4 - Aerodynamic wave rider.*** *Wave rider development will be one of the most important technologies of the New Millennium. Wave riders will be crucial, not only for spacecraft operating between space and the Earth's atmosphere, but also for those voyaging between the planets.*

Trapping the shock wave generates several very important effects: First, the sonic boom is greatly minimized, since the bulk of the shock wave is contained within the trough. Second, compression drag, and its attendant heating of the capsule are significantly reduced. (This will enhance the systems overall efficiency and ease engineering and maintenance problems.) Third, much of the energy in the shock wave goes to generate lift, giving the wave rider a tremendously increased glide ratio. The

Space Shuttle has a lateral range of only 2050 km., compared to a wave rider's potential range of 10,000 km. A wave rider could reenter the earth's atmosphere above the North pole and glide to a landing on the equator.[225]

After the wave rider has rendezvoused with an orbiting space station, or has performed its other mission in orbit, it will fire small on-board retrorockets and reenter the atmosphere. Inside the atmosphere, it will decelerate and glide back to a landing site. To avoid hauling heavy landing gear into orbit, the capsule will be designed to land on water. The capsule will glide in for a landing, not unlike the space shuttle. Instead of using wheels, however, it will settle onto the water and skim along the surface until it comes to a complete stop. After discharging its passengers and cargo, the waverider will be transported back to the launch tube to be readied for another launch.

**No Free Launch**

The present cost of attaining orbit is between $8 and $12 per gram. It is as if every nut, bolt, and lug-wrench taken off the earth were made of solid gold—$11/gm. At this price, we will never become a space-faring people. With the Bifrost Bridge, we can crash these costs, collapsing them to a fraction of their presently bloated rates.

The usage factor for disposable rockets, is one launch per lifetime. The rocket is used and then discarded. All of its capital costs must be charged to the single payload it can carry. The space shuttle was supposed to be more economical because it could be reused. But the shuttles have turned out to be capable of going into space no more than a few times per year, and they require extensive and expensive refurbishing after every trip. Consequently, the shuttle is no cheaper than disposable rockets. The Bifrost Bridge can send up enormous tonnages of material over its lifetime, reducing capital costs per kilogram launched by several orders of magnitude.

A waverider spends about five minutes in the launch cycle: 50 seconds in the launch tube, and just over four minutes on laser burn. Theoretically, we could launch a new capsule every five minutes. As soon as one capsule had entered orbit, we could load and fire another—just like giant machine-gun bullets. In actual operation it will take longer. The capacitors must be recharged and the lasers repositioned for a new launch. On average we could probably launch two or three capsules per hour. At this launch rate, the Bifrost Bridge could put 360 tons of payload into orbit every 24 hours—the equivalent of launching 12 shuttles in the same day. NASA is yet to succeeded in launching 12 shuttles in the same year. Over a 30 year lifetime, the Bifrost Bridge could lift four billion kilograms into space. At a cost of $25 billion for the Bridge, the capital cost per kilogram launched would amount to $6.25.

In addition to capital costs, there are energy and other costs associated with gaining access to space. Together, the launch tube and lasers require 300,000 kwh of power to launch a 14 ton capsule. Aquarius produces this much power every hour. Each sea colony can therefore support 24 launches per day. Three or four marine colonies would be needed to operate the Bridge at its maximum capacity, making multiple launches every hour around the clock. Net weight of payload delivered to orbit will be 5,000 kg. per launch. The energy cost of orbiting payload will therefore be 60 kwh per kilogram. At 5¢/kwh, the total energy cost to orbit a kilogram of payload will be $3. If we assume that operating and other costs are roughly equal to amortized capital costs, then payloads can be sent up the Bifrost Bridge to orbit for something on the order of $15 or $20 per kilogram.

**Snows of Kilimanjaro**

The spaceport should be located as close to the equator as possible. The equator moves faster with respect to space than any other part of the planet, so launches from on or near this line require the least energy. Waveriders at the equator, are already moving 465 meters per second due to the rotation of the earth. By launching our capsules to the east from an equatorial site, we can take full advantage of this free energy.

The spaceport also needs to be adjacent to a high mountain, ideally one more than 5600 meters high. For practical purposes, this should not be a mountain of Alpine conformation. The launch tube must be built in, or along the side of this mountain, so an ideal slope will have a gentle curving profile. The perfect mountain will be a symmetrical volcano—like Mt. Fuji.

What is needed then is a big volcanic mountain located near the equator. There are three mountains ideally suited for this purpose. By odd coincidence, they all happen to be conveniently situated in East Africa, just a few hundred miles from Aquarius.

The best known of these is Mt. Kilimanjaro, immortalized by Hemingway and countless safari epics. At 5895 meters, it is the tallest mountain in Africa, situated just 350 km. from the equator, at Latitude 3° 5' South. Kilimanjaro could hardly be better configured for our launch system. This massive volcano rises in a smooth symmetrical cone from the essentially flat Serengeti Plain, making construction of the launch tube relatively easy. At the top of Kilimanjaro, the rim of Kibo—a caldera two km. across—provides a convenient base for the laser array.

The launch tube will originate near the Ngorongoro Crater, and run due east for 125 km., deep beneath the grasses and grazing herds of the Serengeti. Eventually, the length of the tube will be doubled to 250 km. At one point the extended launch tube will run directly beneath the Olduvai Gorge. (It will be a subtle and cosmic irony to build the Bifrost Bridge under the same terrain where, two million years ago, our infant species took its original first footsteps toward the stars.) Lake Eyasi, 80 km. long and 20 km. wide, just south of Ngorongoro, provides a ready-made landing surface for waveriders.

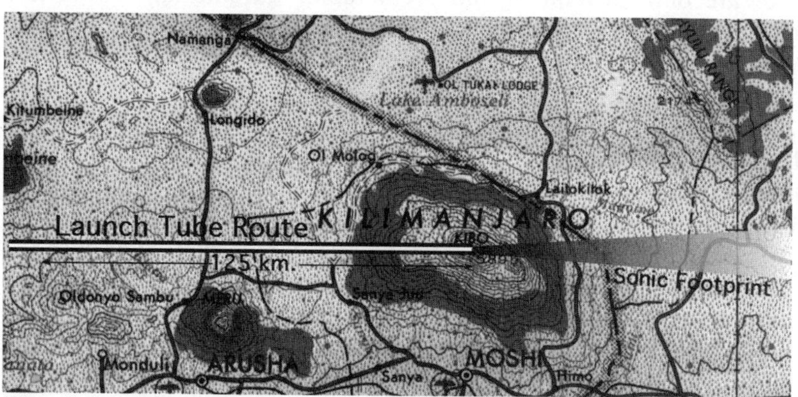


***Fig. 2.5 - Map of Kilimanjaro region, 1:2,000,000.*** *The launch tube runs due east under the broad Serengeti Plain and up the slope of Kilimanjaro. The area affected by sonic booms extends 100 km. to the East. There are presently no communities within that area.*

The Serengeti Plain, home to a multitude of wildlife, is infinitely valuable for its natural treasures, and must not be disturbed in any way. We must and can construct the Bifrost Bridge without doing the least damage to this irreplaceable ecosystem. The launch tube, the capacitors, and virtually every other part of the system will be a hundred meters or more underground. All construction, and even routine traffic will be kept deep beneath the surface. The only surface activity will be at Nainokanoka, a town just north of Ngorongoro. This community will grow into a large and prosperous city as it becomes a global spaceport. All coming and going, to and from the top of Kilimanjaro will be accomplished by airship. The huge spherical ships, supported and fueled by Aquarian hydrogen will carry all construction materials and crews to the peak. This way, no disturbance of the surface will be necessary below the summit. Only the muzzle of the launch tube and the lasers will be above ground. These will be located high atop Kilimanjaro where the only indigenous life forms are lichen and vulcanologists—both extremely tough genera.

There are certain weather restrictions on operation of the Bridge which also favor an East African location. It is not possible to launch our capsules through clouds. The high impact with water droplets causes serious erosion of the ablative heat shields. Similarly, it is not practical to fire a laser beam through cloud cover. Therefore, it is necessary to be in a location where overcast skies are both infrequent and predictable. Equatorial East Africa fits these particulars very well. There are two distinct rainy seasons: April-June and October-December. Between these seasons, rain and clouds are rare. Even though there are snows atop Kilimanjaro—Kibo actually has glaciers—the climatic conditions at the peak are not unlike those of Antarctica, one of the world's driest desert continents. Even during the rainy season, the cloud tops are typically below 5000 M., leaving the launch tube muzzle and the laser array in the clear.

Constructing the Bifrost Bridge on Mt. Kilimanjaro is, of course, contingent on the approval of the people and government of Tanzania. Building the Bridge in Tanzania would transform that country into the central hub of worldwide space activity for centuries to come. The project could catapult this developing country into a position of world prominence. The Bifrost Bridge could provide a source of employment and economic development to Tanzania that would last for generations. Hopefully, this prospect will appeal to the Tanzanian people.

If Kilimanjaro proved impractical or undesirable for any reasons, there are two suitable back-up locations in the same region: Mt. Kenya, which is 5200 meters high, lies due north of Kilimanjaro and sits squarely astride the equator, in the country of Kenya. Margherita Peak, 5100 meters, is located on the border between Uganda and Zaire, within a degree of the equator. (These mountains are also prime candidates for construction of additional launch facilities in the more distant future.)

There are other suitable locations for the Bridge outside East Africa. In South America, Mt. Cayambe, 5800 M., sits directly on the equator, and Cotopaxi, 5900 M., is just a few degrees off. Even New Guinea would work. There, the Nassau Range rises to 5000 meters within 5° of the equator.

The Bridge could even be constructed in international waters if no other options were available. This would mean launching from sea level with the attendant losses to air friction that would entail, but it would guarantee mastery of our own destiny. To be practical, we would have to anchor the launch tube in shallow water. One good place would be on the Mascarene Plateau—a region of shallow water south-east of the Seychelles.

**Tunnel Vision**

By putting all of the mass driver's components deep underground we protect the surface environment, make it easier to maintain vacuum conditions, and gain certain other advantages. However, it does necessitate tunneling on a huge scale. Driving the tunnels for the launch tube, and digging the other subterranean excavations will be some of the biggest and costliest jobs associated with the Bridge's construction.

Fortunately, tunneling is undergoing a dramatic technological revolution. New techniques promise to drive costs down steeply, while rapidly increasing tunneling speed and safety. At the forefront of this revolution are the new tunneling machines. These enormous mechanisms—looking like something out of a Jules Verne fantasy—have made it possible to drive very long tunnels quickly, economically and safely. The Channel Tunnel between England and France is a good example. Even though much of the tunneling was very difficult, the tunnel was nonetheless finished faster than the trains to run inside it could be built.

State of the art tunneling machines typically cut at a rate of 50 meters per day. Costs for a tunnel eight meters in diameter are around $8200 per linear meter.[226] New advances, particularly in hydraulics technology, promise to cut that cost by 40%. The traditional cutting heads on tunneling machines are being replaced with water jets operated at ultra-high pressures. The jets blast pulses of water at the rock face, generating overpressures of a million psi.[227] Since the compressive strength of even the hardest rocks seldom exceeds 30,000 psi, the rock is crushed into fragments and blown away. Other innovative techniques—adding air to the water to cause instantaneous cavitation, for example—enhance the effect of hydraulic cutting.

***Fig. 2.6 - Tunneling machine.*** *Such equipment already brings 100 km. tunnels within technical reach. As this technology advances, such feats of engineering will only become easier.* Photo by The Robbins Co.

With these and other means, the tunnels needed for the Bridge can be cut at a cost of $5000 per linear meter. The 125 kilometer main tunnel could be bored for $625 million. With four tunneling machines, each cutting a different section of the bore, the entire length could be driven through in less than two years.[228]

**Toe-holds**

Work on the Bridge begins at the top of Kilimanjaro and works its way down. Fleets of Aquarian airships will ferry the laser components and drilling machines to the summit. The laser array will be assembled and the first 45 kilometers of tunnel will be bored down through the mountain. This first section of the mass launcher will be equipped with heavy duty electromagnets, capable of accelerating payloads at 30 *g*s. This section of the mass driver will be used to launch cargo and specially trained space cadets. Those who ride the cargo track will have to be in peak physical condition, able to stand crushing *g*-loads for up to 17 seconds.[229]

With the cargo tube in place, we can construct our first outpost in space—Valhalla. This will be a small station in low Earth orbit. Valhalla will never become a colony, but will serve as a staging area for later development. From Valhalla, expeditions will be assembled for departure to outlying bases, and materials and personnel will be marshaled for construction of Asgard—a true space colony in geosynchronous Earth orbit (GEO).

Construction of Asgard will require enormous tonnages of vital materials. Bringing these masses up from the Earth is out of the question. Fortunately space is rich in all the things needed to build space colonies. As a prerequisite to the construction of Asgard and other colonies in free space, we must establish mining bases on the Earth's Moon and on near-Earth asteroids. Expeditionary forces will be sent out from Valhalla to set up these outposts. Once these bases are secure and have built up enough infrastructure to begin sending materials back to Earth, work on Asgard can begin.

# CHAPTER 3

# ASGARD

***The New Frontier is not a set of promises—
it is a set of challenges.***
**JFK**

**Asgard** — Realm of the gods. Kingdom of the sky at the end of the rainbow. In his palace of Valhalla, Odin, monarch of immortals, presides over the endless feasts and battles of the eternal heroes.

Our species is going to move into space. We are going to settle and occupy the raw void as if it were the fertile plains of Canaan. Space is the ultimate environment; if we can successfully adapt to life there, our potential is unlimited. Like Life itself, when it made the move from the oceans to the land, we must learn to live in harsh new surroundings. To thrive as a space-based life form, we must develop new techniques for living. These new ways will incorporate novel tools and skills for dealing with life in space. We need to develop space habitats that are simple to build, easy to operate, and safe to live in. Asgard is a model for one such habitat.

When Asgard is built, we will be free—as a species, and as a phenomenon in the Cosmos. Instead of dying, like the yeast in the bottle, we will blow the lid off our planetary flask. We will escape our gravitational confinement and at last be free in the great outer world of limitless space. Free of planetary constraints, we will rapidly evolve into a space-based civilization. The only boundary to our potential will be the infinite horizon of our own imaginings.

Once we have built our outpost in space, once we have learned to live in harmony with this new environment, our growth will be explosive and unstoppable. The space around our home star is ostentatiously rich in resources. It almost seems as if we have been planted in a well-chosen plot of fertile soil. Everything we need is conveniently at hand, as if placed there by some omnipotent and beneficent creator. We are born heirs to the Empire of the Sun. To claim our inheritance, we need only learn how to live outside the atmospheric womb of our Mother Planet.

Moving into orbit—just a few thousand miles and a scant few minutes from Earth—is only one small step for mankind; it is, nevertheless, one giant leap for Life. When we have built Asgard, and have begun to live self-sufficiently in space, we will have made the ultimate Cosmic transition. Life will have evolved the capacity to propagate itself, independent of the Earth. Thenceforward, there can be no stopping us, neither as a species, nor as a universal phenomenon. Once we have gained this first toehold in space we can go on to spread ourselves throughout the galaxy, and ultimately the universe, virtually without limit.

**Need Another Seven Astronauts**

With our path so clear, and our potential so gleaming, it is heartbreaking to be stuck on Earth watching a hapless bloated bureaucracy like NASA try to wallow its way into space. It's like watching a walrus try to scale a ladder. In its glory-days, NASA was an army of people on a mission. They had focus, they had purpose; with amazing skill and unbelievable speed they tore down every obstacle in their path and catapulted men onto the Moon. But that was over a quarter-century ago. Since then, NASA has been cast adrift. Without a tangible goal, the once potent organization has degenerated into an unholy alliance between politicians and industrialists. In politics, votes are the bottom line; in industry, the bottom line is the bottom line. Neither has much regard for mankind's rendezvous with destiny.

Typical of NASA's present fatuity, more bone-headed even than the graceless Shuttle, is the planned space station—named, with crushing irony, "Freedom". Assuming it is ever built, Space Station Freedom will provide 900 cubic meters of livable space.[230] For this, the American taxpayers will have to spend around 33 billion dollars—36 million dollars per cubic meter, over $1 million per cubic foot. NASA's space station will accommodate a grand total of eight people at a capital cost of around $4 billion dollars per person, give or take a hundred million. It is hard to imagine why it will cost this much, but it's probably the same reason it costs $10,000 to put a toilet seat aboard a B-1 bomber. Four billion each is nothing but the price tag on "impossible".

Like the Shuttle, space station Freedom is emblematic of NASA's failure to come to grips with the "vision thing". Blinded by their perpetual budgetary obsessions, NASA's brass sees the space station only as a means to extract funding from an ever more reluctant Congress.

If we do things the way the military-industrial complex insists they be done, we can just forget about ever moving into space at all. If we are ever going to fulfill our cosmic destiny and carry Life out into the universe, we've got to leap-frog the Pentagon mentality that has paralyzed the American space effort. The aerospace establishment is founded on the proposition that solutions to problems must be complex, cumbersome, expensive, and difficult. We, on the other hand, must find solutions that are simple, elegant, cheap, and easy. Finding such solutions, requires new ways of thinking about space. The Bifrost Bridge reflects a new way of thinking about how to get into space; Asgard should reflect a new way of thinking about how to live in space.

**Bubble Up**

There is a general misconception that space is somehow different; technical bureaucracies like NASA unhesitatingly spend billions of dollars to overcome these perceived differences. Outer space is really no different from the space you are occupying right now. Space is space. The difference is that we live at the bottom of an ocean of air. As oceans go, it is not very impressive. If it were water, it would only be thirty feet deep. Duplicating in space the conditions that persist at the bottom of this shallow sea of air is relatively easy. The proposition is mostly one of confining a few puffs of breathable gas. Following Fuller's design principles we arrive at the simplest of all possible solutions—a bubble.

The design of Asgard is simplicity itself. Instead of a complex arrangement of pressurized cans, Asgard will be a cluster of simple balloons. Asgard's bubbles will be made of clear impermeable silicone membranes, supported by internal air pressure. Temperate breathable air fills these bubbles, providing enormous volumes of habitable space very cheaply. Closed-loop systems, organic to the bubble design, moderate internal temperatures and purify the air, making of the bubbles truly self-sufficient 'ecospheres'.

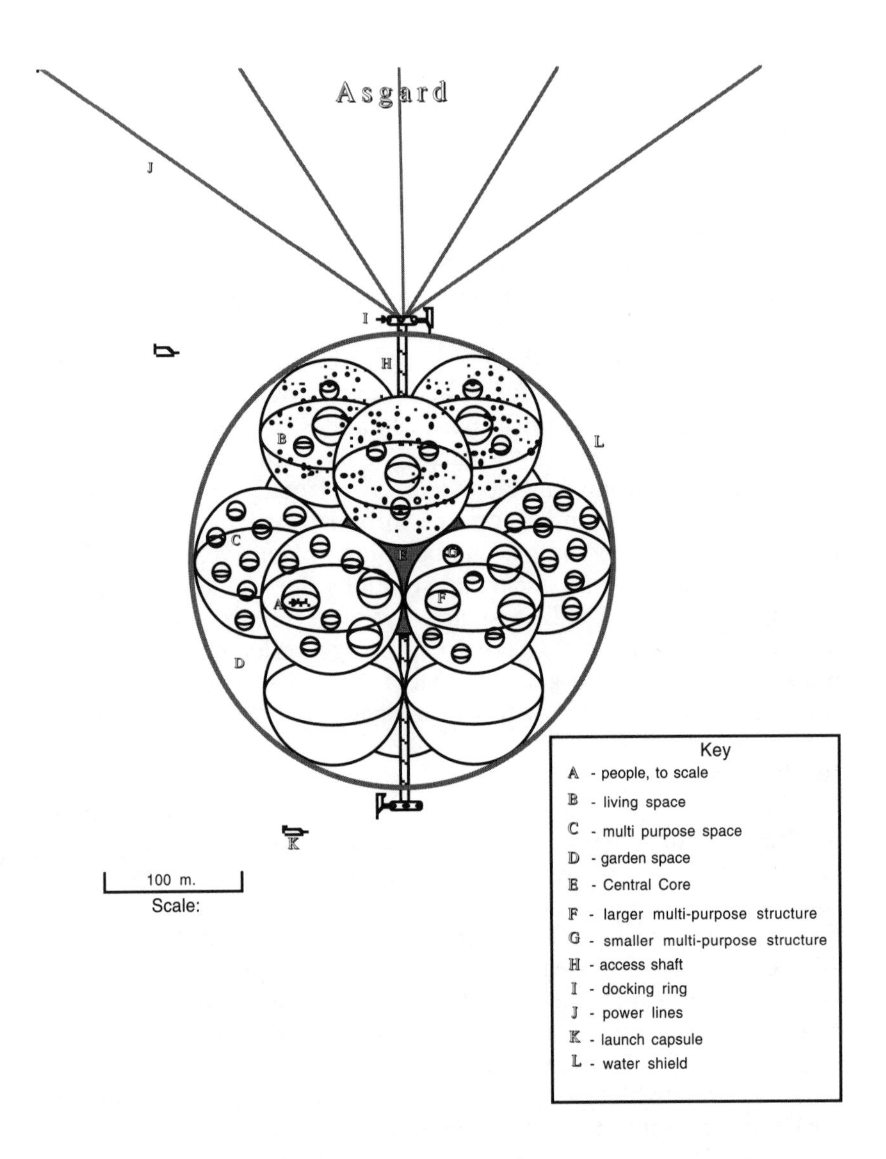


***Fig. 3.1 - Scale diagram of Asgard.*** *Like Aquarius, Asgard is a simple bubble cluster, assuming the same hexagonal symmetry, but in three dimensions.*

The bubble cluster will be in a geometrical arrangement known as 'cuboctahedral packing' (so called because each sphere occupies one vertex of a cuboctahedron). A less tongue torturing name is the 'apostle formation'; so called because there are 12 spheres all equi-tangent to one. The central sphere is surrounded by six spheres in a hexagonal pattern. Three spheres nestle naturally on the top and bottom of the hexagon. This structure occupies the least amount of space and has great dynamic stability. Left to themselves, packed spheres, like lead shot, will automatically assume this formation. This is simply an extension of the geometry that dictated the shape of Aquarius. A bubble float, operating in two dimensions, naturally assumes the same hexagonal arrangement.

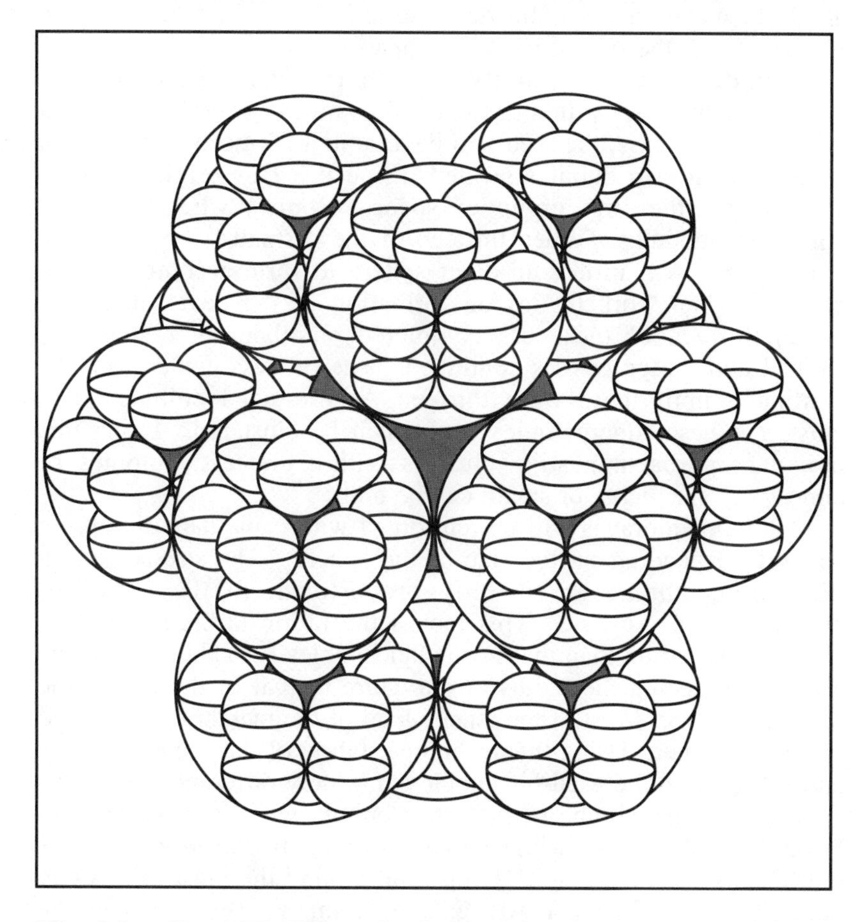

***Fig. 3.2 - Nested bubbles.*** *Both Asgard and Aquarius are built up of clustered modules. Modularity simplifies the construction of Aquarius and lends it redundant security against sinking; in Asgard, nested bubbles protect the inhabitants from a catastrophic puncture of the outer membrane.*

Freed of the gravitational constraints imposed on Aquarius, Asgard is able to take the form of a three-dimensional bubble cluster. The modularity reflected in the largest scale of the structure is repeated over and over again in bubbles nested in bubbles. Each bubble has a cluster of bubbles inside it, down to the smallest individual rooms in the colony. Each bubble in the central cluster holds 13 more bubbles each about 66 meters in diameter; each of these contains another sub-cluster of 13 bubbles, each about 22 meters across. Inside these are the smallest bubbles, 7.4 meters across, the size of individual apartments.

Asgard's fractal modularity is a direct result of the colony's growth process. As with all Millennial colonies, Asgard grows from a seed. In this case, the seed will be an initial 22 meter bubble, with 13 apartment size bubbles nestled inside it. This original colony will shelter the first group of space colonists who will build additional modules. When thirteen of these small bubbles form a cluster, they will inflate an outer sphere to form a complete sub-cluster. Then work begins on formation of the next subcluster. When thirteen of these are completed, another outer sphere is formed and the process is continued. The process could easily be continued until there were thirteen Asgard-sized colonies in a cluster. These supercolonies could then be clustered. Using this modular approach, nesting clusters within clusters, colonies of almost any size, large or small, can be built.

Bubble membranes will be reinforced with cable nets of varying strength, depending on the size of the sphere. The stresses on an inflated membrane are directly proportional to the radius of curvature of its surface. This means the fabric in bigger bubbles faces greater strain than that in smaller bubbles. A cable net has the effect of reducing the radius of curvature to that of the membrane between cables. Most of the tensile load of the membrane is carried by the cable net. Using this technique, fabric of the same thickness and strength can be used for bubbles of widely different sizes.

The clear membrane of the outer bubble will allow sunlight to flood Asgard's interior while providing an unobstructed view of the outside. Asgard is going to be in orbit around the most exquisitely beautiful object in the entire galaxy; being able to look out the window at Earth will certainly be a high priority.

The apostle cluster is enclosed by an outer bubble 600 meters in diameter, with a total volume of 113 million cubic meters. The 13 spheres in the apostle cluster enclose 54 million cubic meters, leaving more than half of the habitable space in Asgard outside the bubbles in the central cluster. Each of the ecospheres is capable of holding its own atmosphere, and could, if necessary, operate independently of the others.

Each of the main bubbles in the apostle cluster is 200 meters across—the same size as the Houston Astrodome—and encloses a volume of over four million cubic meters.[231] The Astrodome will accommodate 65,000 people—sitting cheek-to-cheek, so to speak. Imagine how much more room each person would have if they could all float up from their seats and drift about in the vast unused spaces of the dome.

In Asgard's gravity-free environment, people can float about freely, utilizing all of the space enclosed by the ecospheres. Compared to Earth-bound cities, Asgard is tiny. It is, nevertheless, large enough to accommodate a population of 100,000 people. (See Plate No. 6)

**Elbow Room**

In space, we will at last become truly three dimensional creatures. On Earth we live in two dimensions, stuck permanently to the surface by gravity. On Earth we think of living area in terms of square meters, in space we must think in terms of <u>cubic</u> meters.

How many cubic meters does a person need to live comfortably? There must be enough room to move around freely, room to work, room to live, and room to play. In addition to this 'personal space', there must be room for all the support systems needed to maintain a comfortable environment.

The Soviet Salyut space station, provides only 50 cubic meters per man. A modern submarine has a habitable volume of 10,000 cubic meters, and is occupied by 150 men, providing 70 cubic meters per man.[232] The U.S. Skylab, by contrast, which had roughly the same internal volume of a three bedroom house, provided 100 cubic meters of space per man. This, if the familiar scenes of somersaulting astronauts are any guide, seems fairly roomy.

The residents of Asgard will enjoy spaciousness on a grand scale, 1,130 cubic meters apiece.[233] Living in Asgard will be like having a large mansion all to one's self.[234] Life in Asgard will be a far cry from the cramped existence offered by tin-can space stations. In Asgard, there will be abundant room to swoop and soar in the gravity-free wide-open spaces.

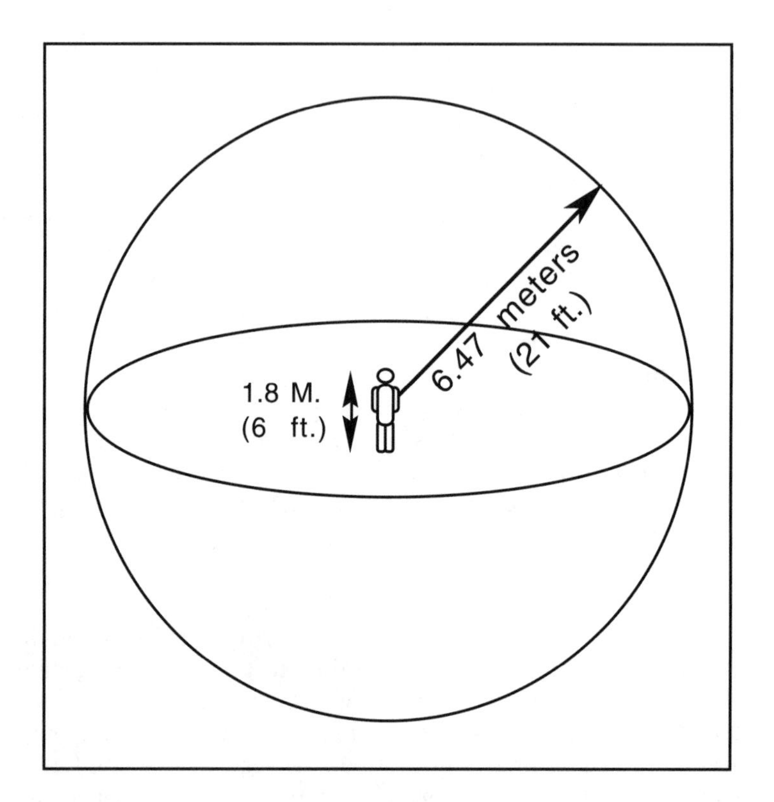


***Fig. 3.3 - Personal living volume in Asgard.*** *1130 cubic meters will provide an abundance of space to meet all needs.*

Such roominess will be a welcome luxury in a place where the childhood dream of flight is not only possible, but unavoidable. All personal transportation inside the colony is accomplished by flying, under one's own power or impelled by turbo-fans no larger than hair dryers. This is practical, since the colony, while spacious in terms of cubic dimensions, is incredibly small in terms of linear dimensions—such is the magic of solid geometry.[235]

Asgard's capacious internal arena is suitable for any of the myriad activities that will occupy us in space. Smaller individual spheres accommodate private living quarters, labs, schools, offices, and workshops. The three lower spheres house utilities and work areas. These spheres are divided into separate spaces for various working groups, not unlike floors and areas in a modern office building, though with decidedly different geometries. The colonists will come here to perform the work of the colony.

Assuming 60% of the population is in the active work force, this provides each worker with 210 $M^3$ of work space.

The six spheres around the central belt of the colony, and the three spheres at the top of the colony are all dedicated to living space. These spheres contain homes, neighborhood services, and family accommodations. The nine living spheres provide each colonist with 377 cubic meters of domestic space—a volume larger than Skylab for each person.

The inner central sphere will hold the common facilities of the colony: school, hospital, auditorium, mall, and recreational amenities. If all the colonists gathered in the central sphere, everyone there would still have almost as much room as each Russian Cosmonaut aboard *Mir*.

**Hanging Gardens**

One of the greatest psychological deficits in space will be stimulus from the natural world. One of the problems with the tin-can approach to space colonization is that it is utterly sterile. Life in such space stations would be very much like life in an Antarctic research station. A few lantern-jawed men and women of iron constitution could probably stand it for awhile, but it would be a dead-end sort of existence. The idea of large populations of humans living that way for years at a time, and trying to raise families, is appalling. An essential aspect of our nature as humans—an aspect that has suffered badly during the industrial era—is our connection to the natural world. We need grass and flowers and trees almost as much as we need food and air and water. Our diminished contact with these sustainers of the human spirit probably has as much to do with the dehumanization of urban life as anything else. To provide these necessities, is perhaps one of the greatest and most essential challenges we face if we are to live happily and successfully in space.

Asgard's voluminous interior will enable us to create a bit of Eden in the void. The region between the big outer sphere and the spheres in the central cluster is wide open. Dotted with free-floating trees and shrubs, there are 580 cubic meters of this open space—essentially park-land—for each colonist; a sure antidote to claustrophobia. (See Plate No. 7)

Open spaces, dedicated to floating gardens, will be thick with living greenery. Colonists will come here to be close to nature and, if they wish, far from people. Even if everyone in the colony went into the gardens at the same time, each person would still have many times more room to glide around in than the Skylab astronauts. The gravity-free environment in the gardens will not be unlike what one would experience if one could fly freely through the canopy of a tropical rain-forest on Earth; a green labyrinth of plants populated mostly by other fliers: pastel butterflies, brightly colored parrots, and iridescent humming birds, all of which, if hatched from eggs in zero-*g* might easily adapt to the new rules of flight.[236]

Many colonists will take their daily exercise in the gardens. Flying with wings or pedal driven propellers will be great for cardiovascular fitness. Clear tracks, each over a kilometer long, will be installed in each hemisphere of the main bubble. Around these raceways, flyers can propel themselves with gusto. A layer of strong, but thin kevlar netting, held out a meter or two from the inner surface, will protect the bubble from inadvertent damage, catching careening flyers and other large objects before they hit the membrane.

Space gardening is as amenable to elegant and simple solution as most of the other problems we will face in our new environment. While we are accustomed to seeing plants firmly rooted in the earth, it is not essential for them. The water hyacinth, for example, floats freely on the surface of a pond, supported by its buoyant leaves, with its roots hanging open in the water. Most plants can live in much the same way, because it is not soil which sustains them. Plants are not dependent for their life on dirt; they depend on water, and the nutrients it carries. If nutrients are present in the water, soil can be dispensed with entirely. Water and fertilizer will flow to the free-floating plants through umbilical cords. The water and nutrients will circulate through a base containing the plant's roots. With such an arrangement, plants can float freely, held in place only by the umbilical cord.

Using this system, we can grow free-floating gardens with flowers, shrubs, trees, and even grass. Individual plants can even be enclosed in their own miniature ecospheres. This way, a wide variety of plants can be grown, even though the general environment has been optimized for human comfort. For example, fruit trees like apple, cherry, and pear require a snap of cold weather in order to blossom. Inside an apple tree's individual ecosphere, the temperature could be adjusted to suit its needs, without giving a nearby banana tree frost-bite.

The gardens will provide the contact with nature that we crave, and will also help to maintain the colony's atmosphere. Green plants not only remove carbon dioxide and produce oxygen, but also cleanse the air of pollutants. Plants actually break down harmful organic chemicals like formaldehyde that can otherwise build up in a closed atmosphere. If activated carbon filters are included in the base with the plants' roots, these pollutants can actually be converted into nutrients. In addition to these purifying services, the plants will also provide fresh fruits and vegetables. While not of real significance as a contribution to the colony's total diet, these products will nevertheless add welcome variety to the bill of fare.

There will also be an abundance of plants in the living and working spheres. Far from living in a dead, sterile, canned environment, we will be surrounded by an abundance of life, color, and natural beauty.

**Domestic Life**

Inside the nine residential spheres, people will usually occupy individual 'domiciles'. A typical family with two children would occupy a bubble with a volume of over a thousand cubic meters.[237]

Living space inside the bubble can be partitioned in a variety of ways. Very little furniture is needed in a weightless household, so there is much more available space in a given volume. Most 'furnishings' will consist of anchored expanses of netting. Velcro will be *de rigueur* in a mind boggling array of applications.

The standard arrangement of 'rooms' will be for the bathroom and kitchen to share space at the 'bottom' of the sphere, where the distance to connect to the utility umbilical is minimized. The umbilical brings all the necessities of life into one's domicile. Along the umbilical come water and electric lines, fiber optic cables ventilation ducts, and pneumatic delivery tubes. As in Aquarius, the pneumatic tube system will deliver most material needs directly to the home. Passing back along the umbilical go the sewer lines and disposal tubes for trash, laundry, and dirty dishes. Along the fiber optic cable comes entertainment, communications, and information. The optical fiber connects directly to the information appliances: one large high-definition flat-plate liquid-crystal color screen, and any number of strategically placed smaller screens. As in Aquarius, color terminals are tied directly into the colony's central computer. An unlimited menu of entertainment selections is available through the data utility. Stereo music or selections from a global library of commercial films and video productions can be piped in. The data terminal provides a window into cyber-space, enabling individuals to communicate directly with anyone in the colony or on Earth, play interactive computer games, or tele-operate remote robots.

In cyber-space the computer will create an alternate reality that is projected on a visor in three dimensions. The image has apparent depth, and covers the entire field of vision, giving the impression of being immersed in the scene. In virtual space, one can manipulate virtual objects, interact with virtual characters, even go adventuring in an imaginary world with real friends. For the next few centuries, virtual space is as close as we are going to get to Star Trek's 'holodeck'.[238] The advent of virtual space is just dawning. By the time we are in Asgard, it will be a highly developed technology, of crucial importance in helping us adapt to our new environment.

**Hard Day's Night**

The apostle formation of 13 spheres is suspended inside the envelope of the outer shield sphere, with some clearance between the walls. A central axis penetrates both poles of the outer sphere. Hard points at either pole provide docking rings for space capsules and access ports to the colony. The inner cluster of spheres rotates around the central axis. The outer envelope does not rotate. The inner cluster turns slowly around the axis, making one revolution every 24 hours.[239]

The slowly rotating spheres present no danger to people gliding about in the garden spaces. The maximum speed of the bubbles at the equator is just over one meter per minute. The cluster spins like the hour hand on a clock. Without watching the bubbles carefully, it will be difficult to tell they are even moving.

The dwelling spheres are grouped in three triplets. Each of the three groups occupies two equatorial bubbles and a 'north' polar bubble.[240] Each of the triads is on a separate 24 hour cycle and forms a distinct work shift. Two shifts are up and operating in the day-lit hemisphere, while one shift is going through its night phase on the dark side of the colony. In this way, the day shifts take advantage of the natural illumination of the sun when they need it, while the night shift is in darkness. Each of the triads makes full use of the work-stations, laboratories, and other facilities in the three working bubbles at the colony's 'south' pole. This way, these assets are in use 24 hours a day.

The central sphere, which is dedicated to providing common services, will be shared by all three shifts and so is open and active all the time. Similarly, the gardens will be open all the time. They will be anchored to the rotating cluster, however, so they will also be passing through a separate phase of the 24 hour cycle. With this system, the colony can be in operation 24 hours a day. Each shift experiences its own day/night cycle, so people don't have to live like vampires, working at night and sleeping during the day.

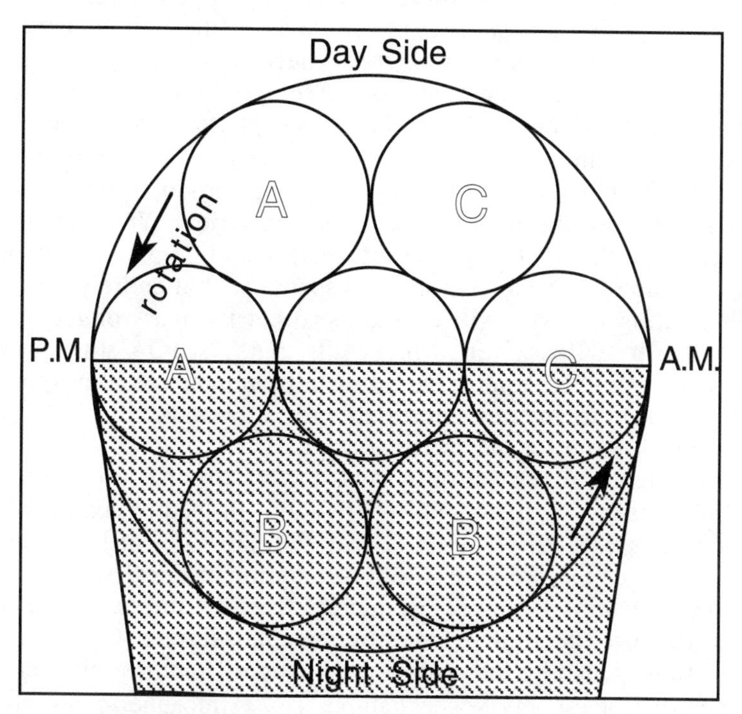


***Fig. 3.4 - Day/night cycle.*** *As the apostle cluster rotates, each bubble passes through a full 24 hour cycle of light and dark.*

**Partial Pressure**

It is Earth's thin skin of atmosphere that makes her a living planet. To live in space we must create a little bubble of this magical fluid. Inside a bubble of temperate air, we can live in space happier than ticks in a pig's ear. Without that bubble of air we would be boiled, roasted, irradiated and freeze-dried in a matter of minutes.

Most space habitat designs call for an internal air pressure close to that prevailing at sea level on Earth. The Russians, for example, have always designed their spacecraft to provide an atmosphere that is Earth normal: 14.7 psi, 80% nitrogen and 20% oxygen. The U.S. Space Shuttle follows the same procedure. While this eliminates a variable from space studies, it is a very expensive proposition.

Organic or 'fixed' nitrogen is a vital micro-nutrient for plant growth, but atmospheric nitrogen is nearly inert, and plays no discernible role in human biology. The only apparent function of nitrogen in Earth's atmosphere is to serve as a 'filler'. It contributes to the total ambient pressure, thereby keeping oxygen concentrations at tolerable levels. We will want to eliminate nitrogen from the atmosphere of our ecospheres. Nitrogen is a heavy gas, and in space, it is hard to come by.[241] Nitrogen for plant growth will come from recycled waste products rather than being fixed from the air by specialized bacteria. We don't need nitrogen in order to breathe comfortably, and if plants can get nitrogen from other sources, we can dispense with it entirely.

Eliminating nitrogen will enable us to cut the mass of Asgard's atmosphere by 80%. Even without nitrogen, Asgard's atmosphere weighs in at 26,555 tons. Including nitrogen in the atmosphere would require lifting an additional 152,000 tons into orbit; a daunting task, even for the Bifrost Bridge.

Eliminating nitrogen not only saves having to transport its mass, but also saves having to contain its pressure. Instead of coping with pressures of almost 15 psi, we can deal with pressures of less than 3 psi. The lower the pressure, the lower the mechanical stresses on Asgard's structural systems. Reducing Asgard's internal pressure allows the use of thinner membranes and lighter cable nets, exponentially reducing the costs and logistic difficulties inherent in the construction of any space habitat.[242] Atmospheric pressure inside Asgard will be three pounds per square inch (155 mm Hg.), equal to the atmospheric pressure on Earth at an elevation of 12,000 meters (39,000 ft.).

One might wonder how humans could breathe comfortably at such low pressure? But the body is not much concerned with pressure; its real concern is how much oxygen is available. This is determined by oxygen's <u>partial</u> pressure, not total atmospheric pressure. In order to exert the same partial pressure in the lungs as normal air at sea level, an atmosphere of 99% oxygen needs to be maintained at a pressure of three psi. If you take a breath of 99% oxygen at three psi, you will have just as many oxygen molecules in your lungs as when breathing at sea level on Earth. Your blood will be richly oxygenated and you will be perfectly comfortable.

Three psi even provides a wide margin of safety. At three psi, the boiling point of water is 60° C. (140° F.). While this might present problems for pastry chefs, it doesn't create any biological difficulties. (If your body temperature rises to 140°, the ambient air pressure will be the least of your problems.) The minimum pressure necessary for breathing is 2.2 psi. (114 mm Hg.)[243] Asgard could lose over a quarter of its atmosphere before people had trouble breathing. The minimum pressure needed to prevent fluids from boiling at body temperature is less than one psi. (48 mm Hg.). With emergency oxygen masks, colonists could survive the loss of up to 70% of Asgard's atmosphere.[244] It is highly unlikely that Asgard could lose this much of its air, because its interior is subdivided into many individual compartments. Even in a great catastrophe, like a strike by a large meteoroid, only a few of the interior bubbles would be punctured.

**Light My Fire**

Excluding nitrogen leaves us with an atmosphere that is almost entirely oxygen. The immediate objection raised to an atmosphere of 99% oxygen is that it will create an extreme fire hazard. But, will it? Fire is just like the human body; it is an oxidizer. The fire doesn't care what the percentage of oxygen is any more than your lungs do; what matters is the number of oxygen molecules present in a given volume at a given time—that is the partial pressure of oxygen. The most familiar example of this is the internal combustion engine in your car. The piston compresses the air in the cylinder which is mixed with fuel and then ignited. The burning efficiency in the cylinder is much higher because the partial pressure of oxygen, together with the total pressure is raised so much higher, even though the percentage of oxygen remains the same. In the highly compressed air inside an engine's combustion chamber, there are many more oxygen molecules colliding within each cubic centimeter than in normal air, and this is all that matters.

The catastrophic fire that resulted in the tragic deaths of three Apollo astronauts, was not caused by 100% oxygen.[245] Those men died because the partial pressure of oxygen was five times its normal value—14.7 psi, instead of 3.1 psi. The same conflagration could have resulted if the capsule pressure had been raised to 70 psi, with normal air.

In a three psi atmosphere of 99% oxygen, fire will burn no more readily or fiercely than it does at sea level. In fact, in a zero gravity environment, it is doubtful that an open flame can be sustained at all. On Earth, because warm air is less dense than cool air, smoke rises. Smoke and other combustion gases like CO and $CO^2$ rise away from the top of the fire, and fresh oxygen-rich air enters at the bottom, creating a convective flow. But convection is a function of gravity; in zero-*g* there are no convective forces, since the difference in density does not cause one gas to be 'heavier' than another. In the absence of gravity, a cloud of smoke and carbon dioxide will quickly build up around any flame. Fresh air will be unable to reach the fire, and, as noncombustible gases accumulate, the fire will snuff itself out. This property, coupled with the relatively low thermal conductivity of a low density atmosphere, will actually make the fire danger in Asgard less than it is on Earth.

**Radiation Hazards**

A simple bubble provides us with most of what we need to survive in space: air pressure and controlled temperatures. There are, however, one or two other provisions we must make for our long term comfort and safety. We need mainly to shield ourselves from radiation. The universe is nuclear powered and space is consequently awash with radioactivity. On Earth, we are protected from most of the harmful rays and particles by the atmosphere. The atmosphere, so seemingly tenuous, forms a radiation shield equivalent to a wall of lead three feet thick.[246] On the other side of that wall is a radiation flux of deadly intensity.

The radiant energies common in space present dangers out of all proportion to their apparent magnitude because they are in the form of "ionizing" radiation. Ions are charged particles—atoms that have gained or lost electrons. When radiation converts atoms into ions at high energy, they can be extremely destructive. The mechanism of ion formation is fairly simple: protons and electrons, moving at high speeds, crash into the electron shells of atoms and knock them to smithereens. High energy particles like protons, packing energies on the order of 100 million electron volts, will tear through millions of atomic shells, ionizing each one. These charged ions can then do tremendous biological damage.

In inorganic materials like air, water, stone, or metal, ionization is not very disruptive; atoms gain or lose electrons with not much harm done. The arrangement of atoms in most materials is pretty random. The molecules in a plastic for example are like strands of spaghetti. Firing bullets into a pile of pasta wouldn't really make a lot of difference. Metals are like piles of bird-shot. Within the pile is a very precise geometric order, but it is based solely on the shape of the little balls and is hard to disrupt. Blasting away at the pile all day, won't disturb things very much. Consequently, inert materials can take huge dosages of ionizing radiation—on the order of tens of millions of rads (radiation absorbed dose)—without suffering much damage.

Biological materials are altogether different. Living tissue is made up of highly complex bits which cannot function if they are rearranged even a little. Ionizing radiation passing through cells in living tissue is like machine gun slugs tearing through a herd of cattle.

Radiation dosages are measured in rads or rems. The rad (radiation absorbed dose) is an objective measure of energy input, amounting to .0000024 calories per gram. The rem (rad equivalent mammal) adjusts the rad to reflect the actual biological damage produced. There is no such thing as "safe" radiation. Even one nucleon passing through living tissues will cause damage, and could theoretically trigger a fatal cancer.

Despite its dangers, radiation is an every day fact of life. On average, each of us will absorb 200 millirem of ionizing radiation every year.

**Table 3.1**[247]
**Natural Radiation Dosage/Yr.**

| **Source** | **mr/yr** |
|---|---|
| Medical X-rays | 75 |
| Natural (from water, soil, etc.) | 48 |
| Bones (natural radioactivity from your own teeth & skeleton) | 34 |
| Cosmic rays (at sea level) | 30 |
| Man-made (A-bomb tests, Chernobyl, etc.) | 13 |
| **TOTAL** | **200** |

We can obviously survive minimal doses of radiation. All organisms have adapted to natural radiation, and possess mechanisms at the molecular and cellular level for repairing the routine damage it causes. As radiation doses rise, however, the body's repair mechanisms can be overwhelmed. Higher rates of cancer, genetic mutation, illness, and ultimately death can result.

Although there is no absolutely safe minimum dosage of radiation, there are reasonably safe levels of exposure. The U.S. Department of Energy sets a maximum of .5 rem (500 mr.) per year for the general population, about the dosage received by people living in Denver. Radiation exposure at this level has no discernible statistical effect; it is lost in the 'noise' of other health factors. For example, people living in Denver—due to the higher altitude—receive twice the radiation of people living at sea-level in New Orleans, but have lower rates of cancer.[248] For workers in the atomic energy industry, maximum safe exposure levels are five rem per year, 5000 mr. This dosage, while ten times that of the average person, is still pretty low. Many stewardesses receive higher dosage levels than that, simply by logging flight time at high altitude. NASA is even more liberal in its radiation allowances, deeming skin exposure to 225 rem per year, 1000 times the average dose, as safe.[249]

The period of time over which radiation is absorbed is critical to determining its potential for causing damage. If more than 450 rem are absorbed in a short time there is a fifty/fifty chance of survival. If, however, rapid dosage exceeds 600 rem, the odds of survival are just about zero.

Fortunately, humans can be "hardened" to withstand higher doses of radiation. It has been shown that taking massive doses of Vitamins A and E can reduce ion damage by up to 30%.[250] Of course, you know where our space colonists get their Vitamins—and it ain't spinach. Blue-green algae generally, and spirulina specifically, are powerful antidotes to radiation. The Russians are using spirulina to treat children who were exposed to radiation during the Chernobyl disaster.[251] Asgardians, living on an algae-based diet, will be highly resistant to radiation hazards. Nevertheless, we will undertake strong measures to protect ourselves from the threat.

**Odin's Shield**

The sources and types of radiation hazards in space are various, and include Van Allen radiation, solar flares, and cosmic rays. (See Appendix 3.6) Protecting ourselves from these hazards, especially cosmic radiation, requires very massive shields. Reducing radiation exposure in free space to under five rem/yr. (the DOE allowance for workers in the nuclear industry) requires a passive shield with a mass of at least 2500 kilograms per square meter. Reducing the dosage to under a half rem per year (the amount absorbed by residents of Denver), will require a shield with a mass of 5000 kg./$M^2$.[252] This will be the shield density we adopt for our long-term habitation of space. Asgard's shield will therefore weigh 5.6 million tons.

It may eventually be possible to eliminate this huge mass with some sort of force field. Electrostatic or magnetic fields could theoretically repel the most dangerous charged particles. But, use of force fields presents huge technical difficulties. (See Appendix 3.7) It may ultimately be practical to build such force fields for later space colonies, but Asgard will have a passive shield which relies solely on mass to protect us from cosmic radiation.

***Fig. 3.5 - Artist's concept of the Stanford Torus.*** *This space colony is designed to be protected from radiation by a layer of slag left over from the extraction of metals from lunar regolith. The strange result is an orbiting cave.* NASA

**Let the Sunshine In**

Most designs for space habitats, like the Stanford Torus, have adopted radiation shields made of rock. Presumably, this would be slag left over from processing mineral ores sent up from the Moon. Slag shields would no doubt be effective, but they create as many problems as they solve. Surrounding a space colony with masses of dirt creates the ironic condition of a free-floating space habitat that is effectively buried underground. Such designs cut themselves off from one of the great advantages offered by life in free space: the abundant and steady availability of the ultimate commodity—sunlight.

In order to illuminate these 'subterranean' space colonies, their designers resort to a variety of ingenious tricks to get sunlight inside—it's all done with mirrors. The Stanford Torus uses a gigantic flat mirror set at a jaunty angle above the colony. This is supposed to reflect light into an elaborate system of angled louvers built into the inside surface of the torus. Gerard O'Neill's innovative designs, both the cylindrical and spherical variety, use similar elaborations of mirrors.

***Fig. 3.6 - Detail of the Stanford Torus.*** *The use of slag as a radiation shield necessitates a complicated scheme to get light inside the space colony. This artist's concept shows installation of the louvers that would allow reflected light to enter the interior of the torus. If made of aluminum, the louvers would need to be very thick to provide adequate shielding. Not only must such components be manufactured, but with this design, they must be supported against tremendous centrifugal forces.* NASA

Some of the difficulties associated with these designs stem from their creator's insistence on producing artificial gravity, but most of the problems are due to the choice of shielding material. Dirt, to say the least, is opaque to visible radiation as well as to ionizing radiation. You can't see through slag. The cybergenic solution is to choose a shielding material that is transparent, but will stop cosmic rays. The clear choice is water.

Hydrogen oxide, $H_2O$, is an ideal shield material.[253] A water shield can provide complete protection from radiation hazards, while allowing an abundance of visible light into the habitat. A layer of water five meters (16 ft.) deep will create a shield density of 5000 kilograms per square meter. This will provide radiation shielding equivalent to a wall of lead a foot and a half thick. Inside Asgard, radiation exposure will be under half a rem per year—the same as in Denver.

***Fig. 3.7 - G.K. O'Neill's space colonies*** *use adjustable mirrors to reflect light into their cylindrical interiors.* NASA

A second outer bubble will serve to contain the water shield. This bubble will be like the one surrounding the apostle cluster, but slightly larger—610 meters in diameter. The two bubbles will form concentric shells, with the layer of water between them. The outer bubble will also serve as a backup to the inner bubble, reinforcing the colony's margin of safety. Air escaping through punctures of the inner bubble will be contained by the outer bubble. If the outer membrane is penetrated, only water will leak out.

Pressure in the water shield is extremely low. The mass of water, being virtually weightless, exerts no depth pressure, and the vapor pressure of water at 20° C. (68° F.) is only .37 psi. Holes in the outer membrane will therefore not create geysers of water blasting out into the vacuum like fire hoses. Rather, if there are punctures, the low-pressure water will tend to just dribble out.

The central sphere, nestled deep in the middle of the colony, will be equipped with its own water shield, also five meters thick. This provides the colony with a storm shelter, where every person can take refuge in the event of a dangerous solar flare (see Appendix 3.6). Inside the central sphere, the colonists will be protected by 10 meters of water, plus the entire mass of Asgard's atmosphere and all of the plants, furnishings, and equipment in the colony. This will amount to a shield density of over 1000 kilograms per square meter, enough to protect the colonists from any foreseeable solar storm.[254]

The outer bubble will be coated with a thin layer of reflective gold.[255] The gold film will screen out harmful ultraviolet and other rays, while allowing visible light through. Gold, a noble metal, is chemically inert, and resists the corroding effects of atomic oxygen which might otherwise degrade the bubble membrane.[256] This may seem an expensive measure, but gold leaf is typically only 3 millionths of an inch thick. At a gold foil thickness of .25 microns, the outer membrane will require five and a half tons of gold, at a cost of about $80 million. (The reflective coating may be bonded to a thin sacrificial film which will form a separate outer layer , see App. 3.1)

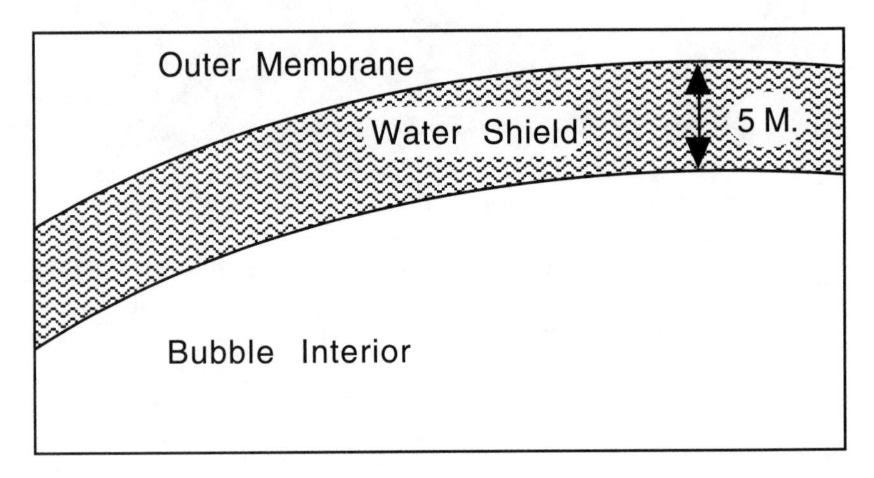


***Fig. 3.8 - Cross section of water shield.***

The reflective coating also helps regulate the amount and type of radiant energy that is allowed into the ecosphere. Maintaining a comfortable environment inside the bubbles is relatively simple. The internal temperature can be regulated with precision by adjusting the amount of sunlight allowed in. If the balloon were entirely opaque, the interior would be pitch black and cold enough to liquefy air. If, on the other hand, the balloon were entirely transparent, the interior would be flooded with raw sunlight and hot enough to boil eggs. A happy medium, between these two extremes, can be reached by adjusting the bubble's opacity to maintain a temperature of 20° C. (68° F.). This approach will enable us to provide interior lighting as well as controlling temperature, without using much electric power.

It would be possible to install an adjustable Polaroid layer in the outer membrane to regulate the amount of energy allowed in. These Polaroid filters can be adjusted electronically so they are entirely opaque or completely transparent. With such a system, we could continuously adjust the amount of radiation entering the ecosphere. Such filters already exist and cost about $8/ sq. ft. We can probably do without them, however, since the colony is in a steady state. Sunlight shines on the colony virtually all the time, and waste heat generated internally should remain fairly constant. Therefore, the reflectivity of the outer bubble can probably be set permanently at some predetermined level.

Five meters of pure water is almost completely transparent. All colors of visible light, except long wave reds, penetrate up to 15 meters of water more or less equally.[257] Consequently, certain shades of dark-red lipstick won't look good inside the colony. Otherwise, things will appear with their normal colors, much as they do under bright indoor lighting.

Water is transparent to visible light, but absorbs infra-red (IR) and ultraviolet (UV) wavelengths. Sixty percent of sunlight's energy, mostly infra-red, is absorbed in the first meter of water. After passing through five meters of clear water, only 30% of sunlight's energy remains, with virtually all of the IR and UV wavelengths screened out. This is a highly desirable property of our radiation shield, since both IR and UV are unwanted inside the colony. Ultra-violet light causes cancer, accelerates aging, interferes with the immune system, and degrades many man-made materials. Infra-red energy causes special problems.

**Stay Cool**

If too much infra-red enters the ecosphere it will cause serious overheating problems.[258] The reflective gold coating and water shield work together to minimize the amount of IR that reaches the interior. The water shield also aids in absorbing infra-red generated as waste heat inside the colony.

One of the colony's main technical difficulties will be in getting rid of its waste heat. Waste heat is generated by every human activity from cooking to making love. The human body is a powerful waste heat generator, as are all types of machinery and electronics. Dissipating waste heat isn't easy because the colony is afloat in an infinitely large vacuum bottle. The only way to dump waste heat is to radiate it away.

A fluid heat shield makes it easier to balance and maintain temperature equilibrium inside the ecosphere. Water in the shield will be separated into cylindrical tubes running horizontally around the latitude of the ecosphere. Water in each tube is slowly circulated from the sunlit side of the ecosphere to the shaded side. The temperature of the water is thereby kept a few degrees cooler than the air inside the sphere. To maintain a shirt-sleeve comfortable 20° C., water in the shield is kept at 18°, on average. The water on the sunlit side will be about 20° and the water on the shadow side will be about 16°. Water in alternate tubes circulates in opposite directions, so that temperature differentials between the two hemispheres are smoothed out. (Figuring out the exact thermodynamics of such a system on the back of a napkin is quite impossible. It may actually be necessary to circulate some of the water through heat exchangers on the dark side to get adequate cooling.)[259]

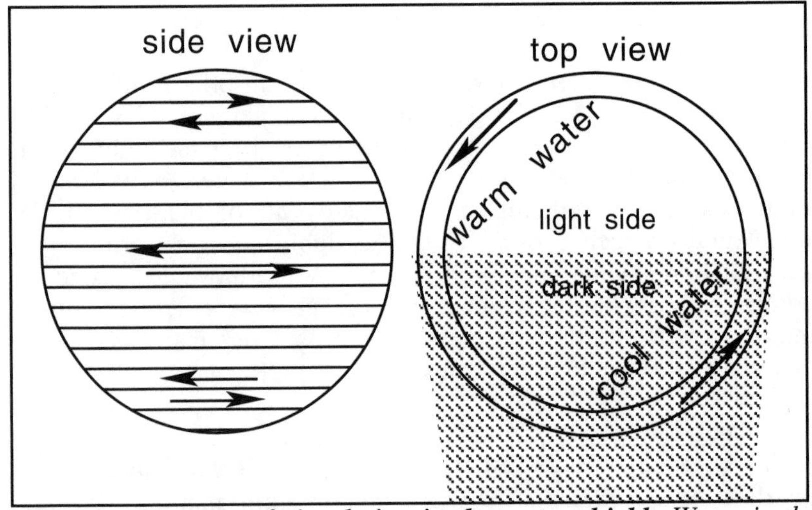


***Fig. 3.9 - Diagram of circulation in the water shield.*** *Water in the radiation shield is contained in tubes five meters in diameter. The water moves in opposite directions in each alternate tube to even out heat distribution. The speed of circulation varies depending on the latitude and corresponding length of the tube.*

A water shield also has the great advantage of serving as a strategic reserve of vital feed stocks. In addition to being a water reservoir, it is also a reservoir of rocket fuel, and of oxygen.

With an abundance of electrical energy on hand, water molecules can be split into hydrogen and oxygen. These are the vital components needed to power LHOX rockets. Asgard can serve as a gas station for wave-riders needing to recharge their maneuvering rockets, and for astro-tugs and shuttles outbound to the Moon and beyond.

By electrolyzing water, we can also generate as much reserve oxygen as we need for the colony's atmosphere. Ninety-nine percent of Asgard's 26,000 ton atmosphere is oxygen, as is ninety percent of the six million ton radiation shield.[260] There is enough oxygen in the water of the radiation shield to replace Asgard's atmosphere over 200 times.

**Attack of the Killer Meteoroids**[261]

In the minds of many people, the idea of a space bubble triggers an automatic reflex: "A meteor is bound to come along and pop it." This is an idea that owes more to Flash Gordon than it does to real space science. Unlike the real and omnipresent hazards of radiation, meteoroids represent only an occasional—and highly improbable—danger to space habitats. The number and size of meteoroids passing near the Earth is well known from ground, lunar, and satellite based observations. Contrary to popular belief, meteors large enough to do serious damage are extremely rare. In space, stuff is sparse. There are lots of bits and chunks of debris flying around, but there is also a lot of space, and the chances of one of these stray pieces coming through your living room in Asgard are remote in the extreme.

Meteoroids with a mass of more than one gram—the size of a pea or larger—are spread through space with a density of one in every <u>trillion</u> cubic kilometers.[262] On average, small meteoroids are separated from each other by over 12,000 kilometers. A given square meter of space will see a one gram meteoroid pass through once every three million years.[263] In terms of raw probability, a person on Earth is five times more likely to be struck by lightning than a person in space is to be hit by a meteoroid the size of a pea.[264]

As the size of meteoroids goes up or down, the chances of being hit by one fall and rise correspondingly, and rather dramatically. A given square meter of space will see a ten gram meteoroid—the size of a large grape—pass through only once in every 320 million years.[265] A large meteoroid, the size of a Volkswagen, will pass through a given square meter once every few <u>billion</u> years.

Conversely, very small meteoroids are relatively common. A sand-grain sized meteoroid, weighing one-thousandth of a gram, can be expected to hit a particular square meter once every 63 years—a once in a life-time event. The really small stuff, microscopic meteoritic dust, smaller than bacteria, is fairly thick. Each square meter will be hit by these micrometeoroids, about three times a year, but these pinpricks are relatively harmless.

Asgard is a big target. With an overall diameter of 600 meters, Asgard presents a strike zone with an area of a million square meters. An object this size has a much greater chance of being hit. Even so, meteor strikes of any magnitude will be the rarest sort of natural disaster. On average, the colony will be struck by a meteoroid as large as a grape once every three hundred years. Over the same time-frame, Los Angeles will suffer through nine or ten catastrophic earthquakes.

Even meteoroids as big as golf-balls will probably not be able to penetrate the colony's radiation shield. To pierce the shield, a meteoroid must possess sufficient momentum to plow through five meters of water. At the speeds involved, hitting water is like hitting solid rock. Since the shield has the same effective mass as a slab of granite six feet thick, it will take a pretty large object to blast through. Realistically, we can expect a meteoroid large enough to actually penetrate Asgard only about once in every five thousand years. An incident that can be expected only once in five millennia—an expanse of time nearly as great as all of recorded human history—is not something we need to lose sleep over.

As for larger meteoroids, their very rarity makes them a risk hardly worth contemplating. A 1000 kg. meteoroid—about the size of a desk—could not be reasonably expected to hit Asgard in less than 100 million years.[266] Larger meteors are able to pass through the atmosphere, and so are just about as apt to hit a given spot on the Earth as they are to hit a space colony. Therefore, the likelihood of a large meteor hitting Asgard is just about the same as the likelihood of one hitting downtown Atlanta.[267]

Even a worst-case scenario, involving a catastrophic strike by a large meteoroid, would not necessarily be fatal to the colony. Asgard's interior space is subdivided into a multitude of spheres nested within spheres, each one of which, down to the smallest apartment, is capable of retaining its own atmosphere. This design gives Asgard's internal space an almost foam-like consistency, which is nearly invulnerable to such disasters. It will be impossible for a single meteoroid to penetrate more than three of the largest inner bubbles. Just so, it will be impossible for a single projectile to penetrate more than three bubbles in any sub-cluster. It would take some time for the air in one of the bubbles to leak out through even a large hole, so people would have time to take shelter in one of the undamaged bubbles. The remoteness of the danger and the countermeasures inherent in Asgard's design renders this space colony nearly as safe from meteoroid threats as any terrestrial city.

**The Kessler Syndrome**

As with so many things, the natural threat is dwarfed by the magnitude of the man-made menace. Orbiting junk is a much bigger danger to us than indigenous meteoroids. The space around Earth is approaching saturation with man's aerospace garbage. Everything from dropped cameras to exploded satellites now orbits the Earth in a deadly swarm. At last count, there were three and a half million bits of rubbish up there, and the numbers are increasing rapidly.[268] The U.S Space Command tracks 7000 objects that are larger than 10 cm. in diameter. Fortunately, most of this debris is confined to two belts at altitudes of 900 and 1500 km. Geosynchronous orbit, where Asgard will be positioned, is still relatively clean, with fewer than 500 large pieces of satellite trash.

Tiny bits of orbital detritus may present the greatest danger. Objects smaller than a centimeter account for 99% of the total number, but only a tiny fraction of the total mass. Despite their tiny size, things as insignificant as flecks of paint can be deadly projectiles at orbital velocities. In 1983, a speck of paint hit the windshield of the Space Shuttle. The thick glass was so badly damaged it had to be replaced.[269] As man's presence in space grows, the cloud of meteoric junk will thicken with potentially disastrous consequences.

It sounds like the title for a Robert Ludlum plotboiler, but the Kessler Syndrome is all too real and infinitely more ominous. Like the China Syndrome, the Kessler Syndrome involves a kind of runaway chain reaction.[270] In the Kessler Syndrome, the chain reaction results from having a critical mass of satellites and debris in orbit. At critical mass, the destruction of a single satellite can create a cloud of debris which, in turn, destroys more satellites, and creates more debris. The result of the Kessler Syndrome would be an impenetrable belt of fragments surrounding the Earth. Such a disaster could make space flight impossible for centuries. By some calculations, the critical mass required is only 70,000 fragments 1 cm or larger. We could reach that density within 20 years.

**Sword of Heimdal**

Natural meteoroids are rare enough that we can take our chances, and trust the gods of odds to protect us. To deal with man-made meteors, and to prevent the Kessler Syndrome, will, however, require a direct application of force. Thanks to the laser array atop Kilimanjaro, we have at our disposal over a gigawatt of power that can be focused with precision on any point within 36,000 km. That enables us to reach out and touch targets even beyond Asgard itself. By positioning additional mirrors in Asgard's orbit, we can reflect beams to focus on any point within an enormous volume of space.

Even very small objects can be tracked by ground-based radar. The deep space radars at Goldstone routinely track particles as small as two millimeters.[271] Radars planned for the Star Wars system will be able to track particles as small as half a millimeter. The Bifrost Bridge will incorporate such radars for precision tracking and aiming of the propulsion lasers. The same radar systems can do double duty, tracking particles of debris and directing the lasers to blast these targets into oblivion.

Hitting tiny high-speed particles at long range is tricky, even with laser beams. A pulse of laser energy can cross the 36,000 km. between Kilimanjaro and Asgard in .12 seconds. In that time, however, a paint chip moving at three kps will move several hundred meters. Therefore, even at the speed of light, it is necessary to lead these targets—like shooting ducks on the fly.

Small bits of debris, up to a few centimeters in diameter, will simply be zapped into vapor—like snowflakes hit by a flame-thrower. Larger chunks, and even whole defunct satellites, can be maneuvered by applying laser propulsion. Parts of the object will be vaporized, thereby providing thrust, just like the ice propellant on the Bifrost launch capsules. This will require precise calculations, factoring in the object's geometry, rate of rotation, and composition; but these are not tasks that will daunt the central computer on Aquarius. By using laser thrust, we can maneuver large objects so that they fall into the Earth's atmosphere; or they can be boosted out of Earth orbit entirely, and sent on a spiral path towards the sun and destruction. With Heimdal's light saber, we can clean up space, protect Asgard from meteoroids of all kinds, and forestall the dreaded Kessler Syndrome.

Only the smallest micrometeoroids will penetrate our laser defenses. Dealing with them will be largely a matter of plugging occasional small leaks in the outer membrane. The exterior surface will be hit by sand-grain sized particles about six times a day. The resulting pinholes will allow water to leak out of the radiation shield. Such leakage would be very minor compared to the mass of the shield, but should nevertheless be minimized. Repair robots will roam the outer surface of the bubble, scanning for leaks with infra-red lasers. Leaking water will instantly evaporate, forming an expanding cloud of water vapor that will look like a smoke signal to an infra-red scanner. The robot will fly to the leak, and then, on its own, or with the help of a human tele-operator, patch the pinhole. One such repair will be needed about every four hours. No more than one or two repair robots will be required.[272]

**Ecospheres**

Once Asgard is constructed, there will be a beautiful golden bubble floating majestically above the Earth. We need only move in to be completely at home in space. The air is breathable; the pressure is adequate; radiation is low; the temperature is a balmy 20° C. (68° F.); and the weather forecast is continued bright and sunny for the next six billion years. If we had a six pack of Coke and some pizza we could stay all week. It takes only this thin bubble to transform the hard vacuum of space into a welcome oasis of life. Such a free-floating ecosphere will be completely self sustaining, with all of its inhabitants' needs for food, energy, and shelter supplied by the colony.

To be a true ecosphere, Asgard must be capable of supporting the lives of its inhabitants without importing any material goods. This is accomplished by raising recycling to its ultimate potential, at which point all wastes serve as feed stocks. This can be achieved with a surprisingly simple process loop, to which the only input is energy in the forms of sunlight and electricity.

There is an experiment presently being worked out in Arizona that is nominally based on the same ideas. Biosphere II represents one approach to creating a completely self-sufficient closed-loop ecosystem. The basic idea behind Biosphere II is to recapitulate the Earth in microcosm. Inside the Biosphere, a tropical rain forest, a desert, a mountain, a savanna, a marsh, a river, and an ocean have been replicated. Each of these "biomes" has been populated with representative species of plants, animals, insects, even microbes. The presumption is that each of the biomes will perform the same ecological function in Biosphere II that it performs in Biosphere I—Earth.

This approach may well be the right one when terraforming on a small scale, as in the construction of ecospheres on the Moon, Mars, or the surface of asteroids. It is too complex and cumbersome, however for free floating space habitats like Asgard. When building space habitats there are three basic tenets to live by: simplicity, simplicity, and simplicity.

There are only two interactive biological components in Asgard's ecosystem: people and algae.[273] Through photosynthesis, the algae convert sunlight, carbon-dioxide, and water into oxygen and nutrients. Humans breathe the oxygen and consume the nutrients to support their metabolism, producing carbon dioxide and wastes in the process. These wastes are broken down into their elemental constituents and are recycled as nutrient feed stocks for the algae. The algae then reconvert the $CO_2$ and wastes to oxygen and nutrients. This closed cycle can reiterate indefinitely, as long as there is sunlight to sustain the algae and energy to run the machinery.

The same oxygen, hydrogen, nitrogen, carbon, and other atoms can pass endlessly around the cycle, changing their chemical formations, but never being expended. Like the Earth itself, Asgard will be an independent living system, capable of spinning through its eternal ecocycle as long as the sun shines.

**Space Diatoms**

Asgard will embody, even more fully than Aquarius, the concept of city as life form. Asgard will function very much like a single celled photosynthetic organism. It will resemble, both in form and function, the Volvox—an elegant spherical colony organism.[274]

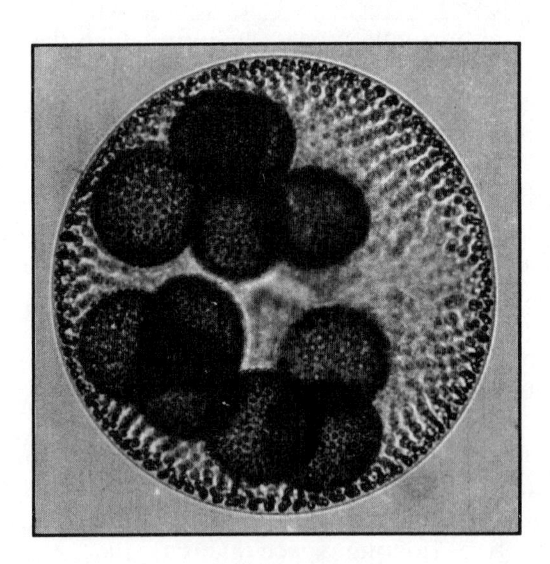

***Fig. 3.10 - Volvox. Asgard will be like a new variety of macro-organism—a space diatom.*** Photo by Eric Grave, © Photo Researchers, Inc.

The internal workings of the colony, much like the metabolism of a diatom, will be solar powered, and virtually self-sufficient. Ultimately, Asgard will be capable of functioning without any inputs from the outside except sunlight. In this regard, Asgard will be like only one other known organism—Gaia.

Comparing the metabolic processes of Asgard with those of Gaia is like comparing a diatom to a blue whale. Nevertheless, the two are alike in their independence from outside material inputs. Asgard will be like a tiny budding spore of the mother planet, microscopic by comparison, infinitely simpler in its functions, but fully capable of supporting itself independently.

Like other life forms, Asgard will eventually replicate itself. In time, space diatoms will come to saturate the oceans of vacuum around our sun. The yellow-white glare of our home star will filter through clouds of these celestial plankton. From a distance, our solar system will glow with their aqua-marine phosphorescence. (See Plate No. 16.)

**Algae Again**

All of our oxygen and most of our food will be provided by our old friends, the blue-green algae. The average human requires 600 liters of oxygen per day. To produce this amount of oxygen, algae will consume 720 liters of carbon dioxide. In the process, the algae also produce 600 grams of food (dry weight).[275]

An average person needs 2500 calories per day, which can be obtained by eating around 540 grams of food.[276] Of this intake, 200 calories, 50 grams, should be in the form of protein. No more than 625 calories, 70 grams, should be in the form of fats.[277] The balance of the diet, 1675 calories, 420 grams, should be in the form of complex carbohydrates. These needs can be supplied entirely from the 600 grams of algae that will be produced as a byproduct of oxygen production.

This is another in a long series of weird coincidences. We need oxygen and we need food, two completely unrelated commodities in our terrestrial existence. Yet, when we contemplate our move to space, we find this strange match between the two requirements.

An ideal algae would consist of 9% protein, 13% fat, and 78% carbohydrate, with an optimal balance of minerals and vitamins. Spirulina, the algae we culture in the marine colonies, is a high protein variety—65%. Other strains predominate in other nutrients. A number of different algae strains will be cultured, each tailored by genetic engineering to produce specific nutrients and compounds to serve as the feed stocks for food synthesis.

In pond cultures, algae populations reach a maximum density of around one gram per liter.[278] The population density of algae in a pond is limited by the simple fact that if the soup gets too thick, the light can't penetrate it, and the algae can't grow. In Asgard, where sunlight is abundant, this problem can be easily overcome. Nutrient-rich water will be passed through coils of clear tubing along the sunlit side of the ecosphere. In the coils, each algal cell is exposed to optimal sunlight. Tubing coils will run around the colony's equator, immersed three meters deep in the water shield. Beneath the gold foil screen of the outer membrane, and sheltered by three meters (ten feet) of water, the algae will be protected from ultraviolet and other radiations that could hamper their growth.

As the algae move along, suffused with light of a controlled intensity and wavelength, they reproduce rapidly. The algae convert $CO_2$ and nutrients into proteins, fats, carbohydrates, and many other desirable compounds. Once the algae have utilized all the available nutrients, they are filtered out, and the water is recycled through the system. Conditions in such an arrangement are strictly controlled, so there is no contamination of the culture by bacteria or predators.

Under controlled conditions, algae can double their population four times a day. Using genetically engineered strains, closed-loop systems have already achieved production rates as high as 100 grams of algae (dry weight) per liter of water.[279] The initial culture needs only 40 grams of algae to produce 600 grams of food, plus having 40 grams left over to start the cycle again.

Each space colonist will require six liters (1.6 gallons) of water passing through the system to produce all the food and oxygen she needs. The system is so compact and efficient that all the oxygen and food needs for the entire colony can be produced from a tiny area. Six helically wound tubes circumscribing the equator are all that are necessary.[280]

The system is closed, except for sunlight which powers the whole process. The gases, nutrients and water can cycle around the loop indefinitely. Algae's genetic material can recapitulate new populations every day, virtually forever without depletion or exhaustion. Algae have been engaged in exactly the same closed-loop process—albeit on a somewhat larger scale—here on Earth for billions of years.

Closed loop systems based on algae have already been demonstrated on a bench scale. Lab rats have been sustained for long periods of time, totally cut off from the atmosphere, breathing oxygen produced by algae.

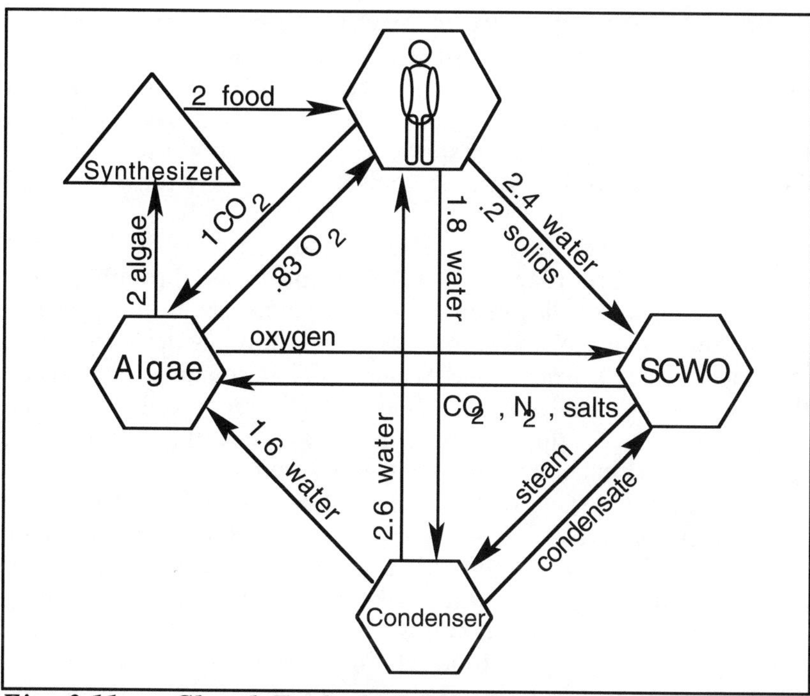


***Fig. 3.11 - Closed Ecological Life Support System (CELSS) process diagram (units in kgs.).*** *Nothing enters or leaves the loop except energy. Algae and humans live symbiotically, providing each other with respiratory gases and nutrients. The process diagram is not much different for the Earth's own ecosphere, where the system also runs on photosynthesis, 90% of which is performed by blue green algae.*[281]

Human beings are biochemical machines. Like other machines, they need fuel and water, and their operation produces exhausts and wastes. The average person needs to take in about four kg. of material every day. Fourteen percent of this mass is dry fuel, 18% is oxygen, and the rest is water. The fuel, in the form of food, is 'burned'—combined with oxygen—to produce energy. Exhaust gases—carbon dioxide, and water vapor—and 'ash' are produced as waste products. Not surprisingly, four kilograms go into the human machine and four kilograms come out.[282] The food provides 2500 kilogram calories of energy and is transformed into 10,000 BTU of heat. The wastes are 20% carbon dioxide, and 75% water; solid wastes—the 'ashes'—amount to only about 2% or 80 grams, dry weight.

Humans actually produce more water than they consume: 2.3 kg. of water are taken in as food and drink, and 3.4 kg. are given off. The extra kilogram of water comes from the oxidation of food, which produces water vapor as a by-product. This is expired through respiration and perspiration. If the build-up of water vapor is not controlled, space colonists can quickly drown in their own sweat.

All of this water needs to be recovered and recycled. Coils of cooling pipes are placed in various locations throughout the spheres. Water that has been chilled by being passed through heat exchangers open to space on the dark side of the ecosphere, circulates through these coils. Water vapor in the air condenses on these cool pipes, and pure distilled water is recovered. The cold water coils serve three vital functions: cooling the colony, recovering purified water, and controlling humidity.

Excretory water and other metabolic wastes are also easily recovered and recycled. Obviously, we do not want to mix such wastes directly with the algae that are going to serve as our food stocks. Nobody wants anything like this anywhere near anything they're even thinking about eating. The idea of directly recycling human wastes into the food chain is repugnant and entirely unnecessary. All wastes, including urine, feces, wash water, and food scraps will be processed and reduced to their basic organic constituents before being sent into the algae tubes as plant nutrients. Essentially, though in a unique way, these wastes are all 'burned', and only the highly purified elements are recirculated as nutrients for the algae. "Sewage", as such, never comes in contact with the algae.

**Super Critical Water Oxidizer**

All organic wastes are completely broken down into their constituent elements. The machine which performs this miraculous alchemy is called a Super Critical Water Oxidizer (SCWO).[283] In the SCWO, organic waste products from the colony are mixed with waste water, forming a slurry. The slurry is heated to high temperature (480° C.) at high pressure (3500 psi). Under these conditions water reaches a 'super critical' state. At super-criticality, there is no difference between the properties of water in its liquid and its vapor phases. In essence, water at these temperatures and pressures becomes "liquid steam". In this phase-state limbo, $H_2O$ can perform amazing stunts, not the least of which is that of sustaining an open flame under water. When compressed oxygen is injected into the hot waste slurry, all organic wastes are completely oxidized in less than a second. In essence, the SCWO 'burns' wastes under water.

A wide variety of noxious wastes, fecal matter, urine, scraps, soap scum, and garbage can be fed into the SCWO, but only pure oxidized compounds come out. The main product is steam, which is recirculated through closed piping coils to pre-heat the incoming slurry. The cooled steam is then recondensed as pure water. The other products are mostly carbon dioxide and nitrogen in the form of fixed nitrates. These refined compounds and nutrient salts then go into the hydroponic solutions supporting the green plants and algae.

**Synthetic Food**

Dried algae will be processed to produce synthetic foods with a variety of flavors and textures. At an average water content of 75%, 600 grams of dried algae can be converted into almost two and a half kilograms (5.3 lbs.) of food. In Aquarius, more than half of our food was still 'natural'; that is, we were still eating fish and clams, bananas and papayas. In Asgard, this will be a rarer sort of luxury. Animal-products will be nonexistent, and even fruits and vegetables will be scarce. Nearly 100% of our daily eating experience will be in the form of synthesized food, derived from an algae feed stock.

It will be a highly efficient and economical system, but it need not be unpleasant. At the mention of eating algae, one imagines a tasteless green paste or something equally unappetizing. This won't be the case. Life is not a zero-sum proposition. Taste and pleasure do not have to be sacrificed to economy. We humans are a clever bunch, and if tasty food is something we want, then we shall have it.

Synthetic food production will be an area of intense study from our earliest days in Aquarius. By the time we are in Asgard, the art of food synthesis will be highly developed. Though all the food starts out as algae, it will end up as something seemingly altogether different. Using recombinant DNA techniques and a growing knowledge base, we will create a wide assortment of flavors; some mimicking natural foods and some entirely new.

By implanting known sequences of DNA into algae strains we can produce virtually any of the 5000 to 10,000 flavors that occur naturally on Earth. This technique is already leading to an explosion in the flavor market. The production of flavor additives already amounts to 25% of the gross value of the world food additive market and is growing at the rate of 30% per year.[284] This technology will not only prove of immense value to us for our own purposes, allowing us to substitute algae for virtually all foods, but will also be an important Millennial industry. Flavor additives sell for up to $4000/kg., and have an average price of $150 to $200/kg. Flavor additives are an ideal export product for a space-based economy. Flavors are light in weight, high in price, and demand compact sophisticated technology to produce.

The genetic laboratories of Asgard will be able to produce any flavor you have ever experienced, and thousands more your taste-buds have never dreamed possible. Just imagine: before the conquistadors returned from the New World, Europeans had never tasted chocolate. What ecstasies of the palate might not await us in the next new world? Celestial Chocolate? Space Spices? Cosmic Caramel Custard? The mind boggles at the mouth-watering possibilities.

Compared to some foods eaten today—TV dinners, microwave pizza, fluorescent orange cheeze spread—our synthetic foods will be gourmet fare. Not only will they taste better, they will provide better nutrition. Foods synthesized from algae will be low in fat, cholesterol, and calories; but high in fiber, protein, and vitamins. Such foods will have the additional advantage of being entirely pure, uncontaminated by pesticides, pollutants, or preservatives.

Admittedly, there is a psychological barrier to be overcome. If we are going to live in space, however, we must make certain adjustments in our basic attitudes. While all the physical pleasures of eating—taste, texture, aroma, and visual appeal—will be available to us in space, some of the psychological pleasures will not be. For those who cannot make this mental adjustment, the new frontier will remain closed. Space is a vast and foreboding place. Unlike pioneers of the past, we can conquer this new domain in relative physical comfort. We may not, however, be able to do it in perfect psychological comfort. Thus, the final frontier will remain, at least initially, the exclusive province of those among us who possess the versatility, the adaptability, the sheer psychic toughness, to boldly eat what no man has eaten before.

**Space Cadets**

Inhabitants of Asgard will either be native Asgardians—in which case everything about life in space will be as natural as "the first breath from a baby"—or they will be Aquarians. A full generation elapses from the beginning of Aquarius to the completion of Asgard. The first inhabitants of Asgard will have lived on Aquarius for years.

Many of the Asgardians will have been born on Aquarius and will have grown up there. These space cadets will have spent thousands of hours under water. For these water babies, dealing with the dangers of life in space will be second nature. Aquarian kids will come and go through air locks without any more concern than mountain goat kids cavorting on a cliff. Space-suits will be as natural to Aquarians as mukluks to Eskimos. Cub scouts in Aquarius and Asgard will earn the same blue-faced merit badge for being able to expel all the air from their lungs and hold their breath for two minutes. People trained to cope with life under water will find life in space a piece of cake.

The great killing danger in both places is the difference in pressure between the inside of the habitat and the hostile environment outside. Underwater, these differences are positive and virtually unlimited. In space the difference is reversed, but the principles remain the same. The pressure gradient between Asgard's interior and the vacuum is what an Aquarian would experience under just two meters (seven feet) of water—a depth any Aquarian kindergartner would sneer at.

All the aspects of life in a space colony, so alien to us, will be familiar and comfortable to Aquarians. Weightlessness, airlessness, pressure gradients, space suits, synthetic food, absence of seasons, boundless spaces, isolation from the terrestrial world—all these things, which might terrorize land lubbers, will seem just like life back home to Aquarians. Asgard holds no terrors for marine colonists.

**Float Like a Butterfly**

Asgard has practically no gravity. It does not spin to produce artificial gravity on its inner surface in the manner of other proposed space habitats. Proposals for space stations, like the classic design from Stanley Kubrick's film *2001*, have a rotating structure. Such solutions require rigid construction to withstand the considerable stresses involved. This necessitates something heavy, and rules out simple solutions like bubbles. Adopting such a system would waste most of Asgard's advantages.

To fly, freed from the clutches of gravity is, for me, a lifelong dream. Fulfillment of this dream will be an attraction which draws many of us into space. In the wide open spaces of Asgard we will be able to fly more freely even than birds. We can swoop and soar with abandon, propelled by jets, propellers, or wings. With a pair of sea-silk bat wings, a flier can propel himself aloft, darting and gliding in ways the Caped Crusader never dreamed.

***Fig. 3.12 - Space station from Stanley Kubrick's* 2001: A Space Odyssey.** *Classic space station design calls for a spinning structure which induces artificial gravity on its inner surfaces. This has certain advantages but is enormously expensive.* Photo Courtesy of the Academy of Motion Picture Arts and Sciences. 

The bicycle can become, with a few modifications—a propeller replacing the back wheel, and ailerons and rudder replacing the front wheel—an aircycle. Can the *Tour de Asgard* be far off? For those who prefer less aerobic means of propulsion, we have a wide array of cheap Japanese flying belts in an assortment of styles and colors. With a small hydrogen powered turbo-fan, one can fly through the air like a lark on the wing.

Artificial gravity has been considered essential for space habitats. Without the continuous stress of gravity, our muscles and bones deteriorate. It's a matter of 'use it or lose it'. In microgravity, the skeleton tends to decalcify, and muscles atrophy. This is not a mysterious process due solely to weightlessness. It is a well documented condition called 'disuse osteoporosis'. Patients confined to prolonged bed-rest, victims of paralysis, even terminal couch-potatoes suffer from the same affliction. The rates of muscle and bone loss are about the same for astronauts as for people confined to bed.

Research seems to indicate that living without gravity for long periods of time is impossible. Long term studies, especially by the Russians, seem to indicate that a person's body will waste away in low gravity. Immobilized experimental rats lose up to 90% of the structural strength in their bones in less than a month.[285] During the Gemini flights, astronauts lost 10 to 20% of the calcium from their bones in just eight days.[286] Loss of muscle and bone tissue has been very serious in astronauts who spend more than a month in space. Rates of bone loss aboard Skylab showed that even with exercise, bone loss in zero-*g* could amount to 25% a year.[287] If we are to become permanent residents of space, we must find a solution to this problem.

***Fig. 3.13 - Bungee jogging.*** *Astronaut Joe Engle, exercising aboard the Space Shuttle to prevent muscle and bone loss. Permanent residence in zero-g will require somewhat more sophisticated countermeasures than these state of the art rubber bands.* NASA

Rather than trying to create gravity, we should attack the problem directly. The problem isn't with gravity; the problem is with human muscles and bones. If muscles and bones waste away without stress, then let's stress them. This is a much simpler solution than spinning the entire colony.

**Sing the Body Electric**

Human muscle is electrically driven. When an electric impulse is applied to muscle tissue it contracts. The greater the impulse, the more powerful the contraction. This makes it feasible to stress skeletal muscles in zero-*g*.

Normally, the impulses that drive muscles are delivered by the peripheral nervous system, but they can also be provided artificially. Contact electrodes, placed against the skin, can deliver an electric charge to a muscle, causing it to contract as if it were being moved voluntarily. Extensive clinical experience has shown that electrical stimulation of muscle can lead to the same increase in strength that is obtained with voluntary muscular contractions.[288] In space, we will use electrical stimulation techniques to maintain and even improve our bodies.

Electro muscle stimulation therapy is still in its infancy, but it has already produced some impressive results. Before the 1988 Olympics, an American weight lifter who had been competing for ten years underwent a program of electrical therapy. His thigh muscles were stimulated by being contracted 10 times, three times a week for six weeks. During the training cycle, his maximum lift went from 330 pounds to 415 pounds, an increase of 25%.[289] The results were so impressive that this particular lifter was tested for steroids three times.

Although electro stimulation therapy is no picnic, neither is weight lifting, running, or any other form of strenuous exercise. There is, of course, a painful sensation, just as there is anytime the major muscles are contracted powerfully and repeatedly. However, there is a happy side effect of the electrical therapy that reduces discomfort. When the voltage is applied in a high frequency current—around 70 cycles per second—the electrical signal cancels pain signals from the muscle which are of about the same frequency. With further experimentation, this analgesic effect could be enhanced, making the therapy relatively painless. Eventually, when fully refined and integrated with a computer to monitor brain waves, this therapy could even be applied during sleep.

Human muscles occur for the most part in pairs. Muscles can only contract, so it takes two opposed muscles on each major joint to obtain a full range of motion. We can take advantage of this arrangement by alternately stimulating opposing muscle groups. While one muscle group is progressively contracted, the other can be relaxed. Floating free in space, we can move the body through a range of exercises, stimulating all of the major muscle groups. It will probably only be necessary to fully exercise the body three or four times a week to maintain a high state of muscular fitness. Electro muscle stimulation (EMS) will make it possible for us to live in zero gravity and yet still maintain a strong musculature.

EMS will not only keep the body from deteriorating, but can also develop it. By integrating the system with a controlling computer, loads placed on the muscles can be incrementally increased week by week, causing the muscles to grow stronger, and, if desired, larger. The computer can be programmed to create a tailor-made body. Working on a video screen, you can custom design your own body, adding mass to shoulders here, flattening the stomach there, and increasing overall tone and definition. People will be able to pick and choose any desired body type; some men might go for the hulking mass of Lou Ferigno,[290] others might opt for the lithe grace of Greg Luganis;[291] women might choose the muscularity of Corey Everson,[292] or the slim toned look of Jane Fonda.[293] Musculature will be a matter of personal choice. Selecting a body will be as simple and routine as choosing a car is today. (For myself, I intend to order a classic '69 Schwarzenegger.)

**Dem Bones**

Electrical stimulation of the body's muscles will have an attendant effect on the skeleton. The skeleton, like the muscles, responds and adapts to stress. On Earth, stress is partly induced by gravity. In space much of this stress can be duplicated or substituted by the action of the muscles. As the opposed muscles are stimulated, they pull against each other and against the bones to which they are attached. When muscles contract, the bones bear the load. Electrostimulation of the muscles will therefore help keep bones from deteriorating in zero-*g*.

The skeleton can also be electrically stimulated directly. Bones, like muscles, are governed by electrical charges, though in a fundamentally different way. The mineral matrix of bone forms a 'piezoelectric' material. The piezoelectric effect is a phenomenon characteristic of certain crystalline materials like quartz. When these materials are subjected to a mechanical force, they generate an electrical charge.

Whenever a mechanical force is placed on the skeleton, the bones produce a small electrical voltage in response. When a person runs, walks, or even stands, tiny electrical charges—on the order of a few tenths of a volt per centimeter—are generated in the bones.[294] In areas where bone is under compression, the charge is negative; in areas under tension it is positive.[295] These piezoelectric charges may be the signals which induce bone growth. Correspondingly, the absence of such signals may cause the generation of bone to cease under conditions of weightless.

The physiology of bone is complex and dynamic. Bone mass is in a continuous state of degeneration and regeneration. In space, the degenerative processes continue, but in the absence of stress-induced electrical charges, regeneration ceases. The result is lost bone mass.

By surrounding the body with electrical coils, and generating an oscillating current in the coils, a pulsing electromagnetic field can be created in the space between them. This pulsing force field will induce an electric charge in the bones, stimulating osteogenesis—bone growth. Immobilized rats given this treatment experience only a fraction of the bone loss suffered by unstimulated rats.[296]

Using an external electrical coil to directly stimulate bone growth would provide a solution to the problem of osteoporosis in zero-*g*. The coils are non-invasive; they can be used during sleep; the process is painless; and, as far as is known, there are no adverse side effects.

In addition to electrical treatments, various hormonal, dietary, and other metabolic therapies can be pursued. There is something about the metabolism of humans that predisposes us to disuse-osteoporosis. Marine mammals like dolphins and whales spend their entire lives in the effectively gravity-free environment of the ocean, yet their skeletons don't decalcify. Hibernating mammals, like bears, are dormant for months, yet emerge from their dens with skeletons intact. Eventually, we will discover the secret of these animal's adaptive responses. Then, perhaps, we can make similar adaptations to deal with the gravity-free conditions of our new environment in space.

Electro-magnetic techniques for stimulating bone and muscle growth could be supplemented by exercising in gymnasiums with artificial gravity. The gyms, equipped with game courts and exercise equipment, might be built along bands running inside a tube, around the inner equator of Asgard's outer sphere. A track could run down the center of the band along the circumference of the bubble, a distance of over 1800 meters. A band 15 meters wide would provide 28,000 square meters of floor space. The band could be reinforced by kevlar cables, running longitudinally down its entire length, obviating any need for it to be guyed from the center. The *g* ring could spin at a rate of two revolutions per minute, around 200 kph. A person on the inside surface of the spinning ring would experience 'gravity' slightly greater than Earth-normal. By speeding up or slowing down the rate of spin, the simulated gravity could be increased or decreased. This would allow a person to continually improve her level of fitness until she could exercise for prolonged periods at 2 *g*s.

Not only would high gravity exercise be guaranteed to halt the deterioration caused by life in zero-*g*, but it would also induce a level of strength and fitness unapproachable by people training on Earth. Space colonists might even be barred from the Olympics as unfair competition. By providing artificial gravity where it is needed, we could avoid the costly and complex measure of having to spin the entire colony.

Electrical muscle and bone stimulation together with high-*g* exercise, will not only counter the effects of zero-*g*, but will also slow the deterioration of aging. Using EMS, a person can maintain physical fitness, even into advanced old age. Elderly space colonists will have strength and muscle tone to spare well beyond the century mark. Without the ravages of ultra violet light, and gravity to wrinkle and distort the skin, or the deterioration of skeleton and muscles to distort the body, white hair might remain as one of the only distinctive markers of advanced age. It may be hard to tell if a given space colonist is 55 or 105.

**Live Long and Prosper**

Living in space can dramatically enhance longevity. Freedom from gravity's tyranny probably won't have any direct effect on the aging process, but it will have a dramatic effect on the life-span of the already aged. For older people, life on Earth can become a daily struggle for survival. Gravity is the arch nemesis. Time takes its inevitable toll: muscles weaken, bones grow brittle, and just moving around saps available reserves of will and energy.

One of the leading causes of death among the elderly is nominally pneumonia. But, pneumonia is only what gets written on the death certificate. What really kills the elderly—in droves—are broken hips. An 80 year old woman falls down, breaks her hip, is confined to bed, weakens, contracts pneumonia, and dies. The real cause of death is gravity.

Extending the human life-span means extending old age. No one is interested in extending years of disability and discomfort. If we can grow old comfortably and conveniently in a gravity-free environment, longevity has much to recommend it. Life in zero-*g* will be substantially easier for someone who has passed the century mark than life at the bottom of a planet's crushing gravity well. In space, one may float from place to place with the effortless grace of a feather wafted on a puff of breeze.

Life in space has other advantages for longevity as well. In terms of appearance alone, the gains will be dramatic. The real symbols of old age, more than white hair or decrepitude, are wrinkles. The wrinkles are not the result of age alone. They are caused by two mutually reinforcing and destructive influences: ultraviolet radiation and gravity. In space, we can live free from both these scourges. Ultraviolet light does not penetrate Asgard's reflective gold surface or the water of our radiation shield. After a life-time of shelter from the ravages of ultraviolet radiation, our space colonists can expect to retain the smooth and supple skin of youth.[297]

Gravity is the other great mutilator of human skin. Fifty or sixty years spent under its remorseless influence produce tell-tale bags under the eyes and chin. The once magnificent temples of our bodies show the gravitational strain. Our once impeccable lines of sculpted marble now sag around the edges. Here in the gravity pit this has the inevitability of time itself, but in space it will not happen. Free of gravity's rapacious pull, our bodies will keep the crisp contours of freshly carven youth, throughout our live long days.

Other factors also contribute to rapid aging on Earth: the sedentary life-style of our modern society, poor nutrition, chronic dehydration, the continual barrage of contaminants in our environment, and many others. Most of these can be ameliorated or eliminated in space.

**Purity of Essence[298]**

We have no idea how detrimental the Earth environment is to our health and longevity. After all, there is no population living elsewhere to measure against. Once we can make such comparisons, we're likely to find that our potential life span is being dramatically shortened. On Earth, our health is continually eroded by pollutants, poisons, and pathogens. Literally every breath we take is loaded with impurities, inimical to life. Our air, our water, and our food are all to some degree poisonous.

We routinely breath—especially in big cities—a witch's brew of carcinogens, chemicals, gases, fungal spores, microbes, viruses, and noxious particles of every kind from asbestos to zinc. It's a wonder any of us are still alive. The body is a remarkable machine, and, for the most part, it is able to cope with this toxic miasma. But no one knows what the life-time cumulative effects are. The planet is a closed bottle, and what goes around comes around. Even relatively remote locations, like the Grand Canyon, are now routinely wreathed in smog.

On Earth, the water supply is often a detriment to health and longevity. Most people in the world drink contaminated water to begin with. Even water that is not overtly fouled, is "treated" with a concoction of chlorine, fluoride, and aluminum ammonium sulfate—alum.[299] Heavy metals, salts, PCBs, things we haven't even heard of yet, all go to season the stew. Water should be the fountain of youth, not the kiss of death.

As the environmental pressures on the Earth continue to grow, these problems will only get worse. Even if we did not have to contend with man's pollutants, we would still have to cope with the stresses of the 'natural' environment. We have only recently learned that radon, a naturally occurring gas produced by radioactive decay in the soil, is perhaps as large a contributor to lung cancer as smoking. Even natural foods, grown organically, without pesticides or fertilizers, are nevertheless virulently toxic. Everyday products like beer, peanut butter, and tomatoes are loaded with substances more noxious and carcinogenic than many man-made pollutants. These natural toxins are present in concentrations hundreds or thousands of times higher than would be allowed for an equally noisome man-made agent. Naturally occurring pesticides comprise up to 10% of the weight of common foods like oranges, apples, even broccoli.[300]

We have evolved the capacity to handle these potent chemicals, but at what cost? The EPA is fond of calculating how many people out of a million will contract cancer from saccharine; but how many will contract it from cauliflower? This is not to say that natural foods are 'bad' or 'dangerous', it is just to point out that our terrestrial environment is stressful, and that those stresses translate into death rates for populations and reduced life spans for individuals.

The present theoretical limit of the human life span is around 120 years. Between the things man excretes into the Earth's environment and the things that occur there naturally, it is not surprising that very few individuals actually survive that long. In space, we can eliminate many of these stresses. In Asgard's ecosphere, water is absolutely pure $H_2O$, produced by combining oxygen and hydrogen. Air is similarly pristine. There are no internal combustion engines, heavy industrial processes, or other such sources of air pollution inside the ecosphere. The atmosphere is produced fresh daily by photosynthesis.

Our food in Asgard, because it is derived from algae, will be entirely free of pesticides and other harmful chemicals, natural or man-made. Algae are virtually unchanged from 3 billion years ago. They have evolved no natural defenses against predators other than their extreme prolificacy. Food derived from algae has none of the toxins and carcinogens that characterize terrestrial produce.

These unadulterated staves of life: clean air, pure water, and toxin-free foods, are bound to enhance our health and longevity. These factors, coupled with the data intensive practices of Millennial preventive medicine, and sophisticated intervention therapies, will all add up to humans who routinely live healthy vigorous lives lasting well over a century.

**Location, Location, Location**

Exactly where Asgard should be located is a complex question requiring delicate balance between advantages and handicaps. If Asgard is placed in a low orbit, just a few hundred miles above the Earth, it will require the least amount of energy to reach, and the view will be spectacular. Low Earth orbit (LEO) is below the Van Allen belts and so offers some protection from radiation (see Appendix 3.6). Beyond these advantages, though, LEO has little to recommend it.

In a low orbit, Asgard will interact with the tenuous and ever-changing boundary of the upper atmosphere. Gas molecules would create aerodynamic drag, which on an object the size of a space colony could be enormous. Over time, the colony's orbit would deteriorate. Without countermeasures, Asgard would eventually plunge into the atmosphere, just as Skylab did in the late '70s. The vision of a whole city hurtling through the sky as an incandescent fireball is as horrific as any nuclear nightmare. Preventing such a catastrophe would require a continual input of energy to maintain the colony's orbit. This would more than cancel out the low orbit's launch energy advantages.

A low orbit would also subject the outer membrane to degradation by atomic oxygen. At low orbital altitudes, the dangers from man-made meteoritic debris are also extreme. LEO would also put the colony in darkness half the time. Even worse, the periods of light and dark would alternate every few hours.[301] Nights and days, following each other so closely, would quickly make cuckoo clocks of everyone's circadian rhythms.

The Lagrangian points would be viable as prospective locations for Asgard or other space colonies.[302] These points, named after the French mathematician who first postulated their existence, are really zones of gravitational stability. These zones are formed when a massive astronomical body is orbited by a lighter one. There are typically five such points. In the Earth-Moon system, $L_4$ and $L_5$ lie on the Moon's orbital path, at points equidistant from the Earth and Moon.

The Lagrangian points have the great advantage of being gravitationally stable—something put there will stay there. Because of this, they will eventually become the loci of large space settlements. But they are impractical as sites for Asgard, simply because they are too far away. L4 and L5 are the same distance from the Earth as the Moon itself.

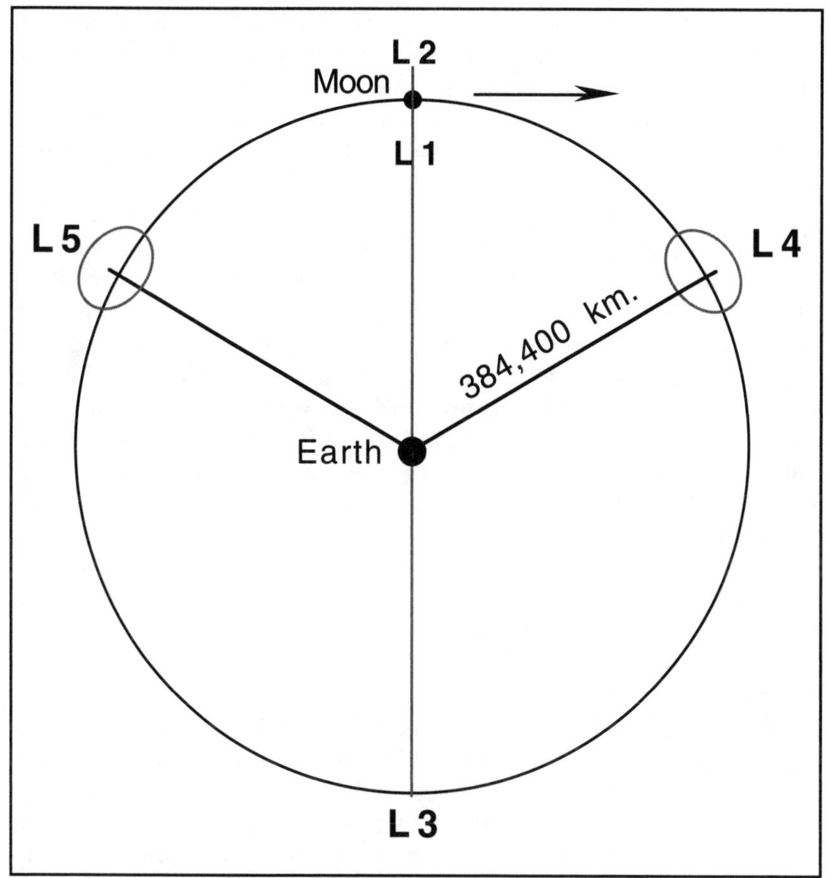


***Fig. 3.14 - Earth/Moon Lagrangian points.***

Asgard will be located in the Clarke orbit. The Clarke orbit, is named for Arthur C. Clarke, who was the first to suggest its importance for positioning telecommunications satellites.[303] Also called 'geostationary' orbit (GEO), the Clarke Orbit occupies a very special region of space, 35,900 km. above the equator. At this distance, a satellite will orbit the Earth exactly once in 23 hours, 56 minutes, and 6 seconds. A "geostationary" satellite goes around the world at the same rate that the world turns under it, thereby appearing to remain motionless in space above a fixed point on the Earth's surface. Such orbits necessarily lie over the Earth's equator.[304]

Our launch facility on Mt. Kilimanjaro, located just a few degrees from the equator, will make it very easy to reach this orbit. Asgard will be positioned so that it sits in space at a spot directly over Aquarius. From Aquarius, Asgard will appear as a brilliant star at zenith. It will be visible even in daylight, and at night it will be the most brilliant object in the sky.

In the Clarke orbit, Asgard will enjoy almost continuous sunlight. The equator is inclined to the plane of the Earth's orbit around the sun at an angle of 23.5°. Therefore, an object in geostationary orbit will very seldom pass behind the Earth's shadow. In Asgard, darkness will prevail for a total of only 48 hours a year—welcome news to children of all ages who are afraid of the dark.[305] Each year, Asgard will first graze Earth's shadow on "Leap Year Day", February 28/29. Then, every "night", we will spend a little more time in the cone of eclipse. On March 21st, we will experience the longest 'night' of the year, lasting 69 minutes. From March 22nd on, the period of darkness will grow shorter, until the nights end on April 11. Asgard then remains in uninterrupted sunshine for five more months. A second eclipsing period, exactly like the first, begins on September 12 and ends October 14. There is no darkness in Asgard except for these two six week intervals. During these twelve weeks, 'night' lasts for at most a little over an hour.

**Calling Planet Earth**

The Clarke orbit is the preferred location for telecommunications satellites for good reason. Satellites in GEO, because they are motionless relative to the ground, are very easy to find and communicate with. By contrast, satellites in LEO zip from horizon to horizon in a matter of minutes, and require elaborate tracking and relay systems in order to maintain communications.

Just as with Aquarius, Asgard will become a tremendous economic engine powering our further expansion into space. People often wonder what there is in space that is valuable enough to create an economic return for space colonists. Solar power, pharmaceuticals, genetically engineered flavors, ultra-pure crystals, tourism, and new technologies will all be important to the economy of Asgard. All, however, will be of secondary importance to the main product of the future—information. Our economy will be based on the creation, management, and transmission of information.

In Asgard, as in Aquarius, data management will be the colonists' main business. The nature of data is such that the physical locale of the person producing it is irrelevant. People can trade in the global stock exchanges, or perform engineering or consulting services just as readily in Earth orbit as they can on Wall Street.

The main business in the New Millennium will be information management, and information means communications. Asgard will be one gigantic telecommunications satellite. From its geostationary position, Asgard will be able to down-link directly to Aquarius. The space colonists in both places are primarily data processing tele-commuters. Therefore, this satellite link will be crucial to our development as a data-based society.

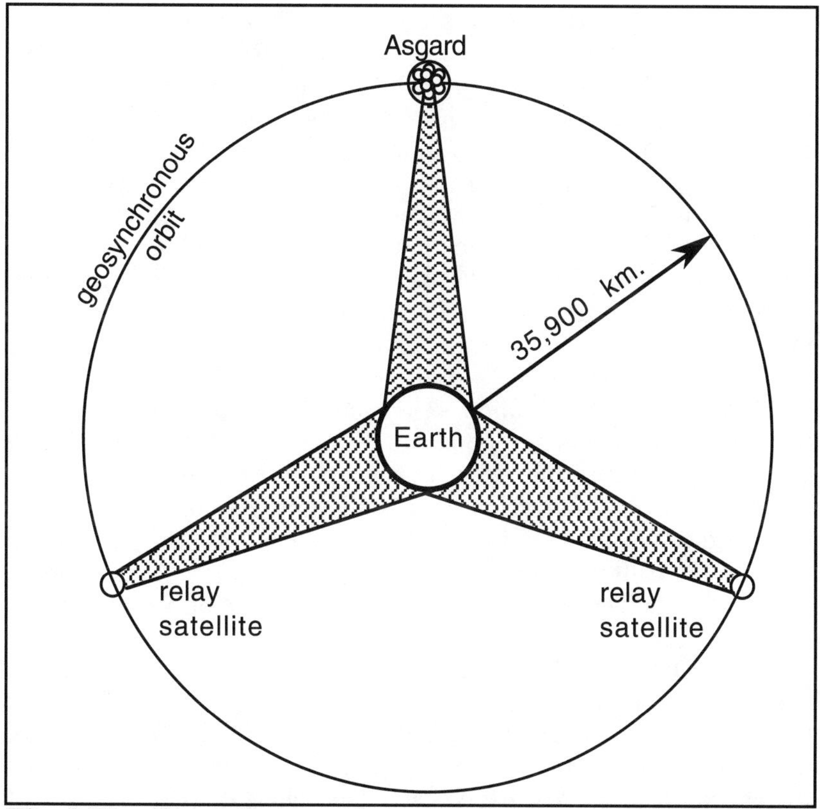


***Fig. 3.15 - Geostationary telecommunications coverage.*** *Asgard and relay satellites in GEO can blanket the planet with powerful radio signals.*

Through two relay satellites—also in GEO—Aquarius and Asgard will be able to communicate directly with any other spot on Earth. The round-trip transit time for a radio message from

Aquarius to Asgard is less than a quarter second, so it will be possible for both colonies to use the Aquarian data utility. Asgard will have its own on-board super computers. The massively parallel computer utility and the global data bank will, however, remain in Aquarius where it can be more easily maintained and protected.

Asgard will be able to function as a central communications hub for the planet. Having an abundance of cheap energy at our disposal and operating from a large space platform, we will be able to generate radio signals powerful enough to be picked up on very small receiving antennas. A typical commercial satellite, currently broadcasting television signals from geosynchronous orbit, operates at a power of five to eight watts—equivalent to a single Christmas tree light. Receiving such a signal requires a back-yard dish, meters in diameter. By contrast, a typical AM radio station on the ground broadcasts at a power of up to 50,000 watts.[306] Asgard's communications facility will transmit at a power in excess of 100,000 watts. At this power, we will be able to broadcast by direct satellite link to tiny individual receivers anywhere on Earth.

The global telecommunications market was a trillion dollar business in 1990. The market is growing exponentially, doubling every 12 to 13 years.[307] In some parts of the world, the growth rate is explosive—24% in East Asia in 1989. In 1970, there were 400 million phones in the world. By the year 2000, that number will have grown to 1.5 billion. The number of calls is growing even faster than the number of phones. By the year 2000, humans will be making a trillion phone calls a year. That staggering volume will double every four years.[308] By the time Asgard is completed, the world's annual expenditure on communications will be between five and ten trillion dollars. If we captured even 1% of that market, we could generate revenue streams on the order of $100 billion a year.

By linking Asgard with two geosynchronous relay antennas we will create a cellular telephone network that blankets the globe.[309] Using a hand-held personal telephone—not unlike Captain Kirk's 'communicator'—our customers will be able to call anyone anywhere on the planet just as easily and cheaply as making a call across town.

Presently, cellular radio is an expensive service; it costs $25 or more a month as a basic charge, plus 25-50¢ a minute, both for calls made and for calls received. Despite its cost, including $500-$1000 for the phone itself, the demand for cellular service is skyrocketing. In 1983, the demand for cellular telephones was zero; by 1987 a million phones a year were being sold; by 1990 the market had doubled to 2 million a year. In many parts of the Third World and Eastern Europe, businessmen wait for years to get a phone line. Cellular phone technology is already leapfrogging the conventional network of land-lines and in these places may become the primary form of telecommunications.

Conventional satellites broadcast at such low power levels, because high power levels demand more weight. At present it costs almost $10,000 to put a kilogram of metal into space.[310] As the power level rises so does the concomitant weight penalty. At $10 per gram no one can afford an overweight satellite. We, on the other hand, having perfected a cheap and reliable launch system will be able to access space for a fraction of that cost. Conventional satellites get their power from photovoltaic cells, either mounted on their rotating surfaces, or held out in wing-like panel arrays. Either way, the power available to the satellite is severely limited. By contrast, in Asgard we will have a super abundance of electric power at our disposal.

Global personal communicators require large space-based antennas. The transmitting antenna must be powerful enough to generate a signal that the communicator's tiny antenna can receive. Correspondingly, the receiving antenna must be large enough to gather and focus the tiny radio signal generated by the battery powered communicator. For Asgardians, these problems of scale are minor. By creating huge reflectors 300 meters across—the size of the Aricebo radio telescope—we can pick up signals generated by a wrist-watch battery.

An essential element in any cellular radio network is the switching system. A large computer is required to monitor the net and coordinate radio signals with the local ground-based telephone system. Ground based systems present complex switching problems, since the radio signal must continually be handed off, from one transmitter to another as the caller moves around. We will only have three antennas in our network so this problem is simplified, but it will still take a mass of computing power to handle the system's volume of traffic. Here, the data utility on Aquarius will be crucial, since we can use the tremendous computer power there to manage the network.

We will have a multitude of on-site personnel to deal with the many problems attendant to having a massive and powerful communications node in space. Our colonists, accustomed to life in space and able to move freely in that environment, will be able to make changes and repairs to the equipment quickly and cheaply. We will even be able to manufacture critical components for the system, like ultra-pure tuning crystals, and high-density computer chips.

**Space Templars**

From Asgard we can dominate the global telecommunications market. The size of our platform and our abundant power supplies will allow us to install a virtually unlimited number of radio transponders. As a space-faring society, based in geosynchronous orbit, our advantages will add up to one thing—in the language of business and the mantra of the 90's—market share. It will not be possible for any other organization—business or government—to compete with us in this realm. Accordingly, we can set the price for our communication services far enough below the going rate to guarantee ourselves whatever share of the global telecommunications market we deem appropriate.

It is not necessary to send people to live in space in order to create a global telecommunications network. If an organization planned to corner the cellular phone market, it couldn't do it economically by manning space stations. If, however, the organization's ultimate purpose is to move Life and civilization into space anyway, then one of the things we can do once we get there is to establish a paying telecommunications network. Our pre-established presence in space will make it possible to build and operate a powerful satellite network far more economically than any competitor.

By way of analogy, consider a situation which might have existed during the Crusades. There was a large and growing trade in spices passing through Palestine. A number of European merchants were involved in this lucrative trade. They faced high risks and incurred tremendous expenses: sending ships across the Mediterranean, establishing defendable trading posts, hiring agents in the Holy Land, and fighting off Saracens.

What if the Knights Templar had undertaken to enter the spice trade? They were already in Palestine; they had already built castles; they had a large organized cadre of literate men in place; they knew the country and the customs; and they were going to fight Saracens anyway. The Templars could have utilized all these assets in the spice trade and operated at a tremendous advantage. There is no way that the European merchants, operating from afar, could compete effectively with the Templars. The Knights could have undercut their competitors and taken whatever share of the market they felt they needed to support their main goals. The Templars were in Palestine to fulfill a higher calling. But, had they entered into business, they could have generated an abundant source of new wealth to finance their main mission.

This is exactly what we will be doing in the communications market, operating from Asgard. We, like the Templars, will have already built our capital infrastructure in pursuit of a higher destiny. Once these assets: launch facilities, power bubbles, and ecospheres are in place, we can then use them to generate a powerful revenue stream.

**Have Space Suit Will Travel**[311]

The Knights Templar would have been well advised to adopt a new panoply before going to fight on the burning sands of the Middle East. Benefiting from their mistakes, we will develop a new type of space suit before under-taking our own crusades in the new promised lands. Like the Crusader's mail hauberk, the conventional space-suit—familiar to everyone as the baggy-pants version from the Apollo era—has a number of serious drawbacks. For astronauts, making occasional, short excursions into space the conventional suit is tolerable. For people who will inhabit space permanently, however, it is not. Conventional suits are just too cumbersome, too uncomfortable, and too dangerous.

The problems of the conventional pressure suit stem from the philosophy behind its construction. The basic idea of the pressure suit is to surround the wearer with a mini-atmosphere. This approach creates a number of immediate difficulties. The suit, because it is pressurized, acts like a balloon with arms and legs. Even at low pressures, it is stiff. Just walking around in an Apollo-type pressure suit is hard work. At a stately pace of two miles per hour an astronaut in one of these suits generates waste heat at the rate of 2250 BTU/hr.—more than an unencumbered man jogging.[312] It takes four times as much energy to move and work in a balloon suit as to move around without one.[313]

To move around in a conventional balloon suit at all, the internal pressure must be kept very low. If a conventional suit were inflated to full atmospheric pressure—760 mmHg.—the person inside it could only stand there with his or her arms and legs sticking out, like some kind of space scarecrow.[314] The pressure in the current generation of space suits used by Shuttle astronauts is 222 millimeters (4.3 psi), about 30% of normal atmospheric pressure.[315] This in itself is no problem, since the human body functions well at this pressure, and, as in Asgard, the astronauts are breathing pure oxygen. The difference in pressure between the suit at 222 mm and the cabin of the Space Shuttle at 760 mm, creates an interesting problem. If an astronaut jumped into his spacesuit and immediately went outside, he could get the bends. In order to survive inside his space suit, the astronaut must first decompress. This must be done in stages, and can take hours. If it becomes necessary to go outside in a hurry, to rescue someone or make critical repairs, this can be fatally inconvenient.

The balloon suit is also extremely, and inherently, dangerous. Like a balloon, it is subject to sudden and rapid deflation if punctured.[316] When this happens, the effects on the astronaut are not pleasant and shouldn't be contemplated by people with weak stomachs. Due to their inherent fallibility, pressure suits must be rigorously inspected and maintained. It takes 80 people 65 days to get a space suit ready to go up in the shuttle, after a single use.[317] Even if the suit is in perfect condition, a simple snag on a jagged piece of metal can mean death. If we are really going to live and work routinely in space, we can't risk dying from a tiny rip in our coveralls.

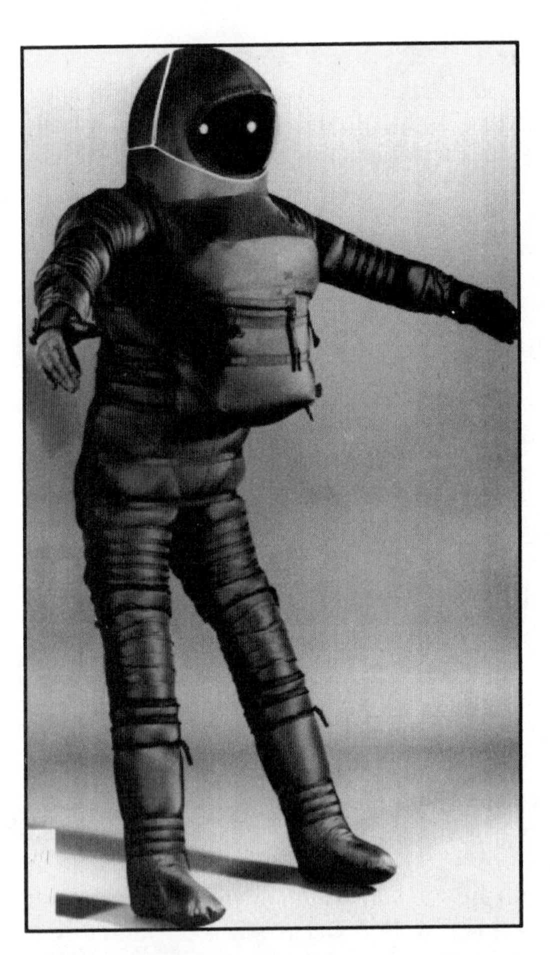

***Fig. 3.16 - Balloon suit.*** *The typical balloon suit encases its wearer in a pressurized bag as hard and inflexible as a fire hose.* NASA photo.

NASA has engineered a solution to the problem. Unfortunately, like most solutions that come from the aero-space industry, it relies on the sheer brute-force of technology. As a result, the right answers are being found to the wrong questions. NASA's solution is a space suit constructed entirely of metal. This space-age suit of armor solves most of the technical problems facing balloon suits. The hard suit, hinged at every conceivable joint, allows the wearer a wide range of motion. The hard suit operates at higher pressures, so an astronaut can put it on and go outside immediately. And the metal construction reduces the chance of puncture to a minimum.

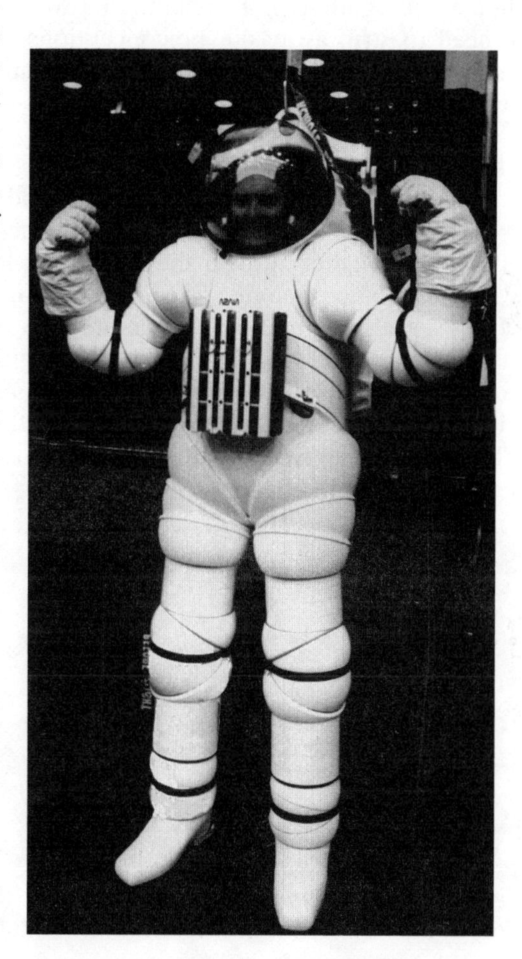

***Fig. 3.17 - All metal space suit.*** *As daywear for space faring people, such suits are too bulky, too heavy, too complicated, and too expensive.* NASA photo.

The hard suit is nevertheless the wrong solution. At 185 pounds, the suit weighs more than most of the people who will wear it. While weight is not a consideration in space, mass certainly is. As well as being expensive to transport, the suit's mass will make it more difficult to control. It is also big and bulky, taking up as much room as a person. The hard suit must be made out of precision-milled aerospace grade aluminum, it has dozens of moving parts, and like any complex piece of machinery it will be expensive.

At a million dollars each, metal space suits won't exactly qualify as leisure wear.[318]

The metal pressure suit's designers are attacking the wrong questions. They assume that the hurdles to be overcome are mere technical difficulties with the conventional pressure suit: how to make the joints more flexible, how to avoid the bends, how to keep the suit from tearing. The right question to address is how to keep a human being alive in a vacuum. To solve the space suit problem, we need to strip away our preconceptions and get down to basics. We have to ask, what happens to a human body exposed to hard vacuum?

**Open the Pod Bay Doors Please Hal**

The typical misconception, used to good gory effect in B-grade sci-fi movies, is that a person exposed to vacuum will blow up like a balloon and splatter all over the set. If we look instead at a Grade-A science fiction movie, like Stanley Kubrick's *2001: A Space Odyssey*, we can get a more accurate idea of the truth. There is a classic scene, involving Commander Bowman's attempt to get back aboard the spaceship *Discovery* from a one-man pod without his helmet. He sets off explosive bolts that blow him into an airlock.[319] For a few moments he is exposed to hard vacuum, but survives with no ill effects.

***Fig. 3.18 - Commander Bowman reenters the* Discovery's *airlock without his helmet, in Stanley Kubrick's* 2001: A Space Odyssey.**

Photo courtesy of the Academy of Motion Picture Arts and Sciences.


The film is often criticized as being unrealistic because of this scene: "Everybody knows your blood will boil in space." But, like meteoroids and many other things people think they know about space, the effects of vacuum on the human body are largely misunderstood.

Most liquids exposed to a vacuum, including blood, will indeed boil at body temperature. But blood inside the body is not exposed to the vacuum. The skin is actually a pretty good space-suit. Imagine this thought experiment: Two samples of water are put into a vacuum chamber; one fills an open beaker; the other fills a tough leather container—a football for example. What happens when the pressure inside the vacuum chamber is reduced to near zero? As expected, the water in the open container boils; but what happens to the water inside the football? The water in the football may try to expand and boil off, but it is held in by the pigskin. Unless the pressure is high enough to rupture the tough leather container, the water cannot boil. The body is just a more complex version of a water filled football.

The human body is surprisingly resistant to the effects of hard vacuum. People's hands have been experimentally exposed to vacuum conditions without ill effects. After about two minutes, the hand shows some swelling, due probably to the decreased circulation caused by pressure of the vacuum chamber's airlock collar. After even 10 minutes exposure to vacuum, such swelling is still minimal, and the hand remains comfortable, flexible, and fully functional.[320]

One's skin makes a good space suit for a several reasons: it is impermeable; it is well attached to the underlying tissue; and most importantly, it applies counter pressure to the fluids of the body. It is this last function of the skin which gives a clue to the solution of the space suit problem.

**Have Space Suit Will Travel**

We can't go into space naked, but we can do the next best thing. Rather than trying to surround the body with a miniature atmosphere, which applies gas pressure to the skin, we can instead apply pressure directly with an elastic garment. Using modern stretch fabrics like lycra and spandex, we can build an elastic body stocking, not unlike the suits worn by Olympic skaters. This suit will really function like a second skin, reinforcing our own skin and applying enough counter pressure to the fluids of the body to negate the effects of vacuum.

According to the researchers who have explored this approach, it is entirely feasible to create an elastic garment which will apply between 150 and 170 mm of counter pressure over the entire surface of the body.[321] This is equivalent to atmospheric pressure at an altitude of 11 kilometers. At 11 km., humans can function normally while breathing supplemental oxygen. Asgard is pressurized to 155 mm Hg (3 psi.), so a suit exerting this much counter pressure could be donned without any depressurization. The absolute minimum counter pressure required for survival is 48.6 mm Hg., (.94 psi.). At this pressure, the boiling temperature of blood is 99.7° F. At any lower pressure, the danger of boiling alive becomes all too real.

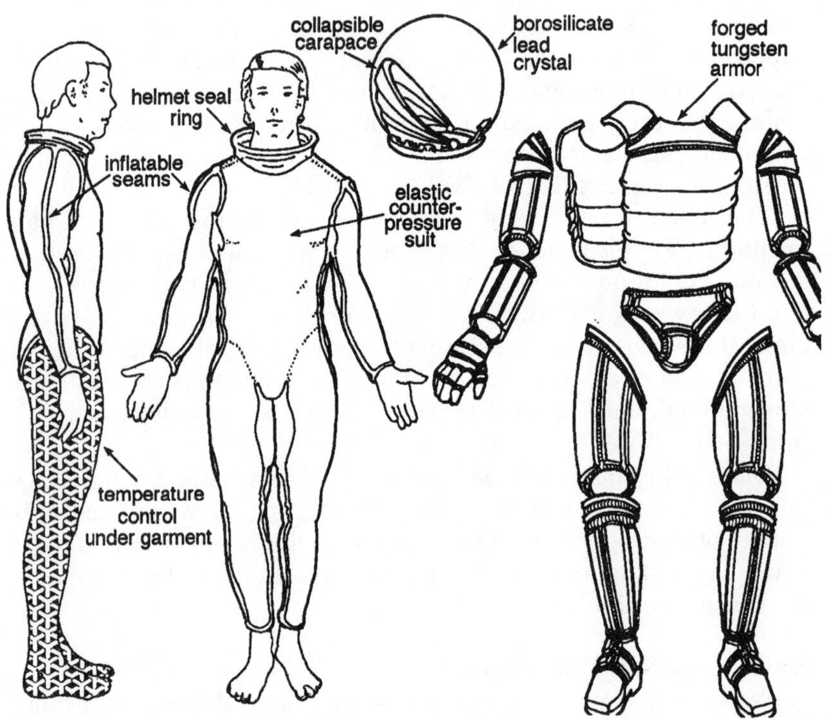


***Fig. 3.19 - Elastic counter-pressure space suit.*** *Fitting like a second skin, this suit will provide protection from the vacuum that is safe, flexible, comfortable, and economical. Stylish too. A panoply of tungsten armor provides protection from Van Allen radiation and micro-meteoroids.*

The elastic counter pressure suit will have a pressurized helmet, so the head can be surrounded by air. The suit itself is made of two pieces: an upper section which covers the arms and torso, like a pullover turtleneck; and a lower section that covers the legs and torso but not the arms. By doubling the torso section, effective counter pressure in vital body cavities is increased without interfering with the mobility of arms and legs. Fluid filled bladders fill in concave surfaces around the armpits, groin, and small of the back.[322]

The counter pressure suit will fit very tightly when it is on, but it can be loose while being donned. Inflatable hoses run down the major seams of the arms, legs, and torso. After the suit is on, these hoses are inflated, taking up the slack in the suit. The same technique was used in the partial pressure suits and *g*-suits worn by early jet pilots. The difference is that the pressure in the hoses of the counter pressure suit need not be very high, since the counter pressure is applied by the elastic material. Therefore, the hoses need not interfere with the suit's flexibility.

A minimal elastic garment, light and comfortable enough to wear as daily clothing, can be developed. When inside an ecosphere, this garment would apply very little counter pressure. Such a suit would not have to be impermeable, and could be equipped with inflatable seams. These measures would make the suit as easy to wear as today's exercise leotards. With a simple oxygen mask and ear plugs, a colonist could survive emergency exposure to space in such a suit for hours.

There will be manifold advantages to the counter pressure suit. Being light and flexible, it will be easy to wear and to work in. It will offer the wearer a full range of motion at very little increase in effort over moving around in shirt sleeves. Experiments indicate that energy expenditures in the counter pressure suit are increased by only 10 to 20%, as opposed to the 400% increase required by conventional suits.[323] The counter pressure suit weighs only a few pounds. When not in use, it can be folded up and stuffed inside the helmet.

The dangers inherent in the pressure suit are largely eliminated by the counter pressure garment. A hole in the garment means only that a patch of skin is exposed to vacuum, with virtually no adverse effect on the wearer. The danger of the underlying skin being ruptured by the body's internal pressure is nil. The tensile strength of adult human skin is 1600 grams per square millimeter; the maximum internal pressure it would face is less than three grams per square millimeter.[324]

Staying warm is seldom a problem in space; vacuum is a near perfect insulator. Staying cool is usually more difficult. An astronaut wearing a space suit is sealed inside a closed bag, making disposal of waste heat difficult. The body's normal heat control mechanism, sweating, doesn't work inside the suit's air-tight environment, so supplemental cooling is needed. This is accomplished by enveloping the body in a network of small water hoses. Water circulates through the tubes and carries away body heat. The same system can be used to warm the astronaut, if the need arises. In the conventional balloon suit, these tubes were threaded through the surface of the astronaut's long-johns. The same system can be adopted, with some modifications to the counter-pressure suit.

The temperature control system could theoretically be eliminated from the elastic style space suit. Unlike the balloon suit, the elastic suit doesn't have to be air tight. The material could be left porous and the astronaut could cool off by sweating, or retain heat with additional insulation. A man doing heavy labor can lose several gallons of water a day through perspiration. Therefore, it will probably be desirable to make the elastic suit impermeable. While not necessary for life support reasons, a sealed suit will save valuable volatiles like water and oxygen for recycling. A system using solid-state thermoelectric refrigeration based on the Peltier effect will both serve for cooling and heating.

Suits can even incorporate the closed-loop algae system which sustains the larger ecosphere. The system, is potentially so compact and productive that it could be packaged in a small enough unit to be carried by a single person. The six liters of algae solution could be contained in less than two meters of helically wound clear tubing.[325] With the algae exposed to sunlight, the closed loop could produce enough food and oxygen from waste products to sustain a person indefinitely.

Early garments will be made of elastic materials similar to those in use today. Later generations of counter pressure garments may become highly complex. Eventually, space garments will be made using nano-technology—millions of micro machines on the scale of single biological cells. Intelligent nano-materials may apply varying degrees of pressure to different parts of the wearer's body. The garment will be a true second skin, actively responding to changes in the wearer's metabolism and the external environment.[326]

**Battle Harness**

Placing Asgard in the Clarke orbit will put us within the outer edge of the outer Van Allen radiation belt (see Appendix 3.6). This is less than optimal, but by no means fatal. There is radiation in the outer belt, but the intensity is 20 times less than in the inner belt.[327] Asgard's radiation shield will stop even high-energy cosmic ray particles, so there is little to fear inside the colony from the relatively low energies of Van Allen electrons. The danger is to people working outside the ecosphere's radiation shield.

Most of the low-intensity Van Allen radiation can be stopped by shields with a density of two grams per square centimeter.[328] Protective armor, worn over the elastic counter pressure suit, will protect colonists when working in space around Asgard. This armor will have an exterior layer of high density ceramic enamel protecting a thin sheet of forged tungsten. The enamel layer, of titanium dioxide, will protect the underlying metal from atomic oxygen, ultraviolet radiation and other potential corrosives. The tungsten sheeting will block Van-Allen electrons and deflect micrometeoroids.

Tungsten, which is three times stronger than steel, almost twice as dense as lead, and has the highest melting point of any element but carbon, is an ideal metal for space armor.[329] On Earth, this metal is very difficult to work with because ceramic crucibles melt at lower temperatures than the tungsten. In the industrial satellites of Asgard, however, metallurgy can be accomplished without any containers. In zero-*g*, tungsten can be melted in an electric arc furnace and then cast and forged as desired. Because of tungsten's great density, the armor plates will be just over a millimeter thick.

The armor's thickness will vary, depending on the part of the body that is being protected. Some parts of the human body are extremely sensitive to ionizing radiation while others are not. The extremities are less vulnerable than the abdomen for example. There are sensitive cells in the extremities, like the marrow cells in the femur, but these are deep in the body and are protected by skin, muscle, and bone. The epithelial cells in the gut are very vulnerable to radiation and relatively exposed, so they require more shielding. The most sensitive areas are the eyes and the reproductive organs. Heavily armored codpieces and girdles will protect men and women's delicate plumbing.

The head will be shielded by a helmet of high density lead crystal, five millimeters thick.[330] Two layers of dense borosilicate glass will be sandwiched between two layers of a clear engineering plastic like Lexan.[331] The middle layer of Lexan will add strength and prevent shattering, like safety-glass in a car windshield; the inner layer will provide a reserve helmet in the event of catastrophe. The outer surface of the helmet will be gold anodized to protect the glass and block out glare, UV, and IR radiations. A nested set of curved armor plates can be deployed to cover part or all of the head to further reduce exposure.

The armor's density will average two grams per square centimeter over the suit's surface. The complete suit of armor and helmet will together weigh around 45 kg.(100 lbs.). On Earth, this might be burdensome, but in the micro-gravity environment of Asgard it will interfere with mobility only slightly.

The tungsten armor will be backed by an inner layer of woven carbon fiber. This layer does not add significantly to radiation shielding, but it is essential to the armor's secondary function as ballistic protection from micrometeoroids. The enameled tungsten is extremely hard and will shatter any small meteoroid that hits it. The woven layers of carbon fiber beneath the outer armor will then catch fragments, and absorb blunt shock. In the helmet, the layers of Lexan perform the same function.

Sheathed in armor, a space colonist can work outside the colony in relative safety. Virtually all of the soft x-rays from Van Allen electrons will be halted by the dense tungsten and lead-crystal shielding. Hits from micrometeoroids up to several millimeters in diameter can be absorbed without fatal consequences.[332] Unfortunately, the armor does increase bremsstrahlung radiation from cosmic rays. (See Appendix 3.6) Cosmic rays are mostly a long-term exposure hazard, and excursion times in the suits will be limited. Accordingly, damage due to secondary radiation can be carefully monitored and minimized. Most outside jobs will be performed by robots and tele-operators, minimizing man-hours spent outside the shields. The huge pool of available workers will help insure that no one ever has to log a dangerous amount of time outside.

**Solar Bubbles**

With enough power we can do anything. On Earth we will tap into the huge thermal battery of the tropical ocean, and extract a vast bounty of renewable energy. We will use the power of the marine OTECs to fuel our preparations on Earth, and then to lift ourselves into space. In space, we must find a new source of power, that, like ocean thermal energy, is bountiful, clean, renewable, and cheap.

Fortunately, space is awash with raw energy. We live—in astronomical terms—astonishingly close to a star.[333] Technically, we are inside the sun's atmosphere![334] Ready access to abundant supplies of free energy is one of the main benefits of living 'inside' a star. Here in our privileged region of space, 1.35 kilowatts of power pass through each square meter every hour. If all of that energy were captured, the electric power needs of an average

American household could be supplied from a patch of space not much bigger than a beach blanket.

Schemes have been put forward for placing huge arrays of solar cells in space.[335] But even very good—which is to say expensive—solar cells can convert only about 20% of the available energy into electricity. Even with a launch system like the Bifrost Bridge, every kilogram we lift into space is going to be costly, and solar cells are heavy. What we need is a solar power collection system that is light, economical, easy to build with the materials at hand, and conceptually elegant. Following Buckminster Fuller's cybergenic design principles, we should seek our solution in the realm of bubble engineering.

***Fig. 3.20 - Echo I.*** *An early telecommunications satellite, Echo I was used to bounce radio waves back to Earth. Power bubbles will be constructed in much the same way as Echo I, but with one hemisphere left transparent.* NASA photo.

The object is to collect large amounts of sunlight, and concentrate it. Looking to the eternal soap bubble, we find our solution ready made. It is a happy law of geometry that a spherically concave mirror will focus all the light falling on it into one area.[336] Constructing a large concave mirror can be accomplished, simply by inflating a large balloon. The balloon can be made out of a lightweight membrane, like mylar. One hemisphere of the balloon is transparent and the other hemisphere is coated with a thin layer of aluminum.

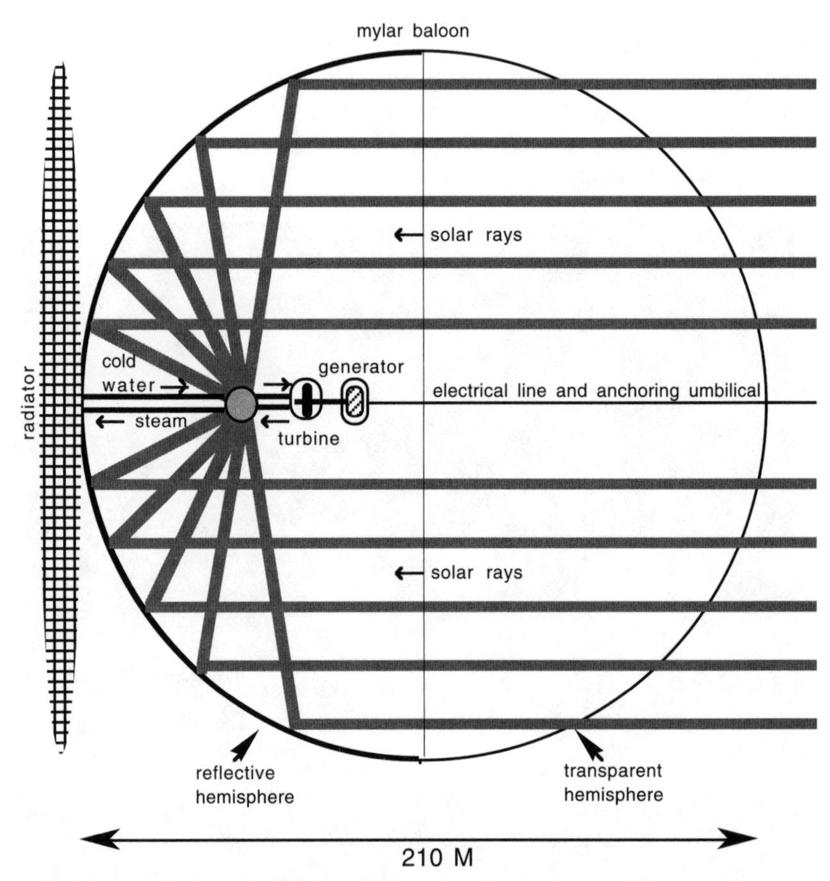


***Fig. 3.21 - Power bubble.*** *This simple system will convert raw sunlight into refined electric power. At 36% efficiency, a power bubble this size will generate nearly 17 megawatts of electricity continuously.*

Once inflated with a small amount of inert gas, the balloon automatically assumes a spherical shape. The inner, reflective, hemisphere will form an optically perfect concave mirror, which can be oriented to face the sun. Light entering the balloon through the transparent hemisphere will reflect off the metalized hemisphere, and converge in an area along the focal axis. The concentrated sunlight can then be used to produce electricity in a steam turbine.[337]

Living in a bubble in space should be much more efficient than living in a terrestrial city. Most light is provided by the sun; heating and cooling are maintained automatically by regulating the absorption of solar energy in the water shield; transportation is minimal and mostly human powered, since no place in the colony is more than 600 meters away; and all manufactured goods, from food and clothing to packaging, are recycled. Therefore, space colonists should need only trickles of power, compared to the amounts required by Earthlings.

Asgard's power needs will be about a billion kwh per year—10,000 kwh per capita.[338] To provide Asgard with an installed power base of 100 megawatts, will require solar bubbles with a total surface area of 200,000 square meters.[339] Six independent collectors, each with a radius of 105 meters, will suffice.[340] Forces acting on the collection balloons like atmospheric drag, tidal effects, and solar wind, are all minuscule, making operation and maintenance of the power facilities relatively trouble-free. If more electricity is needed, for heavy industry or high power telecommunications facilities, the number of power bubbles can be multiplied, almost without limit.[341]

**Onward and Upward**

With the firm foothold in space that Asgard gives us, and the new surge of economic voltage generated by our telecommunications business, we can undertake our next strategic step outward—settlement of the Moon.

With the completion of Asgard, we will have irrevocably crossed the gravitational Rubicon; ours will be a space-based species for the rest of time. With Asgard, the process of space colonization—in the strict sense of colonizing space, as opposed to colonizing planets—will have begun. In Asgard we will perfect those systems and techniques needed to thrive in space. Asgard establishes the basics. Once we have mastered the fundamentals, space will be wide open to us as a new frontier. Power bubbles provide a means to tap the flood of solar energy; water shielded ecospheres provide comfortable, affordable, habitable environments, free of any planetary surface. Given the abundance of energy and raw materials in our solar system, our potential for expansion is truly awesome. We can multiply the number of Asgardian ecospheres until we have fully utilized the resources at our disposal. Then, having transformed our solar system into a living garden, we can turn our attention to the stars.

Our cosmic mandate is to disseminate Life throughout the universe; not just human life, but as many varieties of life as possible. Ecospheres like Asgard provide suitable habitats for ourselves and many forms of plant, aquatic, and marine life, but they aren't well suited to terrestrial animals. Human beings can adapt easily to gravity-free environments, but most animals probably can't. Therefore, we will need to colonize the Moon and Mars, where we can establish habitats able to support a wide variety of other life forms.

"*For I muste into the vale of Avylyon to heal me of my grevous wounde.*"[342] These were King Arthur's last words as he departed for the Mystic Isle. That island is known to be just offshore, swathed in the mists of a mythic tradition. Across a short strait of space awaits the New Avallon, the True Avallon, the once sterile and generic Moon. Like Lancelot and Galahad, we knights of the New Millennium will don suits of shining armor and go forth to battle the dragons of Chaos. A new generation of Merlins will cast their magic spells of science, animating the inanimate. A new people, like Arthur incarnate, risen from his long sleep, will push back the darkness of Mordred.

# CHAPTER 4

# AVALLON

***I am going a long way...***
***To the island-valley of Avilion;***
***Where falls not hail,***
***or rain, or any snow,***
***Nor ever wind blows loudly; but it lies***
***Deep-meadow'd, happy,***
***fair with orchard lawns***
***And bowery hollows***
***crown'd with summer sea...***

**Tennyson**

**Avallon** — The knights of Camelot were the natural successors to the Vikings in the realm of the mythic hero. Like the Norsemen, the Celtic warriors too, had their heaven and it was Avallon.[343]

On the Moon, we will take another giant leap toward fulfilling our cosmic destiny. We will dome over the craters of the Moon, and establish in them self-sustaining pockets of life. Each such ecosphere will contain a particular ecology, imported from Earth. Eventually, thousands of domes will spangle the lunar surface like drops of liquid gold. Each of these jewels will shelter a green bubble of living matter. These corpuscles of life will create a multitude of new domains where animate matter can thrive. We will hereby achieve one of our first great objectives—the dispersal of life to other planets. Once taken, this essential step will assure the survival of Life's diversity.

The Moon is an utter wasteland, a paragon of desolation. It is emblematic of this entire wasted universe. Our first lunar ecosphere—Avallon—will completely transform the iconography of the Moon. That first bubble of Life, sitting alone amongst the ashes of a blasted world, will serve notice to the Cosmos that the Metamorphosis has begun: A dead universe will be coming to life. Outside Avallon's dome, the harsh desert of the ancient Moon will prevail as it has since the dawn of time. But on the other side of the dome shield, a few meters away, flowers a new paradise. There, foliage luxuriates in lush groves; flocks of parrots color the sky in pastel clouds; waterfalls splash into deep pools, and streams trill across the once barren moonscape. Avallon too is like a bud from the body of Mother Earth—a living world in microcosm. This glistening emerald of Life, mounted in a setting of desolate gray regolith, will be the quintessent image of cosmic destiny fulfilled. (See Plate No. 8)

**The World Next-Door**

Asgard provides a comfortable space habitat, but primarily for ourselves. If our only concern was to disseminate humans into space, Asgard and colonies like it could meet that need indefinitely. Our mandate, however, is not that simple. The colonization of space is not an effort undertaken solely for the benefit of *Homo sapiens*. Our intention, and our cosmic obligation is to carry Life to the stars. This means life of all kinds. Therefore we must carry as diverse a cross-section of the biosphere as possible into space and establish a variety of ecospheres to shelter the menageries.

It will be easy for us to adapt to the gravity-free environment of Asgard; adaptation is a human specialty. The same can not be said for many other life forms. Imagine a zoo in zero-*g*, adrift with floating lions and tigers and bears—oh my! The animals we take into space are going to need gravity. To support these other creatures, we need space habitats built on planetary surfaces.

We have been thoughtfully provided with a second world, a lifeboat, placed handily within reach. On this ark planet, the whole glorious rhapsody of Life can be installed and shielded from any conceivable cataclysm. We shall take up "*every living thing of all flesh wherein is the breath of life.*"[344] Only thereby can Life be sustained and protected eternally.

**The Swords of Damocles**

Even now, terrors threaten the very existence of Life. As far as we know, all living matter is confined to the surface of this single planet. One meteor strike could wipe us out. Such a catastrophe might destroy the seed of a living universe. We cannot let that happen.

The universe is a big and dangerous place. Occasionally, one of those dangers comes to Earth in the form of a killer meteor. This is not science fiction; these collisions occur with clock-like regularity. Over the past billion years, the Earth has absorbed dozens of horrific impacts.[345] The last big hit, 66 million years ago, may have wiped out the dinosaurs.[346] The next one could do the same to us.

When an asteroid the size of a mountain slams into the Earth at thirty kilometers per second, the effects transcend the capacity of adverbs to describe. When such a meteorite strikes, it punches a hole 40 kilometers deep, penetrating the Earth's crust and exposing the semi-molten mantle.[347] The instant volcano thus created ejects 40,000 cubic kilometers of debris into the atmosphere. A shock wave as dense as a wall of granite sweeps out from the impact site at the speed of sound. The over pressures crush anything on the surface within a thousand kilometers. Eardrums a continent away are ruptured. Monster earthquakes reverberate across the globe; the whole planet literally rings like a bell. Killer tidal waves three miles high will fill the sky, "the color of boiling blood."[348] A thick pall of dust and debris will settle over the planet like a death shroud, blotting out the sun for months. The atmosphere, poisoned by worldwide fire storms, weeps acid rain for years. Our frail civilization would be utterly annihilated. Humans would be exterminated. All mammals and large animals could disappear entirely. The indestructible cockroach might be left as unchallenged ruler of the planet.

Fortunately, the chance of such a calamity occurring anytime this week is remote. The intervals between such bombardments average 26 million years. We are midway through an interregnum now. The next big bang should be 12 million years off, but we could catch one tomorrow.[349] It has been estimated that a given individual is three times more likely to die from a meteor strike than in a plane crash.[350] A plane crash kills a few people; a large meteor could exterminate our whole species. Given the magnitude of the consequences, we must undertake to escape this possibility, no matter how remote it is.

There is a common, though mistaken, notion that we can easily counter the threat of a meteor strike by nuking it with atomic weapons. There was even a cheesy '70s' disaster movie, *Meteor,* based on that idea. There are several misconceptions attached to the idea of meteor busting which quickly evaporate when exposed to the merciless glare of reality.

It is a mistake to assume that we would even spot a killer meteoroid. Astronomers find meteoroids by taking long exposure photographs. The meteoroids show up as streaks against the background of fixed stars. Any meteoroid headed right at us won't make such a streak. An astronomer could be looking at a photographic plate showing a killer asteroid on a collision course with Earth, and not even know it. The asteroid would just appear as a very faint dot among the myriad stars—completely undetectable. The problem is compounded by the fact that we can't see meteoroids approaching from the sun-side of the solar system. Astronomers looking in that direction will be in daylight, their telescopes useless.

It's often assumed that astronomers have identified all the potential threats. Nearly four centuries have passed since Galileo gave us the telescope. One can be forgiven for thinking that all the meteoroids in our neighborhood had long since been catalogued, and their orbits meticulously calculated. If so, there should be little possibility of surprise. This is sadly far from true. There are probably more than a thousand meteoroids larger than a kilometer in diameter which cross Earth's orbit. Astronomers have determined the orbits of only 77.[351] The rest lurk in the unknown darkness.

Another common misconception is that, after we had spotted a rogue meteor, we would have enough time to prepare for impact. Meteoroids are small dark objects. Not even the largest telescopes can detect them beyond a range of a few million kilometers.[352] Like all bodies orbiting the sun, meteoroids move at high rates of speed. Their velocity, relative to the Earth, can be up to 65 kilometers per second. From the time we spot a killer meteoroid, there may be fewer than 24 hours before impact. The notion that the American and Russian war machines could convert their ICBMs into spacecraft on a day's notice is an idea best left in Hollywood.

In addition to the astronomical hazards, we have to consider our own unwelcome threats to the seed planet. Even as you read this, another unique species is being rubbed out. Life is losing a species a day to the attrition of man's environmental ravages. Without some sort of Cosmic ark, Life could dwindle away to a shadow of its present glory.

We also sit, uncomfortably, on a prickly pile of deadly weapons. There has been a remission in the threat of atomic warfare—at least of "all out nukleer combat, toe-to-toe with the Russkies." But other threats are rising like Hydra heads. Unrest in the XUSSR raises the uncomfortable question of "who's minding the bombs?" It is chilling to realize that accidental discharge of the world's nuclear missiles is prevented only by the same technology that safeguarded Chernobyl. A minimum accidental launch from Russian territory would involve a missile battalion—10 missiles with 100 warheads.[353] The detonation of a hundred H-bombs on American soil would inevitably trigger a spasm of launches in response. The balance of the Russian arsenal would then be unleashed. Cry havoc! Nuclear winter would cool the ashes.

Now, as our old Stalinist nemesis retires, we have to contend with the new kids on the block: Mohamar, and Saddam, and Kim, and who knows who's next. Soon these charming boys will have nuclear toys to play with. They already have ballistic missiles, nerve gas, anthrax, and Allah only knows what else. As nuclear technology proliferates and new nightmares like genetic engineering burgeon, these threats will deepen and darken.

**Moon Mining**

The only sure way to survive such disasters is to disperse Life to other planets. We have been fortuitously provided with a suitable back-up world.

The Moon is a remarkable little planet. It is almost as large as Mercury. Yet, in astronomical terms it is practically in our pocket. If the Earth were the size of a grapefruit, Venus, on the same scale, would be 40 kilometers away, and Mars would be 80 kms. distant. The Moon, however, would be the size of a ripe plum, just 400 meters off. We can get to the Moon in a few days, even poking along with chemical rockets. Getting to Mars the same way would take the better part of a year. The Moon is a conveniently placed handhold in space. Once we get a firm grip on it, we can easily pull ourselves up out of Earth's gravity well.

We will have begun to scratch out a finger hold on the Moon even before the completion of Asgard. Building a city in the sky the size of Asgard is not something that can be done in a single step. It will require the completion of a number of stages of development before the main space colony can be built. First, a foothold must be gained in low earth orbit, then outposts must be established on the Moon and elsewhere to secure raw materials, and finally a network of mass launchers and lasers must be established to move men and materials through space.

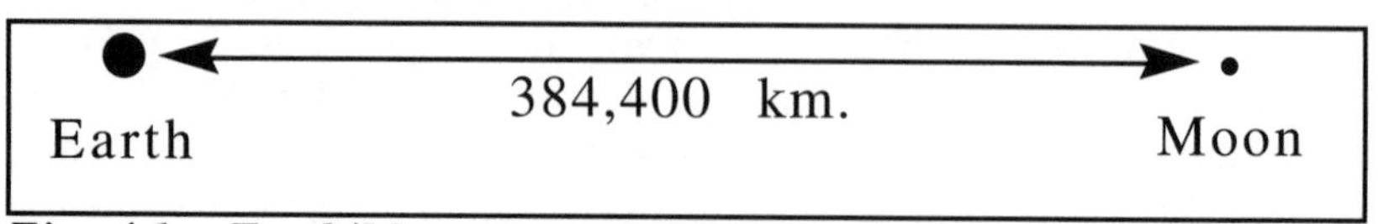


***Fig. 4.1 - Earth/Moon system to scale.*** *In terms of their relative masses the Earth and Moon can be defined as a dual planetary system.*

Valhalla, a relatively tiny and cramped space station, consisting of a single 20 meter bubble in low Earth orbit, will be the first outpost. Valhalla will be shielded by the Earth and Van Allen belts from most radiation, so its water shield can be limited to a meter in depth. This, and its small, size will keep Valhalla's shield mass under 1500 tons. It will be relatively easy to lift the material needed to build Valhalla from the Earth. The Bifrost Bridge can lift 1500 tons of water into orbit in four days. Manned by a crew of up to 50, Valhalla will serve mostly as a way station. Passengers and cargo will transfer from launch capsules to shuttle craft traveling on to Asgard, the Moon, and outlying destinations. Crews will serve in Valhalla for limited shifts, not exceeding six months, so long-term problems will not be critical.

Lifting the entire mass of Asgard from the Earth is not practical, even with the Bifrost Bridge. As reflected in Table 4.1, below, the entire mass of Asgard is over six million tons. To propel this much material into space from the Earth would require over a million launches, occupying the Bifrost Bridge completely for years.[354]

To build Asgard, we must draw on materials from someplace other than the Earth. Fortunately, the solar system is rich in the materials we will need—particularly oxygen. Most of the mass of Asgard is water, and almost nine tenths of the mass of water is oxygen.[355] Of Asgard's 6.6 million tons, oxygen constitutes 5.7 million tons, 86% of the total.

If we find a source of oxygen in space, then we have found most of what we need to build Asgard, both for the water shield and for the atmosphere. There just happens to be a massive deposit of elemental oxygen within easy reach: Moon rocks are 40% oxygen by weight.

**Table 4.1**
**Mass of Asgard by Component**

| Component | Mass - 1000 tons |
|---|---|
| Outer Radiation Shield | 5750 |
| Inner Radiation Shield | 660 |
| Machinery, Equipment, Etc.[356] | 200 |
| Atmosphere[357] | 26 |
| People[358] | 7 |
| Outer Bubble Membranes | 1 |
| Inner Bubble Membranes | 1 |
| Power Bubbles[359] | 1 |
| **TOTAL** | **6646** |

Extracting oxygen from lunar soil could hardly be simpler. When moon dust is heated to 1300° C., by focused sunlight, oxygen is driven off. The precious gas can then be collected and liquefied for transport. To recover the 5.7 million tons of oxygen needed to build Asgard will require processing about fourteen million tons of lunar soil. This sounds like a lot, but it could be mined by scooping up the top seven meters of regolith from just 250 acres—an area not much bigger than a suburban shopping mall.

Heating lunar soil to extract its oxygen requires 450 calories of energy per kilogram of oxygen produced.[360] Heating fourteen million tons of regolith takes a lot of energy—seven billion kwh—but the same energy can be used for multiple purposes. At a temperature of around 1500° C. lunar soil will melt, allowing extraction of various valuable metals and other elements. Processing fourteen million tons of lunar soil would yield an abundance of valuable materials.

The moon is a ready warehouse, stocked with most of the supplies required to sustain a space-based civilization: oxygen for water and air; silicon for glass, fiberglass, and silicone polymers; aluminum, magnesium, and titanium for reflective coatings, machinery, and reinforcing cables; chromium and manganese for metallic alloys; sodium, potassium, and calcium for process chemistry; and even the rare isotope, Helium3 to fuel fusion reactors. Ninety percent of necessary industrial materials can be mined and processed on the Moon. Virtually all data on Lunar minerals comes from surface samples obtained by Apollo astronauts. Rocks from deeper strata are likely to be even richer in metals and minerals.

**Table 4.2**
**Raw Materials From Lunar Soil**[361]

| ELEMENT | % or ppm* | 1000 tons |
|---|---|---|
| Oxygen | 40.8% | 5712 |
| Silicon | 19.6% | 2744 |
| Iron | 12.1% | 1694 |
| Calcium | 8.5% | 1190 |
| Aluminum | 7.3% | 1022 |
| Magnesium | 4.8% | 672 |
| Titanium | 4.5% | 630 |
| Sodium | .33% | 46 |
| Chromium | .20% | 28 |
| Manganese | .16% | 22 |
| Potassium | .11% | 15 |
| Sulfur | [362]540 ppm | 7 |
| Carbon | 200 ppm | 3 |
| Nitrogen | 100 ppm | 1 |
| Hydrogen | 40 ppm | 0.5 |

***Fig. 4.2 - Footprint in lunar soil.*** *The regolith has been pulverized to a fine powdery consistency to a depth of five meters. Lunar soil is like preprocessed mineral ore, crushed, sifted, and stockpiled, ready for use.* NASA photo.

It is more practical to launch these materials from the Moon than from the Earth. Launching a ton of payload from Earth requires 22 times more energy than launching it from the Moon.[363] To get material off the Moon requires accelerating it to a velocity of 2.4 kps, versus 11.2 kps for a launch from Earth.[364] Compare the images of the ascent stage of a LEM taking off from the surface of the Moon to that of a Saturn V taking off from the Earth; it's like popping a champagne cork compared to firing a cannon.

***Fig. 4.3 - Apollo 17 taking off from Earth*** *in a roaring volcano of smoke and flame.* NASA photo.

***Fig. 4.4 - Apollo 17 taking off from the Moon*** *in a silent puff of tenuous gas.* NASA photo (from low-resolution video)

After an object has left the Moon, it requires an additional change in velocity of about one kps in order to enter orbit around Earth.[365] Materials can be sent from the Moon to Asgard for a total change in velocity ($\Delta v$) of 3.4 kps. This requires a total energy investment of around 2.3 kwh/kg.[366] At 5¢/kwh, the energy cost to send a kilogram of material from the Moon to Asgard will be about 11.5¢. The presence of an accessible ore body like the Moon renders construction of Asgard and other orbital colonies practical. Without the Moon, the problems of space colonization might be intractable.

**Light Elements**

About 10% of the materials we need in space aren't available on the Moon. Sustaining our extra-terrestrial civilization will require generous amounts of hydrogen, nitrogen, carbon and other light elements. Hydrogen is essential to form water. Carbon, nitrogen, sulfur, calcium, and other trace elements are essential for biology and for the manufacture of advanced engineering plastics.

Unfortunately, the Moon is almost completely devoid of these elements. Hydrogen concentrations in lunar samples, for example, amount to only 60 micrograms per gram.[367] At this concentration, 16 tons of lunar soil contain but a single kilogram of hydrogen. Out of ten million tons of lunar soil processed for Asgard's oxygen, we will extract only enough hydrogen for 5000 tons of water.

Finding a ready supply of hydrogen and other light elements is one of the big hurdles barring our settlement of space. We could bring hydrogen up from Earth, but that would be an enormous undertaking, requiring 150,000 launches. There is some speculation that there may be water on the Moon, either as permafrost deep beneath the surface, or as surface ice in permanently shadowed craters near the poles. But, the Moon's violent history probably precludes lunar water, so we must seek some other source of light elements in space.

Absence of vital elements on the Moon seems a tragic shortcoming on the part of our celestial gardener. She seems to have so carefully provided our little seed planet with all the things we need to spread our living roots through this fertile plot of space around the sun: the superabundance of clean energy stored in the tremendous heat batteries of the Earth's oceans; the conveniently placed trellis planet of the Moon, put within easy reach of our first grasping tendrils; an abundance of oxygen, glasses, and metals, handily stockpiled and waiting only a burst of concentrated sunlight to liberate them. Such care has been taken to provision us well. Everything else a seed planet needs for quick germination and rapid growth is so deliberately provided; why would we be left to wither, deprived of certain essential nutrients?

Everything we do as we rise to attain our destiny can be done simply, even easily, but only after we have penetrated to the heart of the Cosmic Design and seen its signature in the beauty of soap-bubbles and sea shells. The answers are always at our fingertips, but finding them is an ongoing test of our worthiness. It is a continual challenge to see if we possess the imagination, the intelligence, the vision, to grasp the solutions put before us—like crumbs of cheese designed to lead a hamster through a maze. It should come as no surprise then that there are abundant and easily accessible sources of all the vital materials at hand.

**Asteroid Mining**

We will find the light elements we need on asteroids. The common impression of asteroids is that they are at best a source of iron and at worst a hazard to navigation. Neither impression is really true.

There are three main types of asteroid: the iron, the stony-iron, and the stony. Iron and stony-iron asteroids are the kind most familiar as meteorites in museum exhibits. These are spectacular chunks of more or less solid metal, usually sawed in half and etched with acid to highlight their beautiful crystal patterns. Iron meteorites are fairly common on Earth for the simple reason that they are more likely to survive passage through our atmosphere than are the stony variety. In space, however, the iron and stony-iron varieties comprise only about 25% of the total asteroid population.[368] Most asteroids are of the stony variety, which come in two types: carbonaceous and carbonaceous chondrites.

For space colonists, the carbonaceous chondrites are the most important type of asteroid. Carbonaceous chondrites are incredibly rich sources of water, nitrogen, and carbon; and, at least in the near term, they are the most accessible source of organic materials in the solar system.

Water is of course vital to Life in general and space colonies in particular. As space colonists, we must have ready access to an abundant supply of water, or at least the hydrogen needed to combine with oxygen to make water. Water is the very stuff of life, and space colonies need huge amounts of it. The Moon is, to put it mildly, a very dry place—"a million times drier than the Gobi desert."[369] Hydrogen for water must therefore come from the asteroids. Carbonaceous chondrites contain as much as 20% water.[370] Asteroids will give up this liquid bounty with surprising ease. If a bit of carbonaceous chondrite is heated to 100° C., steam will be driven out of the rock and will condense as pure clear water. It is slightly miraculous to watch the essence of life drip out of what appears to be nothing but a sterile cinder.

Nitrogen is another vital element that the asteroids can supply in abundance. It is more than a little surprising to find outer space liberally sprinkled with bite-sized chunks of rock that are as rich a source of petrochemicals as oil shale.[371] Like oil shale, carbonaceous chondrites are a good source of nitrogen, which composes 4% of its "bituminous fraction" by weight.[372] This is very good news, since the Moon is virtually devoid of nitrogen. Nitrogen is vital in its fixed forms as a plant nutrient, and is crucial to the development of proteins in human metabolism. Nitrogen is a heavy gas, which, if brought from Earth, would be very expensive.[373]

Carbonaceous chondrites, as their name suggests, are also a rich source of carbon. This crucial element comprises as much as 4% of their mass. Carbonaceous chondrites contain an abundance of valuable metals as well.

**Table 4.3[374]**
**Composition of Carbonaceous Chondrites**

| Component | % by Weight |
|---|---|
| Silicates | 76 - 90 |
| Water | 1 - 21 |
| Metals | .1 - 3.5 |
| Carbon | .1 - 3.8 |
| Nitrogen | .01 - .3 |

In addition to its invaluable moisture, this type of asteroid also contains 5% kerogen—a tarry hydrocarbon found in terrestrial oil shale.[375] Over a hundred million billion tons of kerogen are waiting for us in the rich veins of the asteroids.

This material will be precious to us as we go about the ultimate cosmic enterprise of converting lifeless elements into animate matter. For that purpose, there can be no substitute for the stuff of life. We—that is, living things—are mostly oxygen, hydrogen, carbon, and nitrogen, with traces of calcium, sulfur, potassium, chlorine, and many other elements. To sustain carbon-based life like us and the trees, we could hardly ask for a more perfectly formulated nutritive broth than kerogen. Kerogen is like condensed primordial soup.

The evident abundance of these vital light elements on Earth and in the asteroids accentuates the mystery of their almost total absence on the Moon. Luna has apparently undergone some sort of cataclysm which 'boiled off' her light elements. Theories of Lunar formation abound. The current leading hypothesis is that the Moon was somehow spun off from the Earth early in the history of the Solar system. Whatever the true case, we must turn to the asteroids for space-based nutrients like kerogen.

**Table 4.4**
**Elemental Composition of Kerogen**[376]

| Element | Weight % |
| --- | --- |
| Carbon | 75 - 80 |
| Oxygen | 10 -15 |
| Hydrogen | 5 -10 |
| Sulfur | 1 - 3 |
| Nitrogen | 1 - 2 |
| Other | .1 - 1 |

The asteroids represent a substantial material resource. Estimates vary, but the total mass of asteroids in the solar system is certainly somewhere between .1% and 10% of the mass of the Earth.[377] This may not seem like much on an astronomical scale, but consider this comparison: To produce just .1% of the Earth's mass as mineral ore would require strip-mining the entire surface of the planet to a depth of more than a mile. The asteroids contain thousands to millions of times more mineral resource than we can ever conceivably hope to recover from the Earth herself.

Most of the asteroids in the solar system occupy a zone between the orbits of Jupiter and Mars—the asteroid belt. Carbonaceous chondrites make up 50% of the asteroids at the belt's inner edge, and 95% of them at the outer edge.[378] Some of the largest asteroids, like Ceres—940 km. in diameter—are believed to be carbonaceous chondrites.

The asteroid belt will someday serve as a valuable source of materials for our expanding civilization, but it will not help us with our immediate supply problems in Asgard or on the Moon. It is surprisingly difficult to send things back from the asteroid belt.

It is important to remember that distance is a fairly meaningless concept in space. What is really important is energy. It takes a certain amount of energy to get anywhere in space and that, not necessarily its distance away, is the critical variable in determining accessibility. For example, the distance from the Earth to Valhalla is only a couple of hundred kilometers. If you could drive straight up, you could get there by car in a few hours. This, however, wouldn't gain you anything, since once you had driven up to Valhalla's altitude, the space station would come whizzing by at 30,000 kilometers per hour. This is the same situation in virtually any space travel scenario, at least inside the solar system. Everything is moving. Going from place to place means speeding up or slowing down to match the velocity of a moving destination.

Asteroids in the main belt orbit the sun at an average velocity of 24 kps, compared to the Earth's orbital velocity of 30 kps; so an object moving from the asteroid's orbit to the Earth's must first move into a transfer orbit that will carry it between the two orbits; and then it must rendezvous with Earth. Altogether, this requires a change in velocity ($\Delta v$) of around 10 kps. This is almost as high as the $\Delta v$ required to bring cargo up from Earth. To move along a minimum energy (Hohmann) orbit from the asteroid belt to the Earth would take about a year and a half.

So it appears that our celestial gardener has made an oversight after all. Our solar system is loaded with water and the nutrient fertilizers needed to sustain our growth, but they appear to be beyond our immediate grasp. As we have come to expect, though, the caretaker has indeed provided us with an abundance of essential nutrients, well within reach of our first tentative root tendrils. These little bits of ready nourishment are the Apollo and Amor (AA) asteroids.

**Triple AAA**

Unlike the main-belt asteroids which hang out between Jupiter and Mars, the Apollo and Amor asteroids inhabit our immediate neighborhood. In fact, they are so close that one will occasionally 'drop in'. Unfortunately, when they do drop by, they generally come to stay, and are singularly unwelcome. Asteroid strikes routinely wipe out 50-95% of all life forms on our planet.[379] These grim 'planet-crashers' aside, the AA asteroids are the most valuable ore bodies in the solar system.

The Apollo asteroids actually cross Earth's orbit, while the Amors merely graze it. So far, 20 to 30 of each class have been discovered and their orbits calculated. Two or three new ones are found each year. The total population of large AA asteroids, those with a diameter of more than a kilometer, is estimated at around 1600, plus or minus 800.[380] There is a host of smaller ones, over 300,000 with diameters greater than 100 meters.[381] A large proportion of these asteroids are undoubtedly carbonaceous chondrites.

The best thing about the AA asteroids is that they are readily accessible. For example, material can be retrieved from 433 Eros, one of the Amor asteroids, for a $\Delta v$ of less than 4 kps—a transfer energy even lower than that for material from the Moon.[382] Material from some asteroids can be moved into Earth orbit for a thousand times less energy than it would take to bring the material up from Earth.[383]

Eros is an excellent source of vital materials. The asteroid has an appropriately Freudian shape, 24 km. long and 8 km. in diameter. That is a tiny planet, but a pretty good sized mountain. With a mass of three trillion tons, Eros alone could supply enough hydrogen, nitrogen and carbon for hundreds of thousands of colonies the size of Asgard.

**Table 4.5**
**Materials Recoverable from 433 Eros**

| Element | Billion tons |
|---|---|
| Carbon | 60 |
| Metals | 54 |
| Hydrogen | 45 |
| Nitrogen | 5 |

In addition to these materials, millions of tons of sulfur compounds, soluble sulfates, phosphates, and all the other elements and compounds needed to sustain Life can be recovered from Eros.[384]

Transporting materials from the AA asteroids to Earth orbit is relatively easy. The asteroids' gravitational fields are very weak—the gravity on Eros is a thousand times feebler than on Earth. Launching material off Eros is as easy as pitching softballs.[385] A small solar-powered mass driver is all that is needed to propel payloads toward the Earth/Moon system.[386] [387]

Some of the AA asteroids could even be moved closer to the Earth by manipulating their orbits. It may eventually be desirable to nudge smaller asteroids into shared orbits around the sun with the Earth. This could be accomplished by delivering small navigational impulses at key points in the asteroid's solar orbit. By manipulating the asteroid's orbits, and using the masses of the Earth and Moon to provide "gravitational assists", we will be able to bring small asteroids closer to the Earth. Once in the same solar orbit as the Earth, materials can be sent to the Earth/Moon system from these asteroids with very small investments of energy.

Even larger asteroids could be brought into a shared Earth orbit. Smaller asteroids could be used to nudge larger bodies into desired orbits by arranging collisions. There are enormous energies latent in the momenta of these objects. By playing a game of celestial billiards we can tap these huge energies and use them to control the asteroids' orbits. Calculating the bank-shots will take outrageous amounts of computational power, but it is the kind of problem we will be equipped to solve in the next Millennium. Developing expertise at space snooker would also insure that we could cope with any dangers presented to the Earth or the Moon by rogue asteroids or comets.

**Fort Landsberg**

The creation of Asgard depends in large part on establishing a viable base on the Moon. Asgard is mostly water, and water is mostly oxygen. The Moon is in large part oxygen too, so the construction of Asgard hinges on mining oxygen on the Moon and sending it 'down' to geosynchronous orbit around the Earth. Development of colonies on the Moon and colonies in orbit around the Earth must, therefore, proceed in parallel. The first lunar outpost will accordingly be a very important place. It will serve not only as the source of materials for Asgard, but also as the seed for the first lunar colony.

Initially, the lunar outpost will be a rough-hewn mining camp, with the ruggedness typical of any frontier. All the Millennial space colonies share this attribute in the early stages of development. Fully developed space colonies are very civilized places, with every attention paid to human comfort, both physical and psychic. Aquarius, Asgard, and Avallon will all provide pleasant homes for permanent habitation. You can take your children for a lovely, walk (swim, flight, bounce, whatever) through the parks in any one of them. This is as it should be, as it must be. But it doesn't start that way. Each of the colonies begins life as a seed. The vital ingredients that cause the seed to grow—its 'DNA molecules'—are the people in it. They contain in themselves the vision of the mature colony, and the means of bringing that vision to fruition.

These seed people are necessarily a hardy lot. New environments are not exactly pieces of paradise when the first colonists arrive. Paradise has to be carved out of the raw frontier with blood, sweat, and tears. In Aquarius, the initial marine pioneers must be a pack of very salty dogs indeed. They will live for months or years in the middle of the Indian Ocean, wiring rebar forms together 30 meters underwater. The first space cadets in orbit will have to live jammed together in the tight confines of Valhalla, risking daily death from man-made meteoroids and asphyxiation from the smell of each other's dirty socks. Pioneers have always been a tough species and space pioneers won't be any different. Millennialists, sheathed in tungsten carbide, will have to be just as rugged and resourceful as any frontiersman clad in buckskins ever was.

The first 'lunatics' will probably have to be the toughest of the bunch. The difficulties in transporting passengers and cargo to and from the Moon are many times greater than the same problems in Earth orbit. There will be a long time between the establishment of a base on the Moon, and the completion of the first cushy ecosphere. In the interim, the lunar base, Fort Landsberg, will be a bare bones mining camp.

The optimum location for the first outpost will be in a large deep crater as close to the lunar equator as possible.[388] Lunar craters, are in effect, gigantic open-pit mines, already excavated deep into the lunar crust giving us better access to minerals. Landsberg Crater is a good candidate. At over three kilometers deep, it penetrates several different layers of the lunar crust and has a central peak that is probably composed of the moon's primordial bedrock.[389] Unfortunately, no one has yet been on a prospecting expedition to Landsberg. Nevertheless, we can confidently expect to find concentrations of aluminum, titanium, and other metals in the crater's walls and central peak that are far richer than the surface regoliths sampled by the Apollo astronauts.

The first big project we undertake on the Moon will be the construction of a mass launcher—not unlike the launch tube of the Bifrost Bridge.[390] This is a pattern that we will find repeated throughout the Solar System. An early developmental step on each new world will be the construction of a mass launcher, just as construction of the Bifrost Bridge was an early essential step on Earth. With the Lunar mass launcher, we can send oxygen, silicon, fiberglass, aluminum, titanium, and other materials to Asgard.

The Lunar launcher will be about the same size as the Bifrost Bridge's mass driver.[391] Vacuum conditions and the lack of environmental considerations make it possible to build the launch track out in the open. Untroubled by clouds or weather, the Lunar mass launcher will be able to operate around the clock, firing a load into orbit every few minutes.

Landsberg provides good topography for the mass launcher. The floor of the crater is a broad flat plain, 20 kilometers across. The wall of the crater rises 2.5 kilometers over a distance of ten kilometers—a 25% grade. Orbital velocity around the moon is 1.6 kps. At an acceleration of 10 *g* it requires only 16 seconds to attain orbital velocity, over a track length of 13 kilometers. The lunar mass launcher will run from the base of Landsberg's central mountain, along the flat floor of the crater, and up the slope of the crater wall. By tunneling under the central mountain, we can eventually extend the track. This will allow either lower accelerations for passengers, or higher final velocities.

Lunar pioneers will first construct a short cargo track running a distance of six kilometers up the crater wall. This short launch track will accelerate cargo at 25 *g*s, reaching orbital velocity in under seven seconds. The cargo launcher will be suitable for bulk commodities: liquid oxygen for Asgard's water shield and atmosphere, silicone sheeting for bubble membranes, aluminum beams for telecommunications stations, and spools of titanium cable for reinforcing nets. When the liquid oxygen gets to Asgard, it will be used to burn hydrogen from Eros. The water thus formed will be stored in Asgard's radiation shield. The combustion process recovers much of the energy used to extract and liquefy the hydrogen and oxygen.

**Excalibur**

Like the Bifrost Bridge, the mass launcher on the Moon will also use a powerful laser array—Excalibur. But, unlike on Earth, these lasers are not necessary to put payloads into orbit. Because there is no atmosphere to impede them, and lunar gravity is lower, launch capsules can achieve orbital velocity from the impulse of the mass driver alone. To blast the wave riders out of Lunar orbit and send them screaming down to Earth, however, requires an additional jolt of energy. The Moon-based laser array will provide this extra impulse. The Lunar lasers will require larger focusing mirrors than those on Earth, so they can project their power deeper into space. The Moon's low gravity and lack of atmosphere will make it relatively easy to build such large mirrors.[392] In addition to its propulsive duties, Excalibur will, like the Sword of Heimdahl, provide protection from meteoroids.

Payloads en route to Asgard will be carried aboard Valkyrie-class wave riders like those used on the Bifrost Bridge. The wave riders will accelerate along the electromagnetic track atop aluminum sleds. The sleds will be recovered by magnetic deceleration at the end of the launch track.[393] Only the wave rider is propelled into space. The Valkyries carry enough ice propellant to enable them to maneuver out of lunar orbit and rendezvous with Valhalla or Asgard.

The waveriders' ability to surf the top of atmospheres and borrow gravitational energy will make them incredibly versatile. Excalibur pushes the wave riders out of Lunar orbit into free space, where they come under the influence of Earth's gravity. Once 'over the hump' the wave riders will fall towards Earth. The wave riders can then use the Earth's atmosphere and gravity in combinations to achieve a wide variety of orbits. The Valkyries can use the atmosphere for braking maneuvers, and gravity to provide acceleration. By tapping these free energy sources, the wave riders can rendezvous with Asgard in geosynchronous orbit, match Valhalla's low orbit, or even land on Earth without having to expend large amounts of propellant.

On the Moon, Excalibur is free of any atmospheric interference, allowing operation of laser arrays with enormous power. Excalibur can be hundreds of times more powerful than the Sword of Heimdal. Excalibur can thrust its powerful beams deeper into space to decelerate payloads inbound from Eros and other points of origin. Excalibur's range and power, combined with the wave riders' versatility will enable us to send payloads to virtually any destination with speed and efficiency.

Power for Excalibur, the mass launcher, and other surface operations will come from solar power bubbles. The power bubbles will be placed in halo orbits around the Lagrangian points, $L_1$ and $L_2$. These points remain permanently fixed over the same spots on the Moon's surface, much as geostationary orbits do over the Earth. The $L_1$ point hovers in <u>luno</u>stationary orbit over the side of the Moon facing the Earth, and the $L_2$ point is fixed on the opposite side.

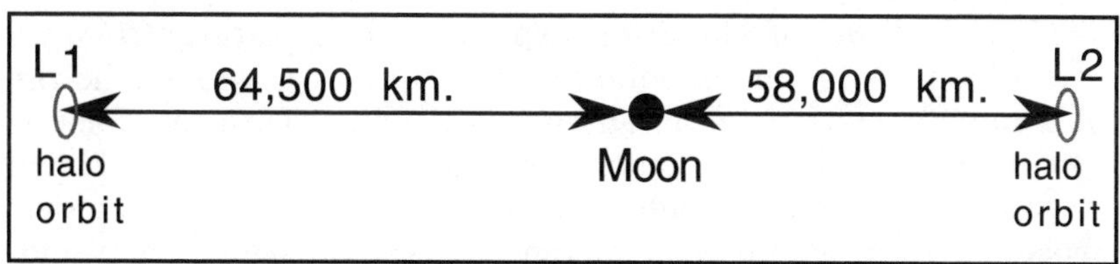


***Fig. 4.5 - Scale diagram of L1 & L2.*** *These points will be enormously valuable assets for a variety of functions in lunar development.*[394]

Because the Lagrangian points are islands of gravitational stability, they can support orbits as if they were planetary bodies. Particularly valuable "halo orbits" are possible around $L_1$ and $L_2$. A power bubble in a halo orbit, 1700 km. from either of these points, will remain in continuous sunlight.[395] In this position, the power bubble will not be eclipsed by the Moon, and will very seldom be eclipsed by the Earth. The halo orbits can hold a large number of power bubbles on each side of the Moon. The power bubbles will convert their electrical output into microwaves which can be beamed to any point on the lunar surface.

**Lunar Ecospheres**

Habitats on the lunar surface will simply be ground-based versions of Asgard. The problems of constructing and maintaining habitable ecospheres on the Moon are really no different than those we faced in free space. In settling the Moon, we will use the same technologies we developed in Asgard: a bubble membrane reinforced with cable nets to contain an atmosphere, a water shield to attenuate radiation, an atmosphere of low-pressure oxygen, electrical muscle and bone stimulation, and self-contained ecological loops based on algae. The major difference between Asgard and Avallon is that Avallon will be a hemispherical dome rather than the spherical bubble of Asgard.

If we set out to build an ecosphere on the Moon from scratch, the first step would be to excavate a foundation for the dome. The circular rim of the excavation could surround a fairly shallow depression. Material removed from the workings would be piled in a berm around the rim, thereby increasing the effective depth, and forming a base for the edge of the dome. It won't be necessary to dig our own excavations. A multitude of ready-made dome foundations already pock the Lunar surface, in the form of meteor impact craters.

Construction of lunar ecospheres is a straightforward proposition. A footing of solid material is first cast around the rim of a suitable crater to form an anchoring ring. Then the dome is installed. The dome is much the same as the outer bubble in Asgard—a gold-coated outer membrane of transparent silicone, separated from an inner membrane by several meters of water. Once the dome material is in place, the bubble is inflated with oxygen to a pressure of three psi. The bubble's water shield will counteract the contained atmosphere's otherwise overpowering tendency to blow out into space.[396]

As in Asgard, the bubble membrane must be overlain with a load-bearing network of cables. Cables of lunar titanium with tensile strengths on the order of 100,000 pounds per square inch will carry the main pressure load. Cable-nets will permit inflation of ecospheres to virtually any size. Even very large craters, like Copernicus, which is 90 kilometers across, can be domed over with a cable reinforced membrane. With the right materials and thick enough cables, the entire Moon could, at least theoretically, be enclosed in a bubble.

**Arthur's Shield**

Avallon is subject to many of the same radiation problems faced in Asgard, though on the Moon, the problems are somewhat less severe. The mass of the Moon blocks half of all cosmic radiation; and the sun, with its attendant flare hazards, is in the sky only half the time. Crater walls further block incoming rays.

To screen out the rest of the radiation hazards, we will use the same type of water shield that serves us in Asgard. Crater colonies require only half the shield mass needed in free-floating space colonies. A water shield 2.5 meters deep would reduce radiation inside the crater to less than half a rem per year.

Avallon's bubble will be even easier to engineer than Asgard's. Avallon's shield, unlike Asgard's, is subject to the pull of the Moon's gravity. Surprisingly, this relieves rather than exacerbates engineering problems. The air inside an ecosphere exerts an upward force of over two tons per square meter. Rather than limiting the water shield thickness to the minimum depth necessary to attenuate radiation, the shield will be made thick enough to counter-balance air pressure inside the dome.[397] This will require a water shield 12 meters deep. The air pressure inside the dome supports the shield mass, and the weight of the shield virtually eliminates pneumatic stress on the dome membrane.

Due to the presence of Lunar gravity, dome profiles must remain extremely shallow. In a hemispheric dome, like that pictured in Plate No. 8, the water shield would exert tremendous hydrostatic pressure on the lower regions of the dome. (Vertical profile of the dome in Plate No. 8 has been exaggerated for clarity. The center of the actual dome will rise only a few hundred meters above the rim.) By adopting a very shallow, almost flat, dome profile, we avoid this problem. The flat dome profile is achieved by expanding the radius of curvature of the dome. Ordinarily this would not be feasible, because the pneumatic stress on the dome membrane is proportional to its radius, but the water shield eliminates these stresses.

As in Asgard, cylindrical tubes, running around the circumference of the dome, will confine the shield water. Such tubes allow water to circulate and maintain an appropriate heat balance inside the ecosphere. In Avallon's dome, the circulation is more complex. The tubes in the dome will describe two continuous interlinked helices. Heated water will move out through one coil, while cooled water moves in through the other. The lowest tubes of the shield dome are buried under the lunar surface where temperatures are extremely low.[398] Hot water, which has traveled the full length of both coils, will dump its heat underground. Then, having cooled, the water will make another trip around the circuit. This, together with active management of the dome's reflective properties, will allow us to maintain desired temperatures inside the ecosphere.

**Day for Night**

The Moon is a good place for space colonization, but it has one serious drawback—the day is a month long. The Moon has been de-spun by the mass of the Earth, so it orbits with the same side always facing us. This means that its day corresponds to the period of its one month orbit around the Earth. Adjusting to the two week day and the fortnight night is one of the great challenges facing Lunar colonists.

The domes of Avallon will have one feature that Asgard lacks: an adjustable layer of Polaroid film. By changing the membrane from clear to dark and back again, a natural day-night cycle will be created inside the ecosphere. With the same technique, we can adjust the length of days and nights and, if desired, even recreate the annual succession of seasons.

During the long lunar day, the Polaroid membrane can provide us with dusk, darkness, and dawn. During the two week lunar night, the colony must somehow be provided with light. Lighting the crater with electric power is a possibility, but not very desirable. Illuminating Avallon's interior for twelve hours a day would require millions of kilowatt hours of power. Some elaborate energy storage scheme, or an alternative power source like a nuclear reactor, would be needed. A huge investment in lighting would be required, and new problems like disposing of waste heat from the lights would have to be solved.

Instead of forcing power into the picture, we should follow Buckminster Fuller's prescription and use design finesse. Space is awash with free power. Our ability to capture and harness that energy will be one of the ongoing tests of our fitness to colonize the galaxy. A means to illuminate the lunar colonies with the available solar power supply must be found.

The answer, as with most magic tricks, is to do it with mirrors. We will place large, slightly curved mirrors, in halo orbits around $L_1$ and $L_2$. When each side of the moon is in darkness the mirror on that side will be facing the sun. These mirrors will be in virtually continuous sunlight and can shine down on the ecospheres during the long Lunar night.

**The Island-Valley of Avallon**

Avallon's crater will measure about 16 kilometers in diameter. Lunar craters are fairly shallow in profile, with a width to depth ratio of 5 to 1, so Avallon will be about 1600 meters deep. The dome profile will be essentially flat. Most of the crater bottom is sloping, with a slightly flatter floor. Once domed over and provided with an atmosphere, a crater eight kilometers in radius will provide over two hundred million square meters of habitable area.

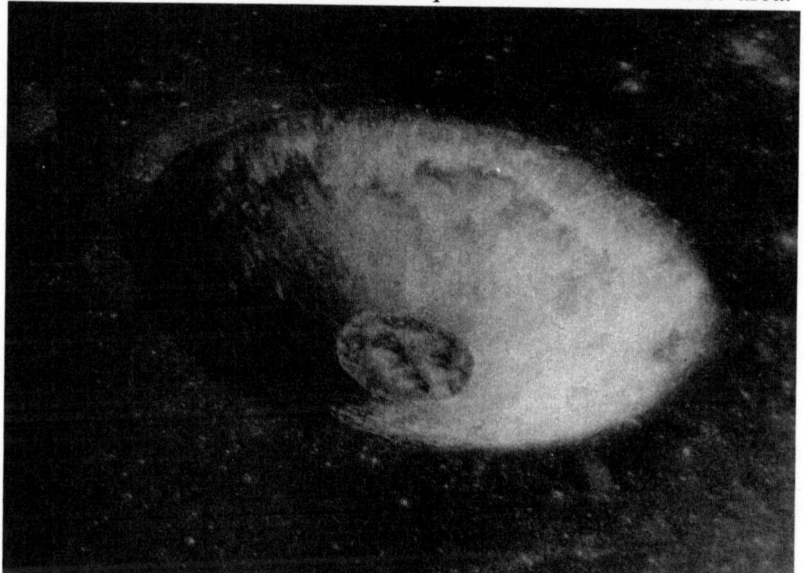

***Fig. 4.6 - Lunar crater.*** *Avallon, and other early ecospheres will be built inside relatively small lunar craters like Messier B.* NASA

It will be a simple matter to recontour the loose material in the crater walls to form broad flat terraces. The terraces will be wide toward the center of the crater and narrower toward the rim, averaging about a hundred meters across. Tubing for drip irrigation will be embedded in the soil of the terraces which can then be planted with vegetation. There are no organic nutrients in lunar soil, so all plants must be irrigated hydroponically. Plants have been grown experimentally in simulated lunar soil with great success. Tropical foliage, of the type common in Aquarius and Asgard, will be grown throughout the ecosphere.

In the center of the crater, we will build an Aquarian city structure. Instead of floating on the surface of the sea, this city structure will rest on the solid surface of the Moon. Rather than growing it out of electrically accreted sea cement, we will build it more conventionally from cement made with lunar regolith, and from aluminum and glass refined out of lunar soil.

On the Moon, the structure does not have to withstand the dynamic forces of weather, nor the crushing grip of Earth's gravity. Inside the ecosphere there will be no winds, waves, or potential earthquakes to contend with.[399] Therefore, we can build it less robustly and much taller.

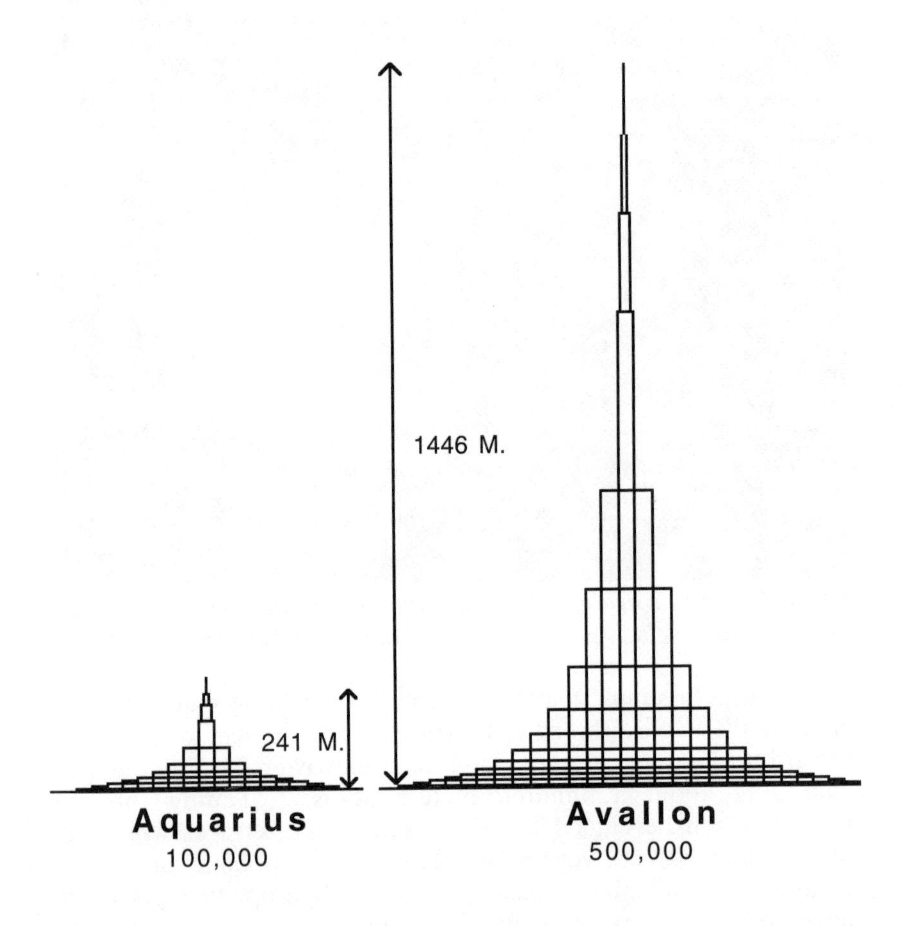


*Fig. 4.7 - Comparative surface structures—Aquarius vs. Avallon.*

The hexagonal tower modules will be the same width as the ones in Aquarius. This will allow us to adapt systems and layouts directly from Aquarius to Avallon. The main difference between the two structures will be the height of the towers in Avallon.

Gravity on the Moon is only one sixth of that on Earth, allowing us to erect towering, spires of breath-taking dimensions. Each tower in Avallon will be six times taller than its terrestrial counter part. Inside the crater, surface area is at a premium, while extra structural height is relatively cheap; just the reverse of the situation in Aquarius, where we have an entire ocean to spread out on, and Earth's gravity to contend with.

The spires of Avallon will have a much higher ratio of surface area to inner volume than their squatter counterparts on Earth. Every tower apartment will have a balcony with a view of the gardens, meadows, and forests covering the crater slopes. An emphasis on tall spires will be important to those Avallonians who prefer to come and go on the wing. Surrounding the central structure will be a small lake. Gardens and orchards will spread up the terraced slopes of the crater walls all the way to the rim. Virtually the entire interior of the crater will be devoted to open park land.

**Never-Trees**

Buildings will not be the only structures able to grow to tremendous heights on the Moon. The forests of Avallon too will be something spectacular. In the low gravity, trees can grow to Brobdingnagian proportions. An oak tree on Earth can achieve a height of 33 meters (110 feet) and a trunk diameter of more than a meter. On the Moon, the same tree could grow to a height of 200 meters (660 feet), with a diameter of six meters (24 feet).[400] Single branches on such a giant would be the size of whole trees. A grove of trees in Avallon will make even the wildest Hollywood fantasy of Sherwood Forest look like underbrush.

Adventuring in a Lunar woodland will be fantasy brought to life. A person who weighs 68 kilograms (150 lbs.) on Earth will weigh only 11.3 kg. (25 lbs.) on the Moon. In Avallon, a fit person can easily jump ten feet in the air. In the jungles of the Moon, you can "make like an ape ape man and swing through the trees with the greatest of ease." Even the most rudimentary wings will allow sylvan frolickers to glide from tree to tree like flying foxes.

Certain of these trees will be hollow, providing an interior landscape enjoyed heretofore only by elves. The interior of a hollow oak might be as wide as your living room. Labyrinthine passages rise hundreds of feet up into the tree, twisting and turning through the hollow branches. The trees in Avallon will be the size of sixty story buildings, and some people will actually live in them.

Complementing this realm of gigantic trees will be our personal powers of flight. In the gravity-free environment of Asgard, flight is almost effortless; in fact, avoiding it requires a strategically placed patch of Velcro. In the lunar gravity of Avallon, flight is a more energetic proposition. In Avallon, flying requires well conditioned muscle power, and good lung capacity.

On the Moon, a human has the strength to body weight ratio of a large bird. Due to the low air pressure—equivalent to that on Earth at an altitude of 11 km.—flying in Avallon will require wings with a large surface area. (See Plate No. 10.) Flying, even in Lunar gravity will be rigorous exercise. To feel what it would be like, hold a dumbbell weighing one tenth of your body weight in each hand—now flap. If you are in good shape, you can probably keep this up for a few minutes. Serious flying will take special conditioning. On Avallon it will be easy to recognize the flyers. Sheathed in form-hugging body suits, their tremendous chests and wide V-shaped backs will be unmistakable trademarks.

For those less inclined to the muscle-beach approach, there will be propeller driven air cycles. These winged bicycles will provide the thrill and independence of personal flight without requiring as much strength and endurance. Even gentle cycling will bear you easily aloft, though you will have to maintain forward speed to stay airborne. Aircycling through the vast internal spaces beneath the dome, skimming over the tops of the gigantic trees, engaging in mock aerial dog-fights, and buzzing hapless pedestrians, will doubtless be popular exercises in the skies of Avallon.

On the ground too, the available forms of personal locomotion will be diverse and entertaining. Jogging, as such, will not really be possible. Running will take the form of either a kind of dream-state slow-motion bounding, or kangaroo hopping. Even the Apollo astronauts, confined in their stiff balloon suits, found they could get about easily by hopping from place to place like cyclopean space bunnies. Out in the grass-carpeted countryside, or along special paths, hoppers can jounce along at high speed, covering three or four meters at every bound.

For transport inside the very large craters, and for those irrevocably committed to vehicular travel, the air car—a dream so long deferred on Earth—will become an everyday reality. With the use of turbofans in swiveling pods, the air car will be able to take off and land vertically, hover, even fly backwards. The turbofans, made of light-weight ceramic and powered by clean-burning hydrogen, will be able to zip across the largest ecospheres in minutes. Another long awaited dream—the personal jet pack—will at last be practical. With a small, quiet, power plant, the jet pack can propel its wearer through the sky with an ease and freedom unmatched outside the pages of comic books.

**Copernican Veldt**

Avallon—16 km. in diameter—is pretty small, as lunar craters go. There are 60,000 lunar craters with diameters of more than a kilometer, and 4000 craters with diameters greater than ten kilometers.[401] After we perfect our construction and ecological systems in smaller craters, we can gradually scale up the size of our ecospheres. Eventually, we can dome over and create ecospheres in even very large craters like Copernicus. (See Plates 8 & 9)

**Table 4.6**
**Habitable Lunar Craters**

| Crater | Depth-m. | Diam.-km. | Area -km$^2$ |
|---|---|---|---|
| Avallon | 700 | 7 | 38 |
| Mosting | 3000 | 25 | 488 |
| Landsberg | 3500 | 43 | 1450 |
| Eratosthenes | 3750 | 60 | 2780 |
| Copernicus | 3850 | 96 | 7320 |

Copernicus alone will provide enough habitable space for several million people. Once a big crater like Copernicus has been domed over and fully terraformed, a person inside will be hard put to tell that he isn't on Earth. The sky is a blue bowl high overhead; plains of verdant grass stretch away to the forest-covered hills on the horizon; giraffes wander by in stately process. Only when you make a spontaneous jump for joy and find yourself clicking your heels six feet off the ground will you know you're not in Africa anymore.

Copernicus will be a self-contained world. A broad ring-shaped lake will surround a large central island which will be set aside as an expansive nature preserve. Herds of animals will graze the lush grass of this lunar savanna. Giant trees will cover the slopes of the mountains in the crater's center.

Big craters like Copernicus will be hybrid compromises, striking balances between pure nature preserves and human habitats. The mix of species may not, therefore, be entirely natural. Large predators, for example, might be less than welcome. Some aspects of the ecology of lunar craters must necessarily be artificial. Even in biomes dedicated solely to ecological preservation, humans must maintain the equilibrium by active and constant management. The lunar craters will inevitably have an unnatural feel. Many native Earthlings may find this disquieting and unsatisfactory. It is nevertheless an inescapable necessity. The Moon is an environment of uncompromising hostility. Animate matter will be able to persist there only through the continual intervention of conscious matter—which is us. More natural environments will have to await the creation of a planetary biosphere on Mars.

Lunar ecospheres will be more like Disneyland versions of ecology, than the 'real thing'. In hybrid environments like Copernicus, this will be especially true. There will be an abundance and variety of life forms in the big crater, but the place is basically a human habitat. There is rain, but networks of subsurface hydroponic tubing sustain the plants. A plant could not derive nourishment from lunar soil any more than a baby could be nurtured on distilled water. The introduction of millions of tons of nutrients into the regolith to create soil may take centuries. The natural processes of decay and nutrient recycling that occur in terrestrial soils will not at first happen on the Moon. Even the simplest sorts of maintenance, like recycling manure, will have to be done artificially.

Fortunately, the technologies at our disposal will be equal to such tasks. On the plains of Africa, animal droppings are a surprisingly rare site. One would think that between the wildebeest and the water buffalo, not to mention elephants, it wouldn't be safe to put your foot down any where. Every night, though, battalions of dung beetles come up from their underground labyrinths and perform a headless dance, rolling up balls of manure and trundling them away underground. On the Moon, the same maintenance chores will be performed by small robots, with about the same intelligence as the dung beetles. Swarms of robotic insects will intervene at every link in the food chain to hold the frail ecologies together.

For natural purists, these artificial worlds, sustained under golden domes, will feel like cartoons of ecology. This may well be, and for people accustomed to the real plains of Africa it may be unacceptable. Native Lunatics, however, are apt to be too busy enjoying Life to spend much time worrying about how their crater worlds compare with Mother Earth. I suspect it would be harder for a native Copernican to adjust to the real world and its harsh struggle for survival than it will be for Earthlings to adjust to the pastoral, controlled ecospheres of the Moon.

The splendor of Life in its glorious variety will be ample compensation for any lack of fidelity to conditions on the mother planet. A riot of Life will romp on the plains of Copernicus, sheltered under the benevolent wing of Lunar man, oblivious to the cruel and arbitrary laws of nature. Flocks of shocking pink flamingos will feed along the shallow margins of the lakes, ignoring the swans gliding through the same waters. Herds of zebra will mingle with mustangs, completely unaware that their ancestors did not live together. Animals in the ecospheres will be chosen only for their adaptability to the climate and their tolerance for other species. Intermingled among the multifarious creatures on the ground and in the air, "in action like an angel, in apprehension like a god," will be that paragon of animals—man himself, living in peace and harmony with his fellow beings at last.

**Metropolis Copernicus**

The floor of Copernicus is 60 kilometers (37 miles) across and covers almost 3000 square kms.[402] The lake and park covering this area will serve the entire populace—human and otherwise. Here, people come to picnic on the close-cropped grass, beneath the spreading branches of giant Acacia trees. It is, on average, another 10 or 15 kilometers from the shore of the lake up the gentle slope of the crater bottom to the steep walls of the crater rim and the atmospheric containment dome.

In the area between the dome and the margin of the lake, sprawling across the slopes of the crater is the city of Copernicus. More of an extended suburb than a city, it covers an area of 4400 square kilometers. Individual homes and community amenities are on the surface, but almost everything else is underground. This leaves the garden-like surface free for dwellings and open space. Homes are in the form of simple tensile structures—high-tech tents of the type perfected in Aquarius. Weather inside Copernicus is limited to occasional showers. Because there is no wind, storms are a physical impossibility. Shelter is more for creating a sense of place and privacy, than for providing refuge from a hostile environment. The great dome overhead provides the real shelter.

In the broad ring of settlements surrounding the central Copernican plain, there is room for millions of individual homes. Habitations will cover only a fraction of the area of the crater slopes. The rest of the land in the ring can be devoted to outdoor recreational facilities: parks, playgrounds, swimming pools, lakes, paths, groves, and gardens. On the lake will float a number of Aquarian structures. Each of these spire clusters can house a quarter to half a million inhabitants. Dozens of Aquarian cities could be sprinkled over the area of the ring lake without crowding.

All of the infrastructure for living: transportation corridors, malls, utilities, offices, schools, warehouses, and factories are below the surface. A labyrinth of mechanical structures, the colony's life-support machinery, will honeycomb the regolith. The subsurface infrastructure provides for all the needs of the crater's inhabitants. Umbilicals connect every home on the surface to the underground utilities. Going into the substructure is like entering the interior of a mall or other large modern building. The difference is that in this case, the building covers several thousand square kilometers.

If colonists drive tunnels into the lunar crust, there is virtually no limit to the number of people who could live 'inside' Copernicus. Although life underground is certainly not as appealing as life on the surface, it need not be too oppressive. The surface is never more than a few minutes elevator ride away. High resolution, flat-panel, video displays could provide views of the surface—or any other views for that matter. The low Lunar gravity will allow enormous caverns to be excavated with high ceilings. These internal spaces could be brightly lit and terraformed to create open park lands deep underground. This combination of countermeasures could make life beneath the surface a tolerable, and even attractive option.

It will be a long time, however, before people have to move underground on the Moon. There are a lot of craters, and there is a lot of room inside them for a lot of people. As we need to accommodate more people, we can dome over more craters, erecting Aquarian structures, tensile surface habitations, and even tree-houses. Eventually, the Moon will be home to hundreds of millions of human beings and a kaleidoscopic array of other life forms.

**Home Sweet Dome**

The number of lunar craters is astonishing. Because our airless companion planet has no weather, there are no winds or spring rains to erase the scars of Luna's violent past. Crater foundations are therefore abundant and come in a wide variety of sizes. The craters on the Moon provide altogether nearly a million square kilometers (386,000 sq. mi.) of potentially habitable space. That is an area larger than California, Texas, and Montana combined.

Some craters are huge. Clavius for example—site of the American base in the movie *2001*—is 235 km. (146 mi.) across. Cut crater dimensions in half, and the number of smaller craters multiplies by a factor of four.[403] There are a quarter of a million craters 600 meters across—the size of Asgard. The number of craters half that size—300 meters across—totals a million.

***Fig. 4.8 - Dark side of the Moon.*** *The far side of the Moon is even more heavily cratered than the side which faces Earth. Each of these craters is a candidate for ecosphere construction. Eventually a large fraction of the Lunar surface will be domed over and terraformed.* NASA.

Small craters will also provide suitable ecosphere foundations. A crater with a diameter of 300 meters encloses an area of 70,000 square meters, too small for an ecological biome, but just right for a cozy human community—a Moon village.[404]

Even single-family ecospheres could be constructed in very small lunar craters. A crater ten meters in diameter will provide 80 square meters (850 sq. ft.) of habitable space. Living and sleeping rooms can be built into the side walls of the crater, freeing up open space for an atrium. There are four billion lunar craters of this size—enough to provide one for every family now living on Earth.

***Fig. 4.9 - Single-family Lunar ecosphere.*** *Living and sleeping rooms are underground, and the atrium is shielded by water.*

Not everyone will be able to afford this lunar version of suburbia, however. Single-family ecospheres will be very expensive. Shield mass per person goes up dramatically in smaller ecospheres. Every time a crater's radius is reduced by half, the mass of its water shield, per person doubles. Therefore, each person in these single-family ecospheres will require more than a hundred times as much water as the inhabitants of Avallon. Costs can be reduced somewhat because the living quarters are countersunk into the crater walls. Time spent under the protection of meters of lunar soil will be virtually radiation free. By sleeping and working underground, a colonist could minimize his time spent in the atrium exposed to higher radiation levels, and the shield mass could be cut accordingly.[405] Lone ecospheres will require individualized utilities, and will otherwise not enjoy the economies of scale available to the inhabitants of larger crater communities. For those who do have the means and the inclination, there will be individual home-sized ecospheres aplenty.

**Arkospheres**

Above all else, the lunar craters will be ecological preserves, providing refuges for Life. In a multitude of special biomes, we can preserve the precious storehouse of Life's genetic diversity. On the Moon, Life will be safe from possible calamities which threaten the Earth—environmental, nuclear, or meteoric. Unlike the vulnerable Earth, there are few conceivable disasters which could wipe out all the lunar ecospheres. Even a giant comet crashing into the Moon would destroy only a few lunar biomes. The same disaster could utterly annihilate Life on Earth. As for man-made threats, it would be a very bad idea to attack the Moon. Positioned on the high-ground, and armed with electromagnetic launchers and the powerful lasers of Excalibur, the Lunatics will be virtually unassailable.[406]

The Moon will be the ultimate safe-deposit vault, securing and protecting the genetic wealth of the universe. The lunar ecospheres will act as living seed banks. Withdrawals could be made if needed for the rehabilitation of a damaged Earth, for the terraformation of Mars, or for the eventual dissemination of Life among the stars.

To tailor-make environments inside individual craters, we can control the amount of energy entering and leaving an ecosphere, and the amount of water inside. Small, self-sustaining biomes can be created to replicate any ecology. One crater might be kept cold and dry and stocked with animal and plant life from the Arctic tundra; another might be kept hot and wet and made a home for a variety of species from tropical rain forests. We can even fill some craters with water to provide environments for marine life. Savannas, deserts, temperate forests, marine estuaries, alpine glades; these habitats and many others can be duplicated. By engineering individual crater ecologies, we can provide a viable home in space for all of Earth's creatures.

Most humans will probably choose ecospheres where the environment and selection of species has been optimized for our own enjoyment. People are likely to prefer the shirt-sleeve environment of the tropics, over the more demanding climes of the Northern pine-barrens or the Nefud Desert. It will, however, be refreshing, to visit such environments when we want to. Some people will even prefer living in one of these harsher climates.

The availability of customized ecozones will engender a wide variety of life styles. It will be possible to preserve and even resurrect traditional cultures. The lunar biomes might then serve not only as preserves of ecological, but also of sociological, diversity. Laplanders, Inuit Eskimos, Kalahari Bushmen,

Yanamamo Indians, Australian Aborigines, Cherokee Tribesmen, Polynesian Islanders—all could reestablish lost or vanishing cultural traditions in ecospheres tailored to reflect their terrestrial habitats. Preserving the diverse roots of mankind's rich cultural heritage might in the long run be as valuable to us as the preservation of the Earth's genetic diversity.

**A Confederation of Lunatics**

Diversity will be the watchword on the Moon. As people find the freedom to explore new modes of social, political and cultural development, the colonies in free space, like Asgard, will develop a tremendous range of cultural options. On the Moon, the same process of diversification will continue, with the added dimension of diverse ecologies. Lunatics will routinely travel from one crater to another, experiencing new climates, new environments, and new cultures. Many crater colonies will probably be more or less normal extensions of established Earth cultures, but some are bound to become bizarre. The colony of New Nippon, for example, will have the elegant style of historic Japan, while the colony of New Nirvana is apt to be a little weird. Some colonies will be dominated by strange splinter cults and special interest groups. Others, like Lothlorien, in the crater Eratosthenes, will evolve into novel societies utterly unique to the Moon. It might be interesting to spend a weekend in New Nirvana though, since the entire population consists of Hare Krishnas—when you arrive at the mag-lev station they will all come to greet you and try to sell you a book. Well, maybe a short weekend.[407] With thousands of colonies available, the options will be limited only by the imaginations of their inhabitants. The crater Mosting B, for example, has been taken over by "The Sons of the Desert", a society dedicated to silliness in general and the films of Laurel and Hardy in particular. Such a colony would certainly be a fine mess. And a lot of fun too.

**Magic Carpet Ride**

The people of the Moon will enjoy certain advantages over other inhabitants of the solar system. They will share many of the conveniences of life in free-floating space colonies like Asgard: closed-loop ecologies, abundant solar energy, and low-gravity; but they will have the edge on space-based ecospheres for a number of reasons. An ecosphere's costliest component is the radiation shield. The shield is 90% oxygen, the most abundant element on the

Moon. Many other valuable industrial materials, especially glass and metals are also available

Other advantages will help accelerate the growth of the Lunar civilization. Transportation will be extremely economical on the Moon. Say you wanted to travel from your home in Kepler, to visit a friend in Eratosthenes—a distance of more than 800 kilometers. You could, of course, take a laser-rocket shuttle. But this expends a lot of energy and precious propellants—an expensive and wasteful proposition. Instead, you will ride the electromagnetic monorail.

Using your universal credit icon, you rent a rail pod the size of a small car. Settling in for the ride, you instruct the on-board computer to take you to Eratosthenes. No direct link between Kepler and Eratosthenes yet exists; so the computer shows the route you will be traveling, which passes through a central rail hub in Copernicus.

The computer screen is available to entertain you during your trip, but you won't be in the pod very long. You settle back and look out through the pod's clear canopy to watch the Moon go by. The pod energizes its magnets and levitates above the superconductive monorail. It is easy to maintain the rail at the low temperatures required for superconductivity since keeping the rail cold involves little more than shading it from sunlight. Under the computer's direction, the pod heads out of the rental garage to merge with other pods on the main track. A pulse of electrical energy traveling along the rail gently accelerates your pod up to a velocity of about one kilometer per second.[408] This is about as fast as you want to go. If the pod went much faster, it would try to launch itself into orbit. The system's computer maintains a spacing of a few tens of meters between individual pods. The high speeds allow people to travel with the comfort, privacy, and convenience of individual vehicles, while enjoying the economies of mass transit.

You watch the lunar landscape whip by at an ever increasing pace. The pod rides the monorail out in the open, accelerating without any friction through the near perfect vacuum. The monorail seldom runs on the lunar surface exactly. It is elevated on trestles in some places, and passes through tunnels in others. Disregarding all surface obstacles, it runs straight and level as a laser beam.

After five or six minutes of gentle acceleration, your seat rotates on its base, turning 180° so you face the rear of the pod to begin decelerating. After a total of ten minutes, you stop in the hub station, deep underground in the center of Copernicus. You transfer to another pod and repeat the process, arriving at the station in Eratosthenes, about eight minutes later. Your 800 km. journey—like traveling from Washington D.C. to Indianapolis—has taken about 25 minutes. If you stop off at the hub station for an ice cream Copernicone, the whole trip might take as much as half an hour.

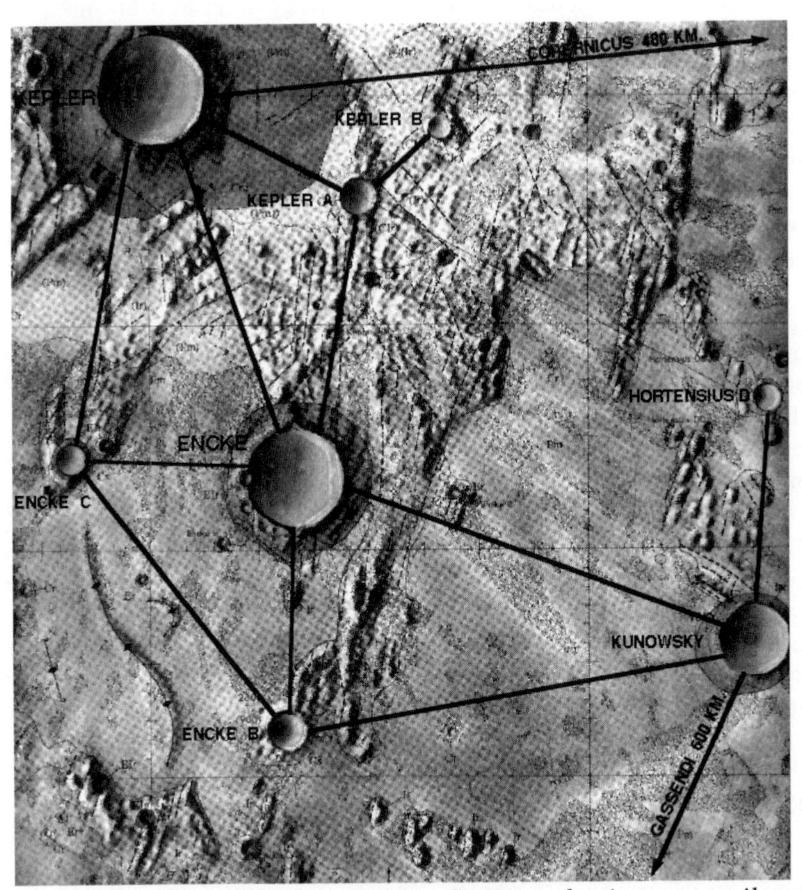


***Fig. 4.10 - Map of Kepler region.*** *Superconductive monorails will link all the major Lunar craters in a network of unprecedented speed and economy.* Based on USGS Map, 1:1,000,000 scale.

Your credit account shows personal pod rentals totaling half an hour, and an energy cost of a few watts. Traveling from place to place at such breath-taking speeds ought to require profligate expenditures of energy. It might, if not for the Moon's special advantages. Traveling through a vacuum on a superconducting electromagnetically levitated mono-rail involves almost no frictional resistance—neither atmospheric, nor electrical. All the energy pumped into the system goes into kinetic motion of the pod. This kinetic energy has to come out of the pod when it brakes to a stop. The brake is applied by reversing the flow of electrical current in the superconductive monorail. Now, instead of accelerating the pod, the pulse of magnetic energy in the monorail acts to decelerate it. Correspondingly, energy flows into the rail instead of out of it. In this way, virtually every Joule of energy that was put into the pod is recovered. Since the entire system is superconductive, the energy losses are minuscule, and so are the costs.

Other costs, like maintenance and capital, are correspondingly low. This railroad is a track engineer's dream: The route is very seldom closed by snow drifts or mud slides, the rails don't rust, the trestles don't decay, and none of the friction-free components ever wear out. The maintenance chief is going to be lonelier than the Maytag repairman. On the time scale of capital projects, the monorails are virtually immortal; so their capital costs can be amortized over centuries and their usage costs shared over billions of passenger miles.

All of the large lunar craters will be linked in a network of superconductive monorails. The wealth of nations has always been founded on transportation infrastructure. Because of the monorails, commerce on the Moon will enjoy advantages that Earth may never have. The Moon's tremendous resource base and its other advantages are likely to make it an economic powerhouse.

**Life Among the Faeries**

(A short digression from conjecturation to confabulation.)

*When you arrive at the station in Eratosthenes, your friend is there to greet you—even though she isn't a Hare Krishna. She has the tall slender build characteristic of native-born Lunatics, and the powerful shoulders and wide back of a serious flyer. Her parents named her Rebecca, but everyone calls her Ariel, her chosen name. She is no Krishna, but Ariel is a bit of a pixie. She greets you with a garland of fresh flowers which she hangs around your neck. Some of the founding colonists were from Maui, and certain traditions linger on. Commandeering an air taxi, you emerge from the subterranean structure of the colony into the mellow air of Eratosthenes—called Lothlorien by its mildly eccentric inhabitants.*[409]

*Every colony on the Moon has its own particular personality, and Lothlorien is quirkier than most. This colony is an artist's haven, and tends to attract a creative population. Flowers are one of the colony's specialties. They grow in riotous profusion on the crater floor—as dense as the tulip fields of Holland. Even as the air cab ascends a kilometer high, the air is perfumed with them.*

*In the distance, you can just see the high peaks of the central mountains, clad in thick cloaks of vivid green. As you approach the mountains, you see individual trees beginning to resolve themselves. They are spectacular giants, even by lunar standards. The cab drops swiftly, approaching the upper branches of what looks like an enormous gnarled oak, half a kilometer high.*

*"Is it real?" you ask your friend. This is your first visit to Lothlorien, and not even the 200 meter ponderosas of Kepler have prepared you for this.*

*"What difference does it make?", she asks, laughing lightly.*

*The cab sets itself down on a platform cantilevered out from the trunk of the gigantic tree; the landing pad's yellow and black striping clashes incongruously with the organic shapes around it. You enter the tree through an automatic door and step into a normal looking elevator. When you step out, you pass through an apartment door, and find yourself seemingly transported to Tolkien's Middle Earth.*

*Your friend's apartment is built into a colossal branch of the tree. Back at home, in Kepler, you live in an expansive pavilion in the highlands of the crater wall. You find Ariel's elven cottage a little claustrophobic. Ariel, however, thinks it is "cozy" and swears she'd never live anywhere else. She shows you into her tiny guest room. There is a single round window set in a knot hole of the hollow branch. Looking out you see other massive branches stretching above and below. The ground is three hundred meters straight down. Although back home you routinely ride an aircycle to work, the precipitous view makes you a little woozy.*

*Your friend has prepared a light brunch and you sit down at a table fashioned to resemble a giant mushroom. You perch precariously on the upholstered toadstools that serve as chairs. Born and raised in Aquarius, you have lived your entire adult life on the Moon, and are completely accustomed to synthetic foods—or at least think you are. But, Ariel's cooking takes even you by surprise. Little tetrahedron shaped bits of purple that taste like lavender are the main course. The meal is washed down with orchid nectar. It is strange to your palate, but delicious in a novel way, and somehow appropriate to the setting.*

*After brunch, Ariel insists on taking you for a flight. She is an experienced and powerful flyer, while you have seldom strapped on wings. At home, you prefer the more straightforward locomotion of aircycles and jet packs. Ariel dons her wings, a gossamer confection of iridescent surfaces, reminiscent of the mythical faeries she seems to embody. She leaps lightly from her balcony, diving a hundred meters before spreading her delicate wings and swooping back up. She beats the air with quick expert strokes, slipping easily through the branches of the tree.*

*"Come on!" she shouts, twisting over at the top of her climb and diving away, out into the open spaces under the wide dome.*

*Taking a deep breath and wishing for a good jet pack, you take the plunge. You are no bird man, but you are a true Lunatic and have flown before. Stretching wide your broad eagle-like wings you soar out away from the tree. It's like riding an aircycle—you never forget. Quickly regaining your confidence, you begin to fly, with slow strong beats that will not tire you, trading altitude for energy. Ariel flits around you like Tinkerbell around Dumbo. All over the sky other flyers are cutting through the air, mostly borne aloft on beating wings.*

*A great flock of flying people gathers, all headed for the same destination as you and your host. Dressed in a riotous array of bright colors and flowing garments they look like a migration of birds of paradise—which in a way they are. You glide down into a broad park land where thousands of people have congregated. There is to be a mass 'lasing' and it is the occasion for which Ariel extended her special invitation. A small crater has been transformed into a natural auditorium. Grass carpets the terraced slopes, and here the people seat themselves. At the crater's center is a raised stage where the lasing band are tuning their instruments. A cylindrical holographic display many stories high rises above the stage, housing and hiding the plasma speakers.*

*As the music starts, the crowd grows silent. Then, as if spontaneously, but actually in response to familiar cues from the holograph, the entire ensemble gives voice. Waves of music and the sweet chorus of a hundred thousand voices wash over you. You are swept away by the tide of harmony, the illusion of identity drowned in a sea of unity. The waters carry you through timelessness, until, all too soon, the great singularity of voices reaches a crescendo of exaltation, and then dissolves in a shower of applause.*

*Blinking your eyes, as if returning from some long inward journey, you look about you at the dispersing multitude. You're a little surprised to find that they are separate people, when a few minutes before you were so sure they had become part of yourself.*

*Ariel is watching you with a bemused and knowing expression.*
*"That's some sing-along you folks put on."*
*She just smiles and says "I'm very glad you made the trip."*
*You smile too, for you're pretty glad yourself.*

**On to Mars!**

For a time, the Moon will serve as Life's main bastion in the wastes of outer space. It will be the ultimate ark. At least it will serve so until we can create whole new planetary ecospheres—our next strategic step toward galactic civilization. It is our policy to enliven this sterile universe. Beginning with our own solar system, we will go about the unlimited cosmic task of coiling lifeless substance up into pulsing spools of living matter. On the surface of the Moon, we will have made Life secure in a multitude of independent ecospheres. Having done so, we will now go on to the creation of the galaxy's second living planet. At this point, we will undertake one of the most colossal tasks facing us in the Third Millennium: the terraformation of Mars.

# CHAPTER 5

# ELYSIUM

***...it is not the gods' will that you shall die and go to your end in horse-pasturing Argos, but the immortals will convoy you to the Elysian Fields...where there is made the easiest life for mortals, for there is no snow, nor much winter there, nor is there ever rain, but always the stream of Ocean sends up breezes of the West Wind blowing briskly for the refreshment of mortals.***

**Homer**

**Elysium**—The Elysian Fields were the Greek's equivalent of the Viking's Asgard and the Arthurian Knight's Avallon. It is a suitable mythic realm for Mars. In fact, the geography of Mars is already replete with landmarks from the Pantheon of Ancient Greece: Olympus Mons, Hellas Planitia, and even the Elysium Plains themselves.

The Earth is a planetary organism; as such, it shares with all life the primal drive to reproduce. A living world can't simply divide like an amoeba, or lay eggs like a chicken; rather, it must send out seeds, like a dandelion, hoping some of them land on fertile soil. Humans are the only life form capable of creating Earth's planetary seeds. We embody the consciousness of the world. It is up to us to fulfill our destiny by reproducing the Earth; not just once, but billions of times; not just here, but throughout the universe. In a million years, there will be a billion new worlds, all living gems endowed with the swirling white clouds and azure blue oceans of their shared Mother Earth. But as in all such things, the many must start with the first; and the first of the new worlds will be Mars.

**The New World**

At this stage of development, we leave the snug confines of our celestial neighborhood, and plunge into the depths of interplanetary space. We have come far since the dawn of the Third Millennium. Now we are a truly space-faring civilization, passing easily from Earth to space and back again. We have fulfilled one of the earliest imperatives: the dissemination of Life. Our numbers and resources have grown prodigiously. There are 100 billion life-seeds (people) now.[410] The population of the Earth has stabilized at 20 billion. Ten billion people live on the land, and another ten billion live at sea. The Moon is home to another 20 billion, and the rest of humanity inhabits free space. Our collective powers are staggering and growing rapidly. The Gross Production of the solar system is 700 times greater than the Gross Global Product of the year 2000. The average per capita income on Earth at the beginning of the 21st Century was less than $4000.[411] In 2125 A.D. it is $128,000. We have numbers, we have wealth, we have power. Now we are ready to make a quantum leap up the Millennial ladder. We are ready to create a new world.

We will use our maturing powers to terraform Mars, transforming that cold desert planet into a living breathing world organism. Consummating this Herculean labor accomplishes many things crucial to the fulfillment of our ultimate destiny. First and foremost, is the reproduction of Gaia. By terraforming Mars, we create a planetary organism that is truly a child of Earth. This first born of the many sibling worlds to follow will inevitably be a son. (In the Roman Pantheon, Mars is the son of Jupiter and Juno; in addition to his martial duties, he is the god of spring—season of pollination; finally, the zodiacal sign for Mars is the universal symbol for 'male'.)

Mars will always be smaller, rougher, and redder than his Mother, but he will be unmistakably a child of Gaia. He will bear his mother's features: swirling cyclonic cloud patterns, sparkling azure seas, meandering rivers, rolling grasslands, and even steaming jungles. The child planet will be a second living world—perhaps the only other one in the entire universe.

By creating a second planetary ecosphere, we virtually insure Life's ultimate triumph over Chaos. Mars will become a self-sustaining planetary organism which, in time, will follow its own evolutionary path. In the end, we will create a unique and infinitely precious second home for Life and for humanity. Space itself provides a habitat for man that is virtually limitless. But the surfaces of living planets will always be magnets to our species.

Our relationship with a living Mars will necessarily be symbiotic. The old parasitic models cannot work on the new world. Terraformed Mars is just another ecosphere—albeit on a planetary scale. The same commandments that apply in closed ecospheres like Asgard and Avallon apply on Mars: pollutants must not be allowed to accumulate; all wastes must be feed stocks; all ecological loops must be closed; all power must come from sunlight; and the whole must be self-sustaining. The inhabitants of Mars will live from the beginning in closed-cycle ecospheres differing little from those of Avallon. The people of Mars must exist in harmony with their new environment. There will be no ecological surplus to devour. Martians must live in balance with their ecosphere as surely as Asgardians. By this time, we will have learned to live in symbiosis with the Earth. If we have not, then the experience of creating and sustaining a planetary ecosphere on Mars will certainly teach us how.

Vitalizing Mars will teach us how to live in harmony with the Mother Planet and how to nurture her celestial progeny. In the process of terraforming Mars, we will learn a million lessons about the ways and means of breathing life into dust. These tools of god-craft will serve us well when we eventually strike out for the stars. When we get to distant star systems, we will probably not find any living worlds. We will find planets aplenty, but all will be dormant ruins. It will be up to us to animate the sterile stones. The worlds we find around other stars are apt to be just raw material. We will have to work these lifeless lumps of clay, laboriously molding them into vessels adapted to hold the green elixir of Life. Having learned our craft on Mars, we will know how to replicate the feat throughout the galaxy. We will perfect the skills needed to go among the stars and transform lifeless hulks into living arks.

**Astride the Steeds of Mars**[412]

The first step toward colonizing Mars is to occupy its inner moon Phobos. It requires only slightly more energy to reach Phobos from the Earth than it does to travel to our own Moon. At first glance this doesn't seem possible. Our Moon is only 384,000 km. from Earth. Phobos can be as much as a thousand times more distant.[413] In space travel, however, distance is more or less irrelevant. Distance translates into time, and time is free. Energy is what matters. Travel in space is measured not by distance moved, but by energy spent. This energy expenditure is expressed as a 'change of velocity' ($\Delta v$). The big question in space travel is not: "How far is it?", but "How much energy does it take to get there?" Changing velocity, either speeding up or slowing down, takes energy. Covering distance takes only time, and is relatively unimportant when going places in space.

In terms of distance, outer space is close to where you're sitting right now—100 km. away. You could drive that far in less than an hour. In terms of energy, though, space is very far away. In fact, once you have escaped Earth orbit, you are "half way to anywhere".[414]

The change in velocity ($\Delta v$) required to escape the Earth's gravitational pull is 11.2 kilometers per second (kps). Once you have climbed out of the Earth's deep gravity well, you are free to go anywhere else with relatively little effort. After having escaped the Earth it requires only an additional $\Delta v$ of 2.4 kps to fly to and land on Luna. Flying from Earth orbit to the Martian moons requires a $\Delta v$ of 3.2 kps. So the energy difference between a flight from Earth to the Moon, and a flight from Earth to Phobos is less than one kilometer per second.[415]

Phobos is more like an orbiting island than a moon. It is only 27 km. long and about 20 km. across, not much larger than the island of Elba. It is perfectly situated for our purposes. Phobos orbits directly above the Martian equator at an altitude of just 6000 km.—about the distance of Lindbergh's transatlantic flight from New York to Paris. Phobos makes one complete orbit in less than eight hours, putting it over every point on the Martian equator three times a day.

Phobos forms a natural space-station sitting at the bottom of a gravity well only five meters deep. Escape velocity is only ten meters per second (22 mph). With aerobraking available in the Martian atmosphere, and Phobos' low escape velocity, the Martian surface is literally only a stone's throw away. You could actually throw a rock off Phobos and hit Mars.[416]

Life inside ecospheres on Phobos will be an interesting hybrid. Conditions will be a cross between zero-gravity habitats like Asgard, and low-gravity ones like Avallon. A person who weighs 75 kg. (165 lbs.) on Earth will weigh just 50 grams (17 ounces) on Phobos —barely more than a loaf of bread.

Phobos is oriented so its long axis points toward Mars, and it spins at exactly the same rate that it orbits. Therefore, it always keeps the same end pointed toward the planet, facilitating ground observations and radio communications.

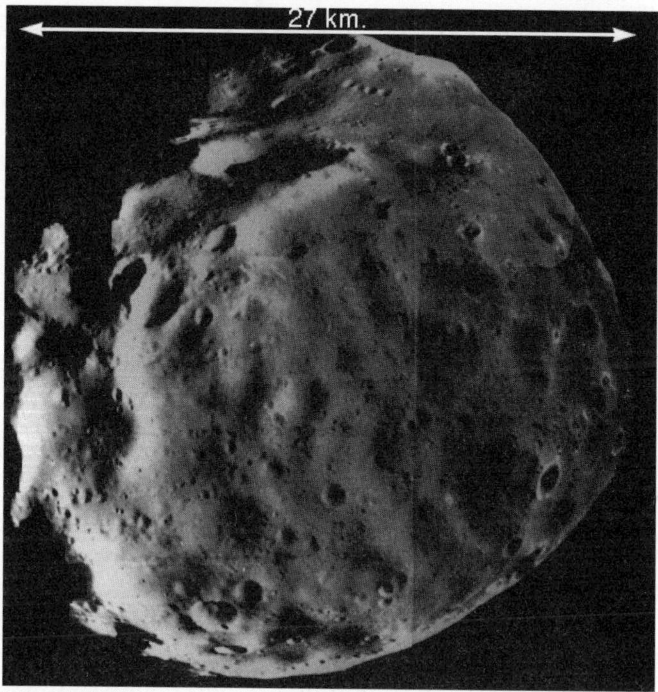


***Fig. 5.1 - Phobos.*** *Mars' inner satellite is tiny for a moon, but gigantic for a space station. Phobos will be one of the assets that makes Mars the epicenter of later human expansion through the solar system.* NASA photo.

There is a large crater on the Mars end of Phobos called Stickney (after the wife of Phobos' discoverer). This crater provides a good site for our base of operations. The crater is more than large enough—ten kilometers across—to accommodate a major base. We will build our first outpost on Phobos in the bottom of Stickney, and then expand the base by inflating ever larger concentric bubbles, until we have domed over Stickney out to its perimeter. Fully developed, Stickney will have an internal area of over 75 square kilometers.

The outpost on Phobos will be well illuminated, even though it always faces Mars. When Phobos is on Mars' dark side, the sun is overhead. When Phobos is on the planet's light side, Mars, which fills a quarter of the sky, is lit up. Mars will shine down on Phobos as brightly as 6000 full moons on Earth. Because of this double light source, Mars on one side and the sun on the other, the ecosphere in Stickney will be in the dark only about 10% of the time.[417] Darkness falls only when Mars eclipses the sun and Phobos is on the dark-side of the planet, a relatively rare circumstance.

By some stroke of cosmic good fortune, Phobos and Deimos (Mars' other moon) are both carbonaceous chondrites.[418] The presence of two such treasure troves in orbit around Mars is unbelievable good luck (if luck it is). Phobos alone will provide us with 16 trillion tons of invaluable supplies, including over 3 trillion tons of water. These key resources are exactly where we need them.

Phobos will be an interplanetary gas station. We can extract water from the Martian moon's carbonaceous ore body, and then break the water down into hydrogen and oxygen. With these elements, we can produce LHOX (liquid hydrogen oxygen) fuel on site. Space craft arriving in orbit around Mars can refuel at Phobos for their descent to the Martian surface. More significantly, passenger and cargo ships coming up from Mars will not need to carry fuel with them.[419] The ships can land on Phobos or Deimos and take on all the fuel they need for flights back to the Earth-Moon system.

Phobos will also serve as a space station for transferring passengers and cargo to the Martian surface. Getting these payloads from Phobos to Mars could hardly be easier. A soft landing on the Earth's Moon requires deceleration by rocket power. Payloads bound for the Martian surface can use the atmosphere as a brake. Atmospheric braking will enable payloads to reach the Martian surface with only a fraction of the rocket fuel they would otherwise require. Martian space pods will use the same wave rider technology we perfected for reentry from Earth orbit. The wave riders will be able to glide payloads to almost any destination on Mars.

Ultimately, we will disassemble Phobos completely. Its orbit is decaying, and without our intervention, Phobos would crash onto the Martian surface in about 100 million years. Our planetary engineering activities might even hurry Phobos' demise. As we mine out this small moon, its relative density will decrease. At the same time the Martian atmosphere will be getting thicker. Together, these effects may hasten Phobos' untimely end. (But I leave the repair of Phobos' creaking orbit to future generations of celestial mechanics.)

To develop Asgard, it was necessary to set up mining outposts on the Moon and the Apollo/Amor asteroids. The development of the Moon will in turn require us to send similar expeditions out to Phobos and Deimos. There is only enough water in the Apollo Amor asteroids to supply a population of a couple of billion people. In 125 years, 80 billion people will be living somewhere other than on the Earth. Supplying such a multitude with water and other resources is an epic challenge. There will be ten or twenty times more people living on the Moon and in space around Earth than we can supply with water from the AA asteroids. Therefore we must scout afield for accessible water supplies.

Bringing hydrogen up from the Earth is a potential solution. There are large supplies of elemental hydrogen in natural gas and other hydrocarbon deposits. The Earth is also amply endowed with water supplies, much of it stored in the form of ice. Eighty billion people, each with an allotment of 60 tons, will require five thousand cubic kilometers of water. This may seem like a lot, but it could be skimmed off the Greenland ice cap without making much more than a scratch. Greenland's glaciers cover an area of nearly two million square kilometers and are on average two kilometers deep. Enough water for 80 billion people could be obtained by shaving off just the top three meters of ice. This water is not part of the active biosphere and would hardly be missed. Even though there is water to spare on Earth, there is still a compelling reason to go elsewhere for our supplies—cost.

The cosmography of the solar system actually favors the export of materials from Mars to the Earth-Moon system. To send a kilogram of material from the Earth to the Moon requires a total change in velocity ($\Delta v$) of 13.6 kilometers per second (kps). The energy cost is, at the very minimum, 92.5 megajoules.[420] By contrast, moving a kilogram from Mars to the Moon requires a $\Delta v$ of only 9.4 kps.[421] Energy requirements are very sensitive to total $\Delta v$, due to the square function involved in the kinetic energy equation; it therefore takes less than half as much energy to send a kilogram of material from the surface of Mars to the Moon—44 megajoules—as it takes to send the same payload from the Earth to the Moon.

Hydrogen could be sent from Mars to be recombined with plentiful lunar oxygen to form water on the Moon. Only one tenth of the mass of water is hydrogen. Therefore, the transport cost for water on the Moon, if produced with Martian hydrogen, would be 4.4 megajoules per kilogram—about 6¢ per liter.[422] Transporting enough hydrogen from Mars to produce 60 tons of water for each Lunar colonist would involve an energy cost of around $3600 per person.

There is an interesting dynamism at work in our spread out into the solar system. It is like scaling a cliff. Every time we want to step up to the next level, we must first get a good grip on a hand hold above us. To build Asgard required us to establish outposts on the Moon and AA asteroids. Settlement of the Moon will require the establishment of outposts on Mars. The lunar colonies grew up from the frontier mining settlements in Fort Landsberg. In the same way, the Martian colonies will grow out of ice mining settlements on Mars.

**Slow Boat to China**

The journey to Mars requires relatively little energy, but it takes a long time. A minimum energy trajectory to Mars takes eight months. (Such flight paths are called "Hohmann Orbits") Even by the standards of Columbus, this is a long voyage. It can be done faster of course; it is only a matter of expending energy. Fusion rockets, generating a continuous thrust of 1*g*, could propel a space ship to Mars in a couple of days.[423] This, however, would require an outrageous amount of energy, and energy's alter ego—money.

All cargo, and most travelers, will take the slow boat to Mars. It is relatively easy to accelerate payloads to Mars. The change in velocity from Earth orbit to the transfer orbit is less than three kilometers per second.[424] The harder task, is to provide life support and radiation shielding during the eight month trip. Providing these necessities aboard a space ship requires accelerating and decelerating many tons of water and machinery for each passenger, consuming enormous amounts of energy in the process.

Instead of sending spacecraft, it will be more economical to set up a network of sun orbiting habitats that travel between Earth and Mars.[425] Using gravitational assist maneuvers, we can place a habitat in an orbit that will carry it back and forth between the two planets. Once placed in such an orbit, an ecosphere can serve as a transport vessel.

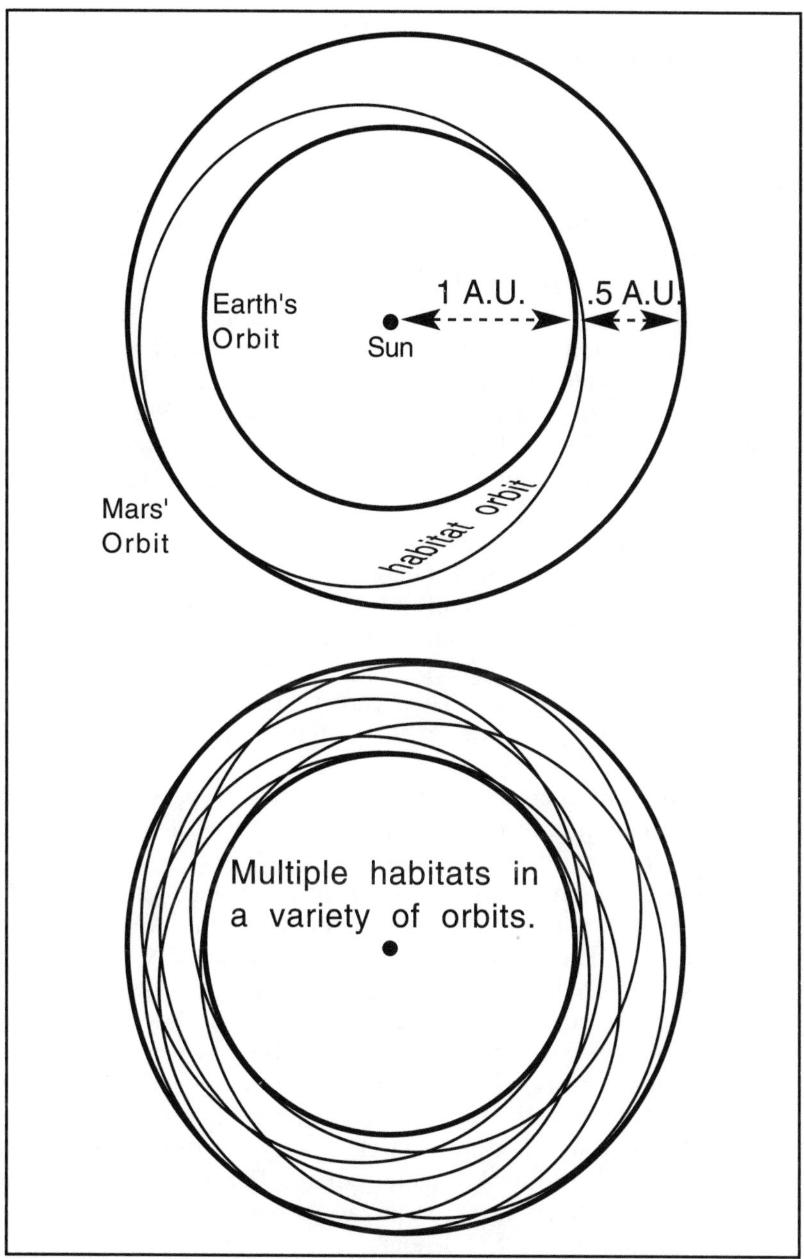


***Fig. 5.2 - Transfer habitat orbits to scale.*** *Gravitational acceleration and deceleration maneuvers are made at each planet.*

Eventually, we will establish a chain of small ecospheres in transfer orbits between the two planets. The transfer colonies will take an average of 26 months to cycle between Earth and Mars.[426] With a couple of dozen such cycling habitats in place, waiting times between outbound and inbound ships could be as little as 30 days.[427]

The passenger and his baggage still have to be accelerated and decelerated to the same relative velocities, of course. There is no such thing as a 'free ride'. The transfer habitats do, however, save having to repeatedly accelerate and decelerate the extra mass required to make long-haul space travel safe and comfortable.

As Millennial colonists, we will have developed the ability to live freely in space, not tied to any planetary surface. We will have adapted fully to zero-*g* habitats like Asgard. These adaptations will enable us to overcome some of the biggest obstacles impeding our settlement of the solar system. We will be true spacemen.

Passengers in a transfer habitat will live very much as they do in Asgard or Avallon. They will inhabit a spacious ecosphere with all the comforts and amenities of home. Outbound colonists can continue to live and work, more or less normally, for the eight months of the journey. People can go about the productive business of their daily lives. Their careers, families, and life-styles will be minimally dislocated, even though they are en route to a distant world.

When the long journey is nearly complete, passengers will transfer to a fast shuttle craft. The fast shuttle will then make a short hop to the final destination.

By the time we are ready to undertake large scale settlement of Mars, we will already have colonized many of the Apollo Amor (AA) asteroids. There are about a hundred known AA asteroids and perhaps hundreds more yet undiscovered. So there will be numerous orbital colonies hurtling through space between Earth and Mars. The presence of these colonies will be a great asset. The AA's orbits periodically take them close to both planets. When one of these asteroids is in the right position, a fast shuttle can reach it from Earth or Mars in a few days. AA asteroids sometimes approach within a million kilometers of the Earth. A shuttle pod catapulted off the lunar mass driver at 10 or 15 kps, could reach such an asteroid in less than 24 hours.

Once on the asteroid, the passengers will move in with the permanent inhabitants. The asteroid colonies function much like the cycling orbital habitats. The passengers go about their lives, working and living normally, until the asteroid reaches a position close to Mars. The passengers then make the hop to their destination. A given asteroid colony might only have a favorable conjunction with both Earth and Mars once in a decade, but when it did it would carry passengers. With a hundred asteroid colonies cycling between Earth and Mars there will usually be one or two with favorable orbital windows. The AA asteroid colonies will together act like a continuous conveyor belt, moving people out to Mars, through celestial osmosis.

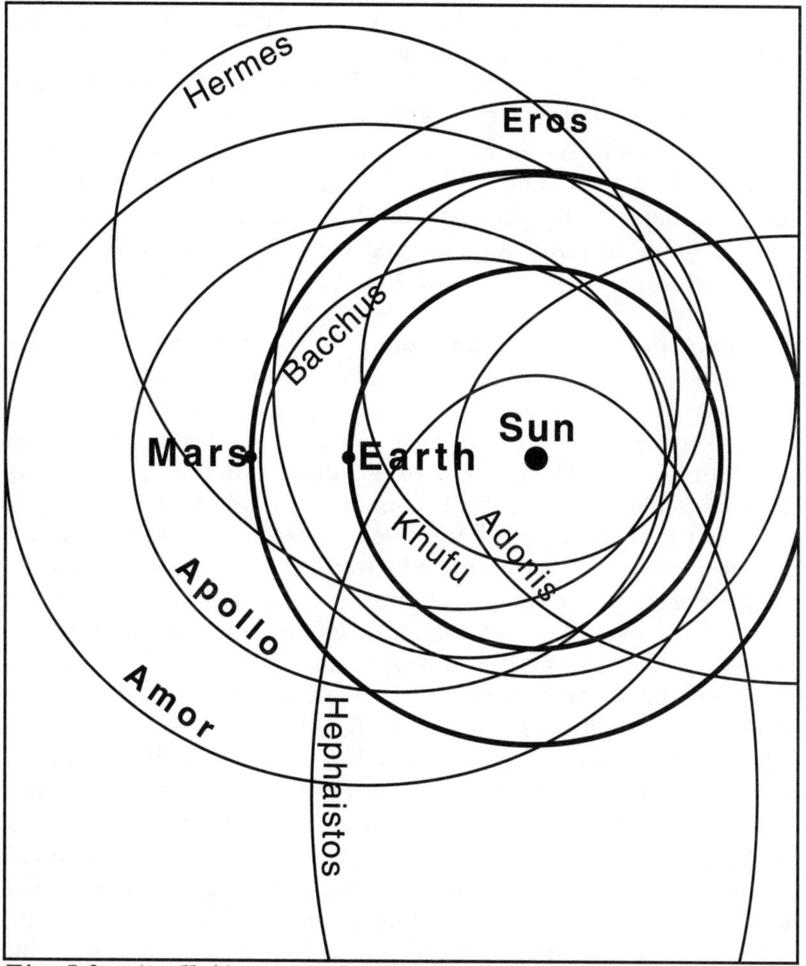


***Fig. 5.3 - Apollo/Amor Asteroid orbits to scale.*** *Only a few of the thousands of available orbits are shown here. The orbits plotted above all have inclinations with respect to the ecliptic of less than 12º.*

**Fast Fusion**

To this point in our development, solar power alone has fueled our expansion. Solar energy in sea water powers the floating islands of Aquarius. The same power source fuels the Bifrost Bridge. Orbital solar bubbles provide the electricity for Asgard and Avallon. Once we begin the push to Mars, however, we will enter a new energy realm. To terraform Mars, we are going to need a more concentrated form of energy than sunlight.

Fusion has been with us for several decades now—in the form of H-bombs. Fusion powers the universe. It is a force we must eventually harness to the chariot of our Cosmic purpose.

Fusion has two main attractions: First, the energy density in a given mass of fuel is truly amazing. A single kilogram of deuterium contains as much energy as a train-load of coal.[428] A kilogram of coal contains 5500 kcal of energy—enough to boil 40 liters of water.[429] By comparison, a kilogram of deuterium—the most common fuel for a fusion reaction—holds 86 billion kcal. That is enough energy to boil the water in 935 Olympic-size swimming pools.

Second, deuterium is abundant. It is a randomly occurring heavy isotope of hydrogen. Hydrogen contains on average .02% deuterium. Five thousand kgs. of hydrogen contain one kilogram of deuterium. The deuterium in a gallon of sea water contains as much energy as 300 gallons of gasoline; so the Earth's seas represent an energy resource equivalent to 300 global oceans overflowing with high octane fuel. Deuterium occurs in all hydrogen, and can be extracted from any source of water or hydrogen gas. (Extraction requires only a tiny fraction of the available energy.) We can therefore include the hydrogen in the giant gas planets in our energy supply calculations. The hydrogen in Jupiter alone contains enough deuterium to produce as much energy as the sun itself for a period of 2000 years.

The preferred fusion reaction combines deuterium and helium3. This reaction is the easiest to initiate, produces the greatest amount of energy, and creates no dangerous flux of neutrons as a byproduct. The problem with this reaction is that it requires helium3—an isotope of helium containing one extra neutron in the atomic nucleus. Helium3 is virtually nonexistent on Earth. The only terrestrial source of helium3 is the fusion of deuterium with deuterium. Production of helium3 is a chicken-and-egg proposition: It requires fusion to produce helium3, and helium3 is required to produce fusion.[430] In space we won't face this quandary. The Moon is a rich source of helium3. This isotope is common in the solar wind, which has infused large quantities of it into the lunar soil. The presence of helium3 on the Moon is a boon to space colonization, and its extraction will be a major Lunar industry.

Ultimately, we will require tremendous concentrations of energy in areas of the solar system where the solar flux is exceedingly weak. This creates a compelling need for fusion. At the orbit of Jupiter, the solar constant will be only 4% of its value on Luna. Power bubbles in orbit around Callisto—one of the Jovian moons—would have to be 25 times larger than power bubbles in Earth orbit to collect the same energy. For energy intense operations, like powering mass launchers on Callisto, or guidance lasers in the asteroid belt, solar power is too diffuse. For these applications we need the concentrated power of fusion.

We also need fusion power for speed. Without fusion, distances, even in the inner solar system, are daunting. Voyages to Mars last over half a year, and trips to Jupiter can take several years. For crews going to Triton, a moon of Neptune, the trip out could take decades. Fusion rockets will shrink the solar system to the point that interplanetary flight times look like continental bus schedules.

Traveling to Mars in one of the cycling habitats will take five to eight months. Traveling between planets on one of the AA asteroid colonies can take years. With fusion power, we can get to Mars much faster. A fast fusion clipper service will carry people on urgent missions. The clippers can avoid long elliptical Hohmann orbits and travel between Earth and Mars on straight line routes. The 'straight line' orbit to Mars requires a velocity of 32 kilometers per second (kps). At this speed, a ship can fly directly between the two planets. Stopping at Mars requires another 32 kps. The total change in velocity of 64 kps is more than ten times the $\Delta v$ required for a minimum energy transfer.[431] At 64 kps, the minimum transfer time to Mars is 20 days. The maximum transfer time could be as long as 144 days, depending on the relative positions of the Earth and Mars in the solar system.

There is no limit to how fast a ship can travel on a straight-line orbit—up to 300,000 kps, the speed of light. A continuous acceleration of just one-tenth of a *g* would enable a spaceship to reach Mars in just a few days.[432] Achieving 32 kps requires accelerating at one *g* for only about an hour.[433] This appears to be a modest requirement, but it is not. Such accelerations are beyond the capacity of any conventional rocket fuel in even the most colossal step-rockets.[434] Such accelerations will be outside the reach of even our Lunar-based laser propulsion systems. To attain velocities like these, we must make the quantum leap to fusion power.

Fusion rockets generate combustion temperatures of a hundred million degrees, and exhaust velocities as high as 37,000 kps.[435] The products of fusion between deuterium and helium3 are helium4 atoms and protons.[436] Both types of particle carry a charge, and the fusion reaction takes place within a magnetic field. Thrust is produced by expelling these superheated ions through a magnetic rocket nozzle. Single stage fusion rockets will be able to achieve velocities of 30,000 kps and more.

At even a fraction of this velocity, the solar system will be a very small place. At 1000 kps, flight time from Earth to Mars will be less than a day. A fusion rocket with a mass ratio of only 1.4, could attain 10,000 kps. At this velocity—equal to 3% of the speed of light—a ship could fly from the Earth to Pluto in less than a week.

With fusion power and the network of cycling habitats in place, we will be ready to undertake the transformation of Mars into a planetary ecosphere.

**Bifrost II**

A mining settlement served as the locus for further colonization of Luna, and the same model applies on Mars. Long before we are ready to undertake the terraformation of Mars, there will be mining settlements on Deimos and Phobos and the Martian surface. The growing colonies on Luna and in the realm of Asgard will have a powerful thirst for Martian water. Some time early in the 22nd century, the first serious expeditionary forces will head for the red planet. Major settlements will already exist on Deimos and Phobos, and these will provide support systems for the ice miners on Mars.

Getting people and materials back and forth from the Martian surface is essential. One of our first items of business on Mars, therefore, will be the construction of a bridge between the planet and space.

On Earth, we looked for a high mountain near the equator, where we could build the Bifrost Bridge. We found Mt. Kilimanjaro. On Mars too we need an equatorial mountain, and as you might expect by now, we find exactly what we are looking for.

Olympus Mons, an extinct volcano of prodigious size, is the most monumental mountain in the known solar system. Its base would cover the country of France. Its summit towers 27 kilometers (88,600 ft.) above the planet. Everest, Kilimanjaro,

McKinley, and Aconcagua, all stacked on top of each other, wouldn't reach that high.

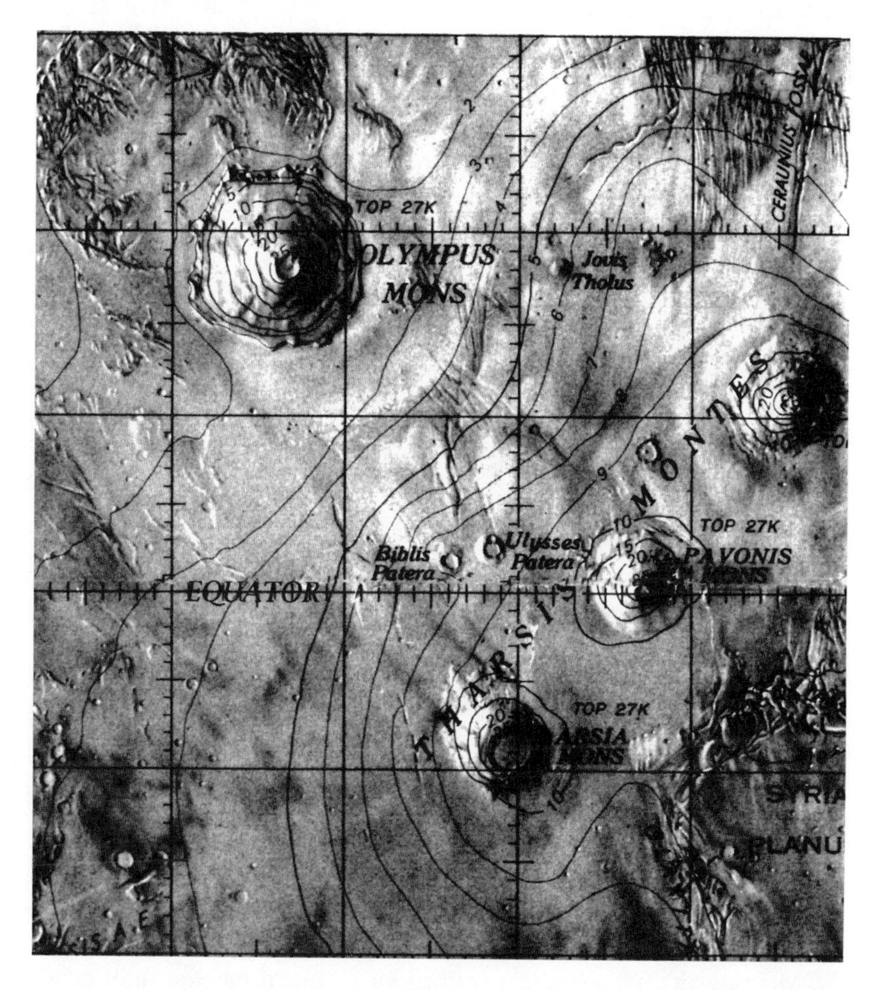


***Fig. 5.4 - Map of* Tharsis Montes *region, 1:25,000,000 scale.*** *Pavonis Mons is one of a quartet of giant Martian volcanoes, all of which top out at altitudes of 27 km. Pavonis' equatorial site makes it an ideal location for a five* g *launch tube.* USGS map.[437]

Olympus Mons gets all the press, perhaps deservedly, but it is by no means the only great mountain on Mars. There are three other giant volcanoes, each nearly as tall as Olympus Mons. Olympus Mons is 18° North of the equator, and so is not ideal as a launch site. Pavonis Mons, though, sits exactly astride the equator, providing a perfect site for the Martian launch facility. On Earth, the Bifrost Bridge's launch tube accelerates payloads to five km. per second—less than half of escape velocity. On Mars, five kps is escape velocity. Orbital velocity around Mars is only 3.5 kps. On Earth, we needed a battery of lasers to propel our payloads through the atmosphere and on into orbit. On Mars, we won't need the lasers. Payloads will emerge from the end of the launch tube at a high enough velocity to attain orbit. Also, atmospheric drag at an altitude of 27 km. will be minimal, even after we have thickened the Martian atmosphere.

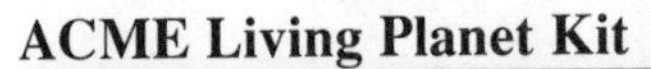

**ACME Living Planet Kit**

Mars awaits us; a living world in kit form. It has the right orbit, the right seasons, the right day; it has a ready-made atmosphere; it even has hidden oceans. Mars needs only a touch of magic catalyst and it will explode into life. That catalyst is us.

Mars is like a life raft, stored within easy reach, just in case we should need a spare planet. It is a pre-manufactured, instant ecosphere—just pull the rip-cord and it inflates itself.

This solar system looks suspiciously like an artifact, engineered for the rapid and convenient expansion of Life into space. Mars seems to have been put in place with us in mind. Its orbit—228 million kilometers from the sun—is only half again as far out as

Earth's. This puts Mars well within the sun's 'ecosphere'—the region around the sun where liquid water, and so Life, can persist on a planet's surface. The solar flux at the orbit of Mars is about half of that at the Earth, still an abundant enough supply of sunshine to support life. On Mars, as on Earth, the sun is unquestioned king of the skies. Seen from Mars, the sun is about two-thirds of its apparent diameter as seen from Earth. Even on Mars, the sun brings the light and life of day.

By contrast, Jupiter, the next planet out, is five times further from the Sun than the Earth, and receives only one twenty-fifth as much light. At Jupiter, the sun is hardly more than a very bright star. The Jovian system, and all planets beyond it dwell in relative darkness.

Only Venus, Earth, Moon, Mars, and the asteroids orbit close enough to the sun to support water-based life.[438] The universe is a limitless expanse of frozen space. If Life is to persist within its frigid wastes, it must huddle close to the celestial hearth of a burning star.[439] Mars is close enough to bask in the sun's warm glow. This privileged place makes Mars a good candidate to become the second living world in the Cosmos.

Mars is uncannily well suited as a home for Earth life. One of Earth's peculiarities—and a driving force in our evolution—is that she has seasons. If the Earth's axis was not tilted relative to the sun, then the global climate would have long since settled into a stagnant, stratified system. Temperatures and climates would be more or less uniformly distributed from the equator to the poles and would not change over the course of the year. A tilted axis endows a planet with dynamism that may be essential to a diverse biosphere. The annual succession of seasons is the driving engine of Gaia's circulatory systems, both atmospheric and oceanic. A living planet needs a tilted axis, but like Goldilocks' porridge it must be just right. It is more than a little surprising then to find that the axial tilt of Mars almost exactly matches the Earth's: 24.5° on Mars vs. 23.5° on Earth.

The strangest 'coincidence' of all is the length of Mars' day—24 hours and 37 minutes. Earth's day is 1444 minutes long, Mars' is 1477. Mars is barely half Earth's diameter and a mere one-tenth Earth's mass; it spins at half the speed of the Earth, yet Mars' day is the same length. Venus, Earth's near twin in size and mass, has a day eight months long; Jupiter, 11 times Earth's diameter and 320 times its mass, spins a complete revolution every 9 hours. The Earth, flies in tandem with a moon so large it is virtually a companion planet. Mars' moons put together are no bigger than the island of Maui. Why should such different planets have such similar days? That Mars and Earth should have almost identical days is just too weird.

**Space's Oasis**

Conditions on Mars are so hospitable that Earth-life could persist there, unsupported, even now. Space is almost wholly a radiation-blasted, frozen waste. Within that context, one develops a keen appreciation for those rare and precious places where Life can survive. At last count, there were exactly two: Earth and Mars. The high dry valleys of Antarctica are the quintessence of a hostile uninhabitable environment—compared to everywhere else on Earth. Compared to the surface of the Moon, however, the dry Antarctic is a tropical oasis. Conditions in the dry valleys of Antarctica and at the bottom of Valles Marineris—a deep canyon along the Martian equator—are comparable.[440]

The only forms of life that can persist under such harsh conditions are our old friends, the algae. In Antarctica, hardy strains of these primitive plants live beneath and inside the surface of sedimentary rocks. These are pretty tough customers. (One sample of dried algae was revived after spending more than a century in a museum display case.)[441] Lichens, (forms of fungi, living in symbiosis with algae) also cling to life in the high Antarctic. Some lichens can even tolerate temperatures as low as -200° C.—equivalent to immersion in liquid nitrogen. Any form of life exposed on the Lunar surface, however, including the toughest strains of Antarctic black lichen, would die within minutes.

There is the exceptional case of a bacterium that survived for years inside a camera lens aboard *Surveyor 3*, but that is the exception that proves the rule.[442] The bacterium, and it was only one, survived in a microscopic bubble of air trapped in the lens. Within an air bubble, even a very cold and dry one, life can persist. Without an air bubble, life has no chance; and that is exactly the point about Mars—like the Earth, Mars is surrounded by a bubble of air. If that bubble is thickened, Mars too can become a living planet.

Most schemes for terraforming Mars involve planting seed organisms and waiting for biology to create an ecosphere. This approach may work, and while cheap, it is only for those with patience—a lot of patience.

As we have come to expect, the key life forms that would terraform Mars are the algae. But even algae would be hard pressed to fulfill such a task alone. Under present conditions, algae could photosynthesize for at most 3-5 hours per day at the equator. To produce a mass of oxygen equivalent to the $CO_2$ in the Martian atmosphere would take algae 7000 years.[443] To produce an atmosphere with the minimum density required for breathing (100 millibars) would take another 133,000 years. In 133,000 years we'll be terraforming worlds around stars in the center of the galaxy! We must intervene to accelerate the transformation process.

There has been much speculation about whether or not life already exists on Mars. There may have been life in the past, but there almost certainly isn't now. James Lovelock—father of the Gaia Hypothesis—points out that living worlds are easy to recognize.[444] Living worlds, like any other living thing, modify themselves and their environments in dramatic and unmistakable ways. If life ever did take hold on Mars, it would transform that planet into an Earth-like world—complete with oceans—in just a few hundred thousand years. If life-forms on Mars were similar to our own algae in age and function, then these microscopic planetary engineers should have transformed Mars into a living planet long ago.

Mars is a skeleton planet whose evident barrenness leaves us with only two possible conclusions: Either it has always been a dead planet, or it has died. One way or the other, we still have to accomplish the same thing—bring Mars to life.

**Pull Tab to Inflate**

In the anatomy of a living planet, the rocks are its bones, and the atmosphere is its skin. Mars still has a bare skiff of skin left on him. Atmospheric pressure at "sea-level" on Mars is 6 millibars.[445] That equates to the pressure of the Earth's atmosphere at an altitude of 40 kilometers—almost outer space. To transform Mars into a living world we must increase the mass of his atmosphere. Earth has an atmospheric mass of 5,200 trillion tons. Mars' atmosphere weighs a scant 26 trillion tons. Mars has less surface area than the Earth, but it also has less gravity; so overall, it takes more mass to get the same surface pressure. Adding a thousand trillion tons of gas to the mass of Mars' atmosphere will create a mean surface pressure of 263 millibars (3.8 psi).

But, which gas? Earth's atmosphere is 78% nitrogen. To create an Earth-like atmosphere on Mars, would require 3000 trillion tons of nitrogen. Other than Earth's own atmosphere, there is no ready source of bulk nitrogen in the solar system short of Triton (Neptune's largest moon). Stripping our own planet of half its atmosphere is out of the question, even if it was feasible. Importing thousands of gigatons of nitrogen from the orbit of Neptune is equally impractical—at least in the near term.

Fortunately, nitrogen is not essential in the atmosphere. Oxygen and carbon dioxide are the essential gases. Mars' atmosphere is now virtually devoid of oxygen—.2%. Its free oxygen was locked up as oxides in the crust eons ago (these oxides give the Red Planet its color). There is already a relative abundance of carbon dioxide. Mars' atmosphere is basically all carbon dioxide—95%. There is actually a greater mass of $CO_2$ in Mars' atmosphere than there is in the Earth's.[446]

We could conceivably create an atmosphere out of either one of these gases. An atmosphere of mostly oxygen at 4 psi is a possibility. As in Asgard, a low pressure, high percentage $O_2$ planetary atmosphere presents no insuperable combustion problems. There is plenty of oxygen available in the rocks and soil. Liberating it, however, requires heating the rocks to over a thousand degrees. To release the volume of oxygen needed would require baking over a half million cubic kilometers of rock. The energy cost would be over two million terawatt hours. That is about the same amount of energy as cumulative industrial energy use on Earth since 1850.[447] With fusion energy at our disposal, the human race could certainly take this approach to creating a Martian atmosphere, but there is a better way.

Mars shares characteristics with Earth that turn out to be crucial to our purpose: well-defined polar ice caps. The difference is that Mars' ice caps are composed mostly of 'dry' ice—frozen carbon dioxide. The Martian polar caps contain 3000 trillion tons of $CO_2$.[448] If we liberate all of this frozen gas, Mars will gain an atmosphere with a surface pressure in excess of 11 psi. That is the same pressure as at an elevation of 2250 m. (7380 ft.) on Earth. To liberate oxygen from rocks requires raising their temperature to almost the boiling point of lead. To liberate carbon dioxide from dry ice requires raising its temperature to only sixty degrees below zero. Creating a planetary atmosphere from the

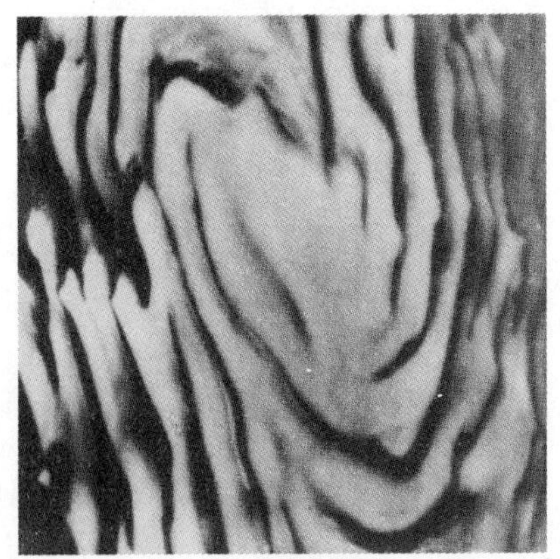

***Fig. 5.5 - North Polar cap of Mars.*** *Cosmic coincidence or thumb-print of the gods?* NASA

frozen stores in the Martian ice caps is by far the easier option.

It is more than a little strange that Mars has a tailor-made atmosphere waiting in the freezer ready to defrost. It seems too perfectly arranged to be coincidence. Looking at the swirling whorl patterns in the ice caps of Mars, one can almost see the fingerprints of destiny.

**Terraforming Mars**

Humans are the ultimate catalyst in the universe, animating the inanimate. As a living catalyst, we can quicken the evolution of Mars, from lifeless dust ball, to living world. We can compress what would ordinarily take hundreds of thousands of years into a period of mere centuries. The means of this metamorphosis are straightforward. Like reconstituting freeze-dried soup, we will just add water.

Mars is a freeze-dried planet, pre-packaged and ready to go; all we have to do is thaw it out. If we can thaw Mars, then its latent atmosphere will be liberated and the planet can come to life. So, how do you defrost a planet?

Let's look at the brute force approach first. It's usually the worst option, but almost always the easiest to figure out. Suppose we put Mr. Mars in the microwave. If we set up gigantic microwave antennas over the Martian poles and beamed energy at the dry ice caps, it would take 200,000 terawatt hours to vaporize them. This is a big improvement over extracting oxygen from rocks (2 million twh), but is still too energy intensive. Producing energy on such a scale would be difficult and expensive. A thousand hundred-megawatt power bubbles would take a quarter of a million years to vaporize the polar caps. Even with a million power bubbles it would take centuries.[449]

Brute force is not the answer. We'll need to use a little finesse. The solution might be to substitute mirrors for microwave antennas. Vaporizing the polar caps would require the entire solar radiation incident on Mars for three years. A mirror just 1% of the size of Mars' surface area would have to be over 1300 kilometers across. Even if we did build and deploy a structure the size of Alaska, this giant mirror would still require 300 years to do the job.

We could get more subtle, and change the polar cap's albedo (the amount of light it absorbs), then we could let the sun do the rest. This would only require lowering the ice cap's albedo from .77 to .73, which could be accomplished by sprinkling dust on the ice caps.[450] However, to get the $CO_2$ to vaporize would still mean increasing the total insolation at the poles by 20%. It would then take hundreds of years for the ice caps to vaporize.

All of these schemes have a fatal flaw, aside from their unworkable time frames. Even if we do add a pure $CO_2$ atmosphere to Mars, it won't really solve the basic problem: Mars is too cold. Despite its reputation as a greenhouse gas, $CO_2$ is not that good at holding in heat. If all the carbon dioxide in the Martian ice caps were vaporized, it would raise the mean surface temperature by only 15° C.[451] The result would be a less than balmy eight degrees below zero.

The champion green-house gas is water vapor. Adding enough water vapor to increase the pressure of the Martian atmosphere by just 10% would have the same effect as increasing the density of carbon dioxide by a factor of 100.[452] Adding enough water vapor to Mars' atmosphere to produce a water vapor pressure of 6.1 millibars will raise the red planet's mean temperature above freezing.[453] It will require adding 24 trillion tons of water vapor, doubling the mass of the atmosphere. Once this change is made, the dry ice, locked in the polar caps, will begin to vaporize.

Admittedly, 24 trillion tons is a lot of water. It is, however, only a fraction of the 1200 trillion tons of gas the total atmosphere requires. Twenty four trillion tons of water is a volume of 24,000 cubic kilometers—equal to twice the amount of water in Lake Superior.[454] Such a quantity is just a drop in the planetary bucket by Earth standards, but it is a lot of water on a desert planet like Mars.

Mars is dry. The Martian atmosphere presently holds less than a single cubic kilometer of water.[455] If all the water in the Martian atmosphere were condensed, it would form a body no larger than the Great Salt Lake in Utah. To transform the arid red desert of Mars into a sapphire jewel like Gaia will literally require oceans of water. First, we need enough water to liberate an atmosphere, and then enough more water to fill streams, lakes, and seas. Mars is a small world as planets go, only half the size of the Earth. Nevertheless, it still takes an abundance of water to saturate a ball of rock 6800 km. across.

The Martian moons, Deimos and Phobos, contain some water. But if we chewed them both up and digested them completely, they could provide only 600 cubic kilometers. There is more water in the carbonaceous chondrites of the Asteroid Belt, and the Moons of Jupiter are encrusted with oceans of ice. Tapping those reserves means going very far afield at enormous cost in time and energy. There is, however, a more convenient source of water at hand.

Once again, we find our solar system well provided with what we need—this time in the form of comets.

**Catch Me a Comet**

We can get the water we need from a single comet of surprisingly small size. Comets are 60% water by weight.[456] Just one comet, with a radius of 21 kilometers, could provide all the water needed to reconstitute Mars.

There are abundant cometary resources available. Comets comprise by far the greatest mass of the solar system apart from the sun. The gross cometary mass in the solar system may total an amount equal to as much as 1% of the mass of the sun.[457] This may appear to be a small fraction, but it is a mass ten times greater than all the planets combined. Unfortunately, the bulk of this mass orbits in the far away Oort cloud. This cloud of comets begins beyond the orbit of Pluto and extends out to a distance of 100,000 A.U.—nearly half the distance to the next star system.

The situation with comets in the development of Mars is analogous to our situation with asteroids in the development of Asgard. When we were developing Asgard, the asteroid belt contained the bulk of the carbonaceous chondrites we needed. But the asteroid belt was out of reach. Now we need cometary material, but the Oort cloud is out of reach. The solution now is the same as it was for Asgard. To develop Asgard, we exploited the Apollo Amor asteroids which orbit closer to the Earth. To hydrate the Martian atmosphere, we will turn to those comets which inhabit the inner solar system.

There are 121 known 'short period' comets, trapped in orbits no further out than Saturn.[458] Of these, at least 105 have orbital periods of 30 years or less. All of these short period comets are in 'direct' orbits. That is, they pass around the sun in the same direction as the planets (counter-clockwise as viewed from the Earth's north pole).[459] At least 60 comets are known to belong to the 'Jupiter Family'. These are comets whose orbits have been altered by Jupiter's gravitational field. All the comets in the Jupiter family, and most orbits of other short period comets, lie within the plane of the ecliptic of the solar system. The Jupiter Family comets have very short orbital periods of from three to nine years. On average, six comets a year cross the orbit of Mars in the plane the ecliptic.[460] All we need to do is arrange a rendezvous between some of these comets and the surface of Mars.

Most comets are less than 15 km. in diameter, with an average diameter of three to four kilometers.[461] Halley's Comet is typical.

It measures 15 kilometers long by eight kilometers wide. (Halley's is not a good candidate for capture, however. Its orbit is highly inclined to the plane of the solar system, and it only shows up once every 75 years.)

We could use strong-arm tactics to get comets to Mars. We could build a mass launcher on a comet and use the launcher as a rocket motor to push the comet out of its orbit. A mass launcher expelling ice chunks at 50 kilometers per second (kps), could impart a change in velocity of 3 kps to the comet, while consuming just 6% of the comet's mass as propellant.[462] Such a change in velocity would be sufficient to maneuver the comet onto a collision course with Mars. The energies involved, however, would be colossal. Accelerating two trillion tons of ice to 3 kps would require something on the order of 30,000 terawatt hours of power. Less than vaporizing the polar caps with microwaves (200,000 twh), but still expensive.

With fusion energy at our disposal, we could certainly produce that much power. Deuterium concentrations in cometary hydrogen are relatively high, at one part in 5000.[463] A comet 10 kilometers in diameter could provide 35 billion tons of hydrogen, yielding seven million tons of deuterium. The Deuterium to helium fusion reaction yields 100 million kilowatt hours of power per kg. of deuterium. To power the mass driver we would need to consume 300,000 tons of deuterium fuel, out of the seven million tons available on the comet. While this may be inside the realm of possibility, it is probably not practical or desirable. If it took ten years to change the comet's orbit, it would still require a power plant with an output greater than three hundred gigawatts operating continuously. By comparison, the combined output of all the world's electrical power plants today is less than 2.5 gigawatts.[464]

**Low Rent Rendezvous**

Luckily, it isn't necessary to force the comets into Mars' orbit. Most comets orbiting the sun inside the solar system cross Mars' orbital path anyway. All we need to do is adjust its timing so the comet is crossing Mars' orbit at the same moment Mars is there to meet it. The nature of orbital mechanics makes this a relatively simple proposition.

One of the interesting characteristics of elliptical orbits is that an infinite variety of orbital shapes can all share a common point on their paths. Typically this point is at perihelion—the point of closest approach to the sun. A very small impulse delivered to a comet at perihelion can put it on a radically different orbital path.

Instead of having to change the comet's velocity by three kilometers per second, we can alter the path of its orbit by just a few arc seconds. This tiny change at perihelion can set the comet on a collision course with Mars. With 60 candidate comets on a ten year closed track, we can afford to take our time. Each time the selected comet passes through perihelion we can give it a nudge, until we have it in exactly the orbit we want. We might adjust the orbit of a five year comet several times before finally locking it on target for its cataclysmic meeting with Mars.

We can deliver these navigational impulses by any number of means. The simplest approach will be to detonate several Hydrogen bombs in the right spot on or near the comet. Other energy saving options will be available: laser propulsion from solar powered FEL arrays in orbit around the sun, solar sails, even space billiards—arranging collisions between comets and asteroids.

**Sudden Impact**

When a comet several kilometers in diameter crashes into the Martian surface, the results will be spectacular. It will probably be desirable to rig the comet with explosive charges so it breaks up into many chunks just before reentry. Otherwise, the intact comet will blast a crater miles deep in the Martian crust. Detailed seismic surveys of the comet will be conducted to determine its inner structure. The identified fault zones will be drilled and explosive charges planted in the weak areas. Shortly before impact, the charges will be detonated, breaking the comet into a number of smaller fragments. These fragments will then impact the surface over a wide area of the planet. By breaking the comet up, each of the impacts will be exponentially less severe than a collision with the whole comet would be. This will minimize the amount of dust thrown into the upper atmosphere and will reduce the amount of water vapor that is blown back out into space.

On impact, all the ice in the cometary chunks will instantly vaporize. At the same moment, a large amount of ice held as permafrost deep in the Martian soil will flash into steam. In fact, the comet will vaporize an amount of permafrost equal to hundreds to thousands of times the mass of the comet itself.[465] Moisture released from the soil can reduce by scale factors the amount of cometary material we need to bring in. Comets will therefore be directed to impact in higher latitudes, where permafrost is concentrated. Water vapor, liberated into the atmosphere by just a few small comets, will generate a greenhouse effect powerful enough to warm the Martian surface above freezing.

**Meltdown**

The comets will not by themselves have contributed enough gas to create an acceptable surface pressure. The comets will raise the water vapor pressure to six millibars, and total atmospheric pressure to 12 millibars. While this doubles the present atmospheric pressure on Mars, it will still be far too low to support life. At this pressure, the boiling temperature of water is only 10° C. To render the Martian surface hospitable to Life, we need to raise the air pressure to 500-600 millibars, half the pressure that prevails on Earth at sea level. There is enough carbon dioxide in the dry ice glaciers of Mars' poles—3000 trillion tons—to do just that. There is even more dry ice trapped in Martian permafrost—enough to increase atmospheric pressure to three times that of the Earth at sea level.[466]

The greenhouse effect on Mars, as on Earth, will have its most pronounced effect at the poles. As the greenhouse effect warms Mars, the dry ice in the polar caps and the permafrost will boil away, adding its mass to the atmosphere. Surface pressures will quickly rise to livable levels. The new thicker carbon dioxide atmosphere will also contribute to the Martian greenhouse effect, adding 15° to the average temperature of the planet.[467] Once all the $CO_2$ has been liberated from the polar caps and the permafrost, Mars will have an atmosphere thicker than the Earth's, and a balmy mean surface temperature of 15° C. (60° F.).[468] These dramatic changes can be triggered by the collision of just two good sized comets. Eventually temperatures and pressures will be high enough to allow liquid water to flow openly on the Martian surface. As water begins to flow, Mars' metamorphosis will accelerate rapidly.

**Water Works**

To survive as a living world, Mars needs oceans. Earth's oceans hold over a billion cubic kilometers of water. If it were spread evenly over the surface, the entire Earth would be submerged two kilometers deep. Inundating a planet under kilometers of water is the kind of extravagant gesture best left to the original creator. We ersatz gods should content ourselves with just enough water to do the job.

The water budget for Mars is conservative, compared to the vast oceanic basins of the Earth. The surface area of Mars is less than a third that of the Earth. An amount of water sufficient to cover the entire surface of Mars 300 meters deep, 40 million cubic kilometers, will be ample. Such a supply will provide Mars with

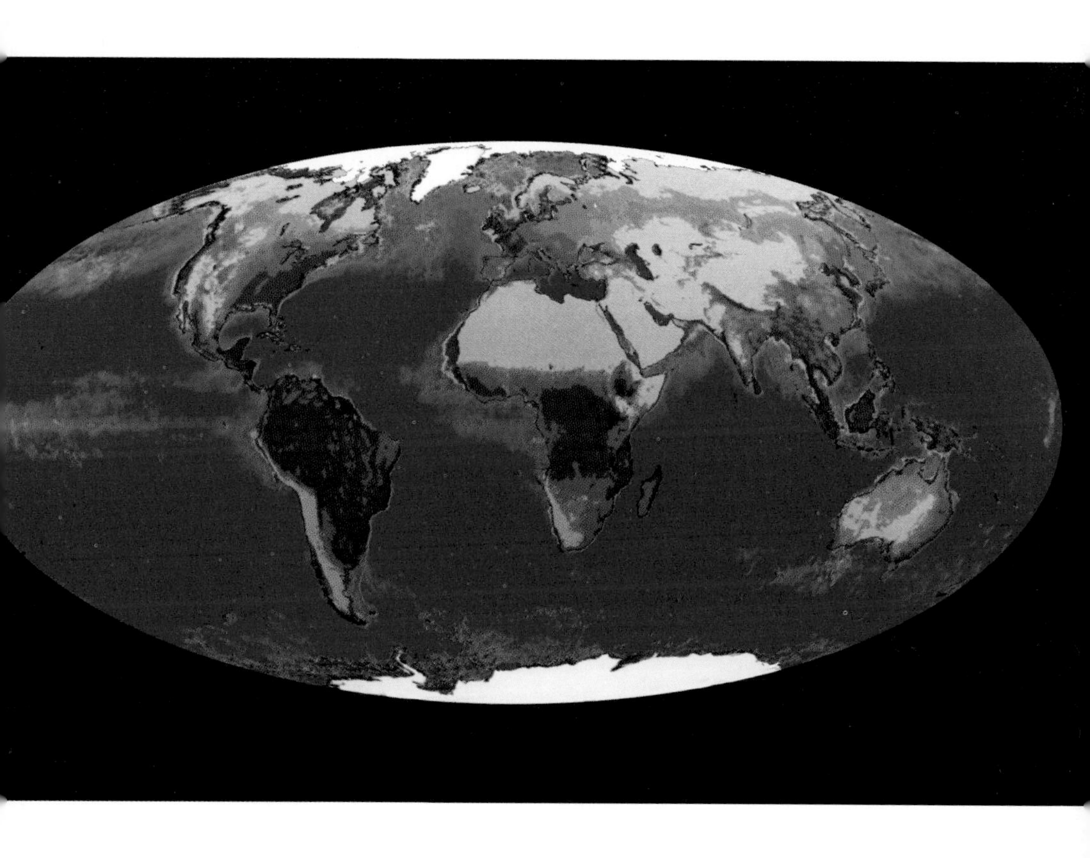

*e **1** - World Productivity. This composite satellite image shows concentrations of planetary biomass. and, forests and grasslands are indicated in green; at sea, plankton are shown in yellow and red. areas of highest oceanic productivity are concentrated along coastlines and in regions of upwelling. -ocean zones are nearly devoid of algae and most other life. These vast oceanic deserts, indicated in blue and purple, can be colonized without disturbing any preexisting ecologies.*

***Plate 2*** - (opposite) *Aquarius is a cybergenically grown floating island—a space colony at sea.*
***Plate 3*** -(above) *Parklands, atop the hexagonal towers of Aquarius, are terraced down to the beach*
***Plate 4*** - (below) *Aquarians learn to deal with life in space by living in bubble habitats under wate*

***Plate 5*** (above) - *Powerful laser beams of the Bifrost Bridge converge to propel a wave-rider skywa*

***Plate 6*** (opposite) - *Asgard is a city in orbit, home to thousands of space colonists.*

***Plate 7*** (below) - *Sheltered inside Asgard, life thrives in a gravity-free oasis in space.*

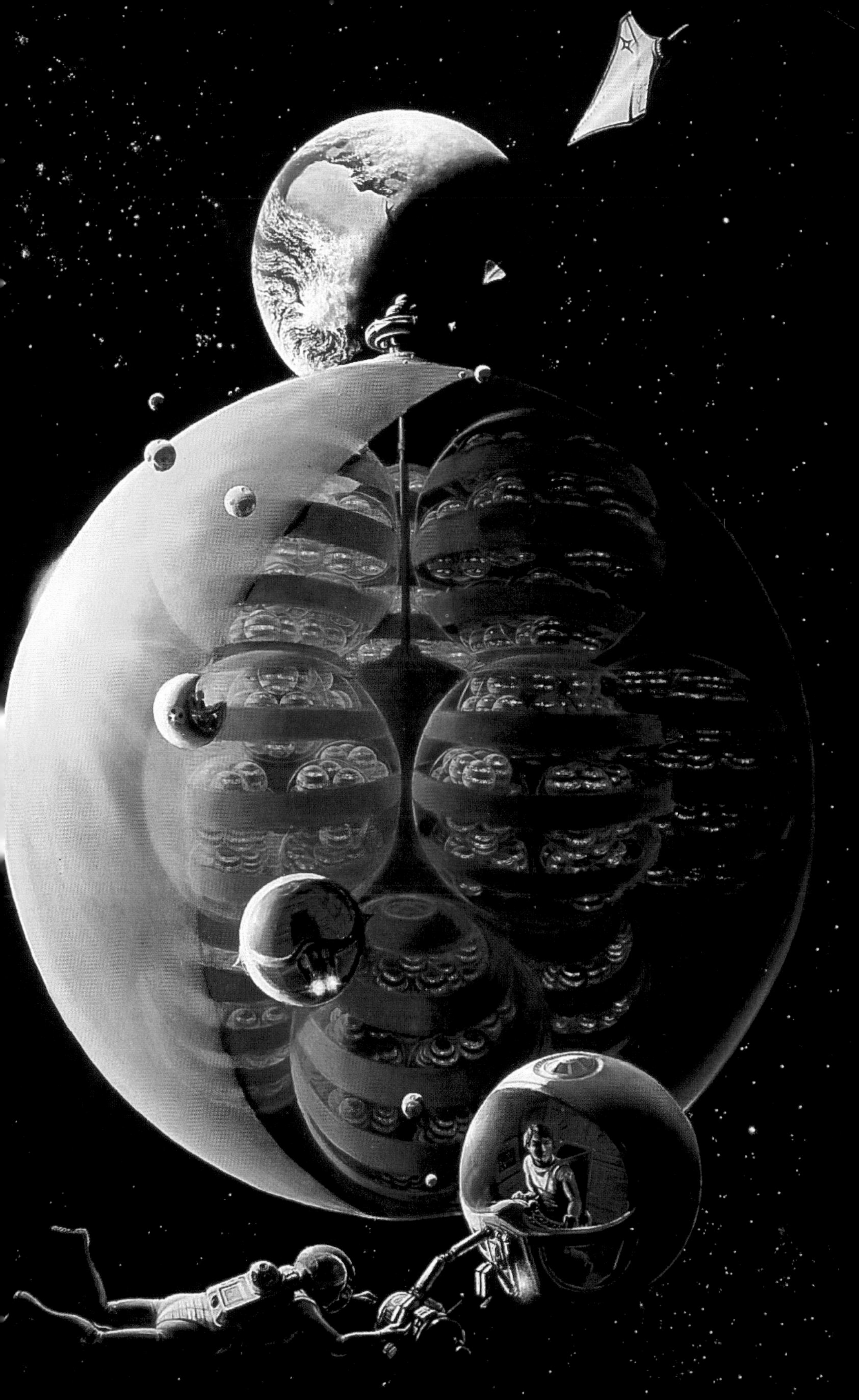

***Plate 8*** - (above) *The Lunar crater Copernicus, domed over and terraformed to create an ecosphere.*
***Plate 9*** - (below) *The huge crater, 90 km. across and 6 km. deep, awaits the touch of Life.* NASA ph
***Plate 10*** - (opposite) *In Avallon, people fly under their own power, cavorting among the giant trees.*

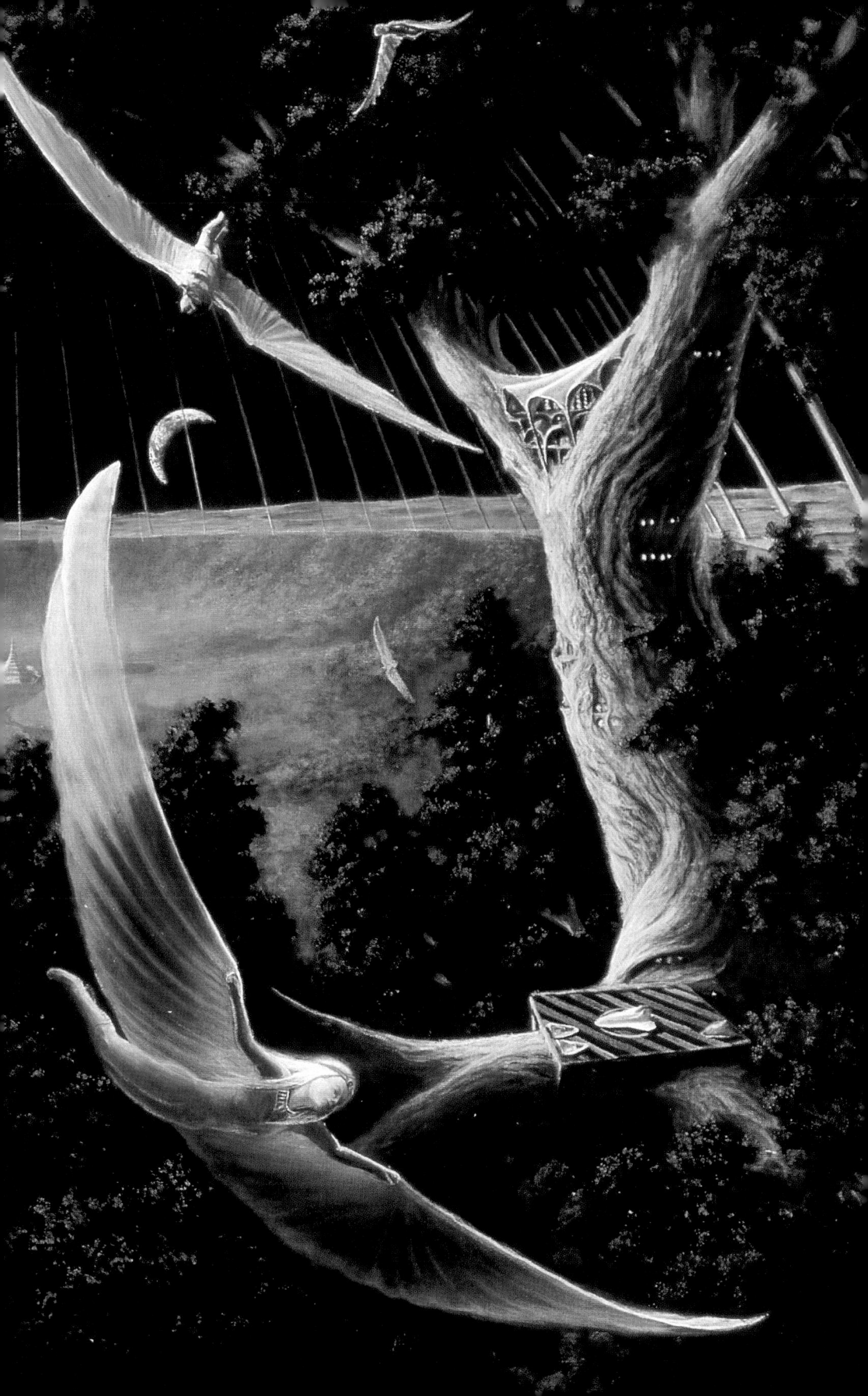

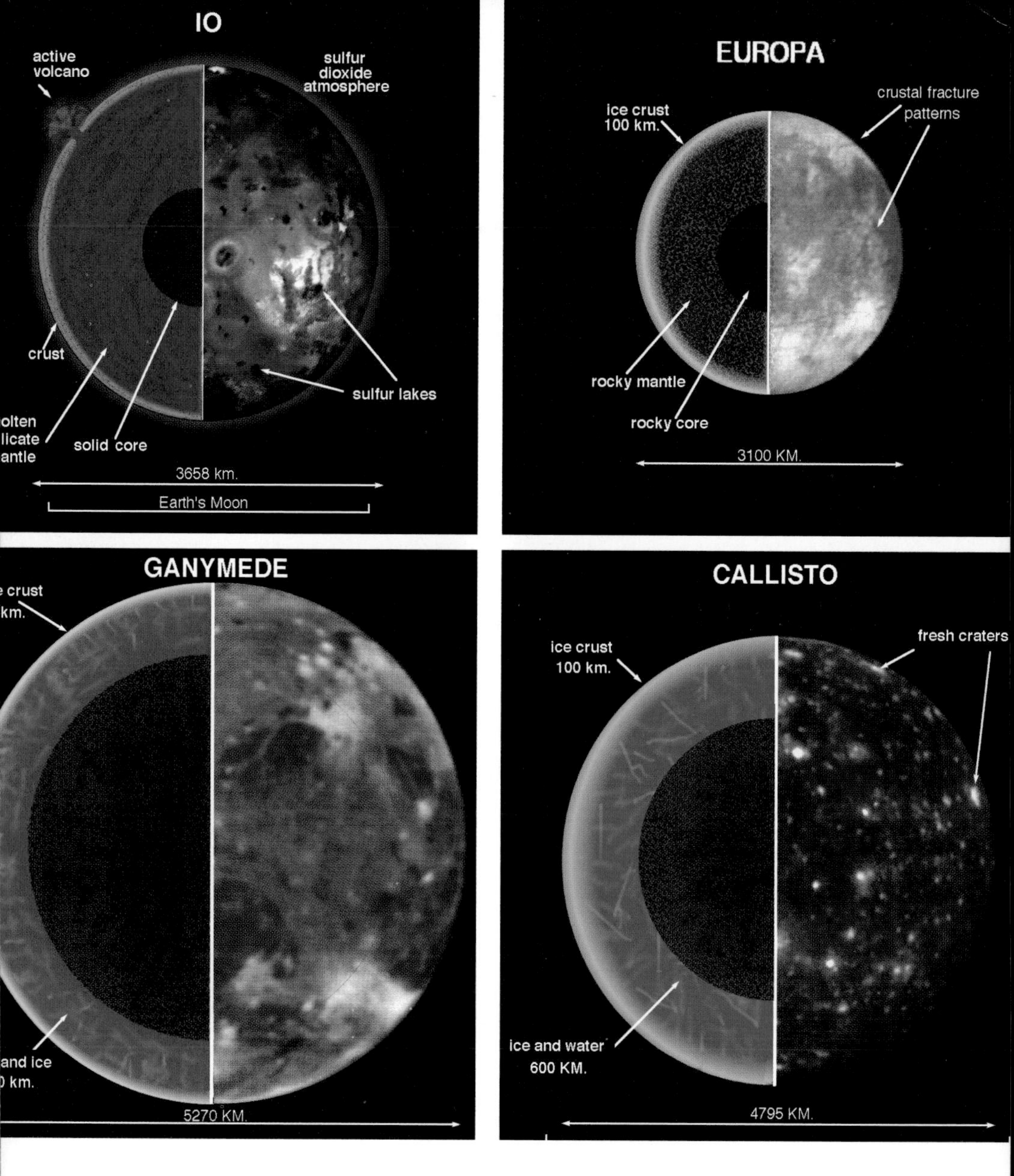


***late 11*** - *Terraformed Mars as seen from one of its moons. The sleeping planet's oceans have been berated, fleecy clouds nourish the desert landscape with life-giving rains, and verdant forests spread cross the once sterile globe.*

***late 12*** - (above) *Cutaway views of the four largest Jovian Moons.* NASA photos.

***late 13*** - (overleaf) *Lasers carve out 100 M. blocks of ice on Callisto. The mass driver launches the unks into Jupiter's gravity-well which slingshots them toward the inner solar system.*

***late 14*** - (overleaf opposite) *The asteroids are hollowed out and surrounded by bubble membranes so oth their inner and outer surfaces can be terraformed.*

*ate 15* - (opposite) *Billions of billions of bubble habitats surround the sun in a ring of gold.*
*ate 16* - (above) *The Dyson cloud of our mature K2 culture, as seen from the orbit of Jupiter.*
*ate 17* - (below) *Cometary nomads slowly cross the dark gulf between stars.*

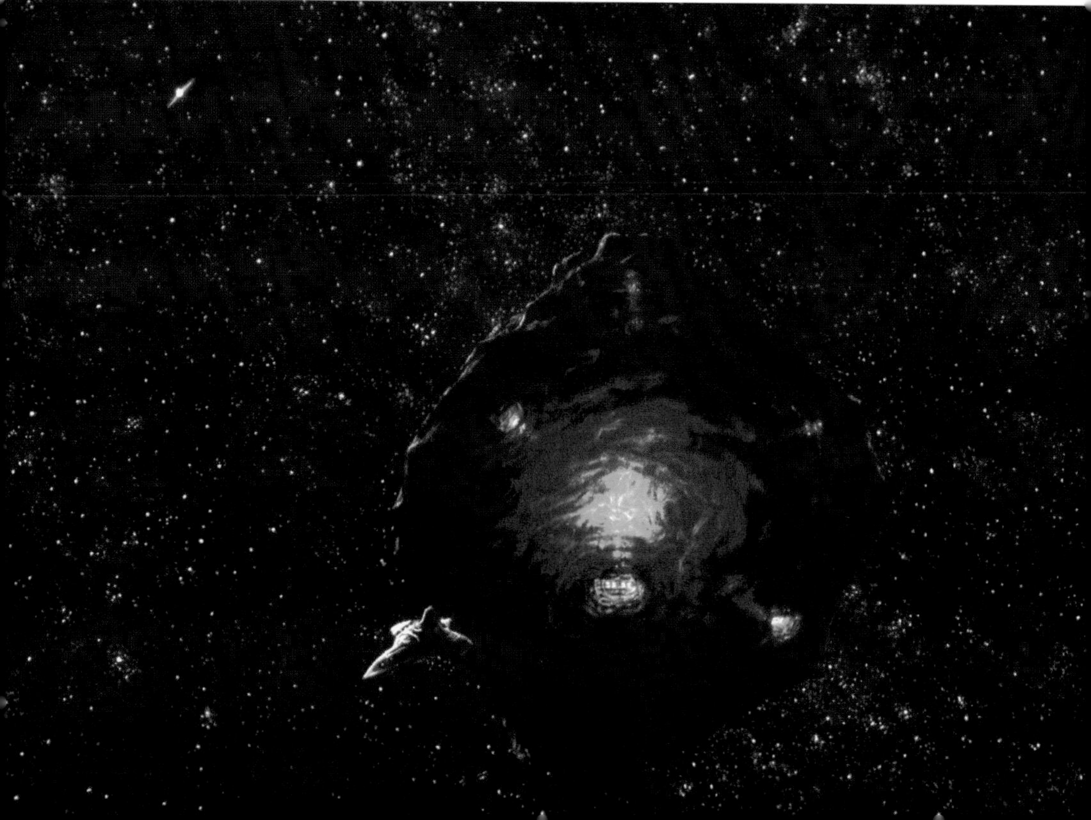

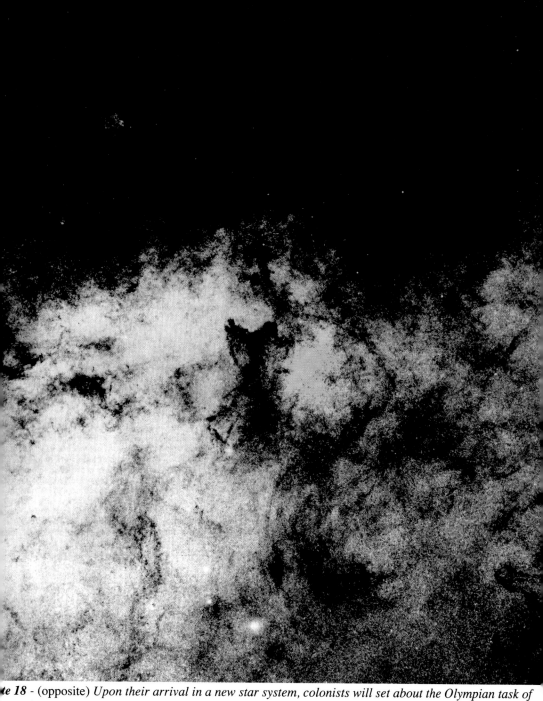

*te 18* - (opposite) *Upon their arrival in a new star system, colonists will set about the Olympian task of ısmuting raw materials into living worlds. Here twin seed ships have parked in orbit around a giant planet and the space colonists are transporting ice from the planetary ring to one of the small moons 're terraforming is well underway.*

*te 19* - (above) *Stars roil in their billions. Each point of light is a star, not unlike our own Sun. ween each particle of the cloud yawn the unfathomable chasms of interstellar space. Ten thousand rs from now, our seed ships will have penetrated deep into the Milky Way, and our region of the ıxy will be suffused with the glowing green fireball of Life's exponential explosion.* Photo courtesy of : Observatories.

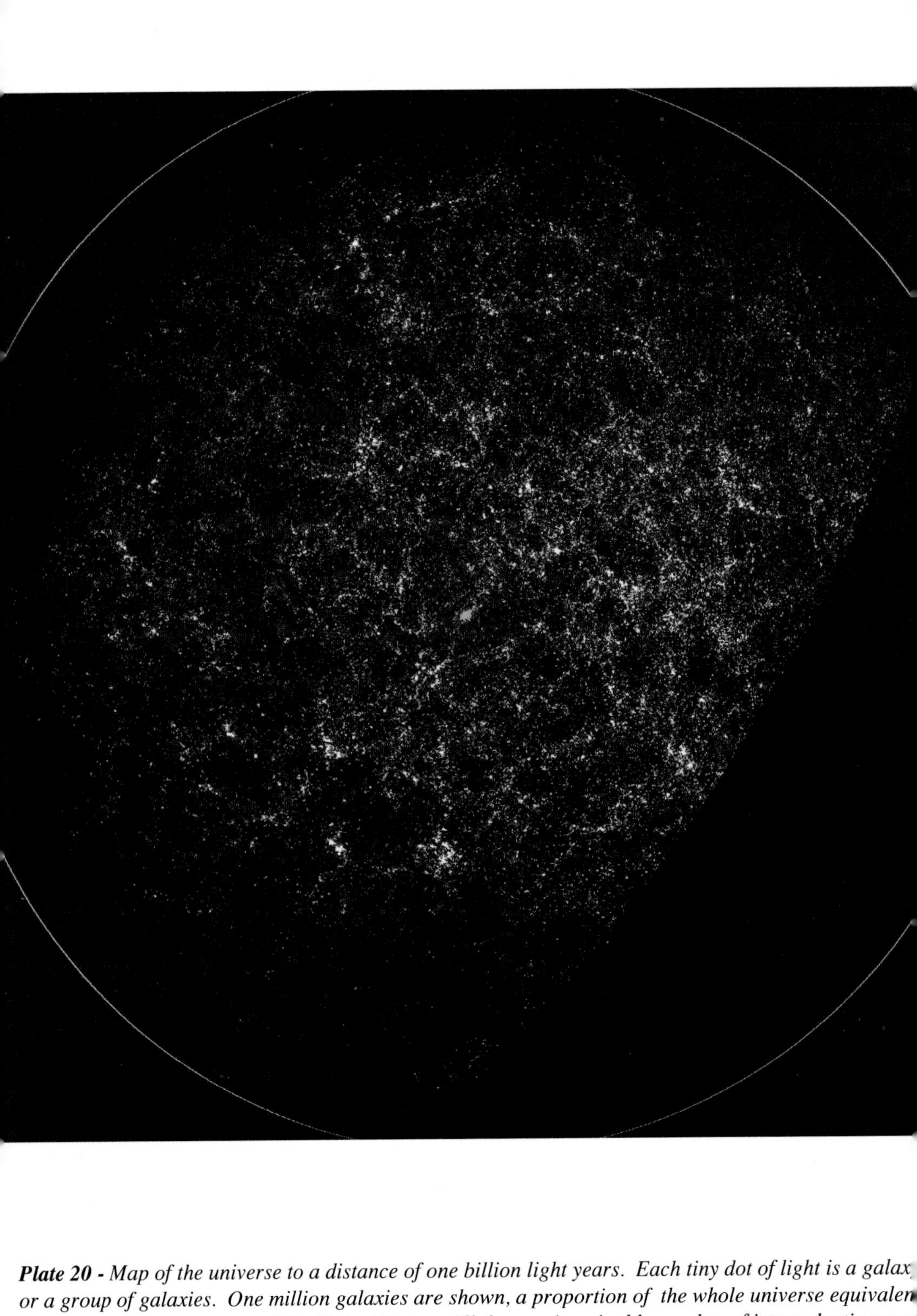

**Plate 20 -** *Map of the universe to a distance of one billion light years. Each tiny dot of light is a galax or a group of galaxies. One million galaxies are shown, a proportion of the whole universe equivalen to Yuma, Arizona on a map of the United States. All these unimaginable reaches of intergalactic spac are utterly barren; but soon, Life's unquenchable conflagration of trees will explode across this desola starscape, transforming the very nature of the universe forever!*

Photo courtesy of P.J.E Peebles, M. Seldner, B.L. Siebers, and E.J. Groth, *Astronomical Journal*, 82, 249, 1977.

thousands of small crater lakes and ponds, a number of large lakes, several inland seas, and one planet-girdling ocean.

Where we can lay our hands on 40 million cubic kilometers of water is a big question. Obviously, Mother Earth could spare it. She could give us that much or more and hardly miss it. Getting out of the Earth's gravity-well requires accelerating to a speed of 40,000 kilometers per hour. Reaching Mars requires speeding up another 22,000 kph. Accelerating forty thousand trillion tons to 50 times the speed of sound is a daunting prospect for even the most ambitious would-be terraformer.

Bringing in additional comets isn't the answer. The 60 short-period comets in the Jovian Family, altogether contain less than a thousandth part of what is needed. There are more comets of course; billions in the Oort cloud, but those are too far afield. It might take centuries for a comet from the Oort cloud to even get to Mars.

It is obvious to even the most casual observer that Mars has had water running over its surface in the past. Water eroded features like Valles Marineris and depositional basins like the Acidalia Planitia are apparent on a continental scale. Fine scale features like dry river-bed channels can be seen many places. Mars may now be dry, but it once was wet.

Mars possesses some of the most dramatic volcanoes in the solar system, and volcanoes are nothing less than planetary water wells. Like artesian geysers, volcanoes bring steam to the surface as well as lava. Calculations indicate that volcanic out gassing on Mars may have released $3.5 \times 10^{17}$ tons of water.[469] This huge amount—350 million cubic kilometers—is enough to have submerged the whole of the red planet to a depth of 2.4 km. (1.5 mi.). Mars cannot have lost all this water. Geologic evidence suggests that Martian water has been liberated, perhaps many times, in the past.[470] As Mars cooled, ground water must have frozen and been permanently trapped. Ice on the surface might eventually sublimate and be lost into space, but underground ice, compressed by the pressure of overburden would be preserved. A tenth to a third of the total water released by Mars's great volcanoes must remain in the soil as permafrost, or underground lenses of ice.[471] The remaining 35 to 100 million cubic kilometers of water is ample to meet our needs. Spread evenly over the Martian surface, this water would form a planet-wide ocean at least 200 meters deep.

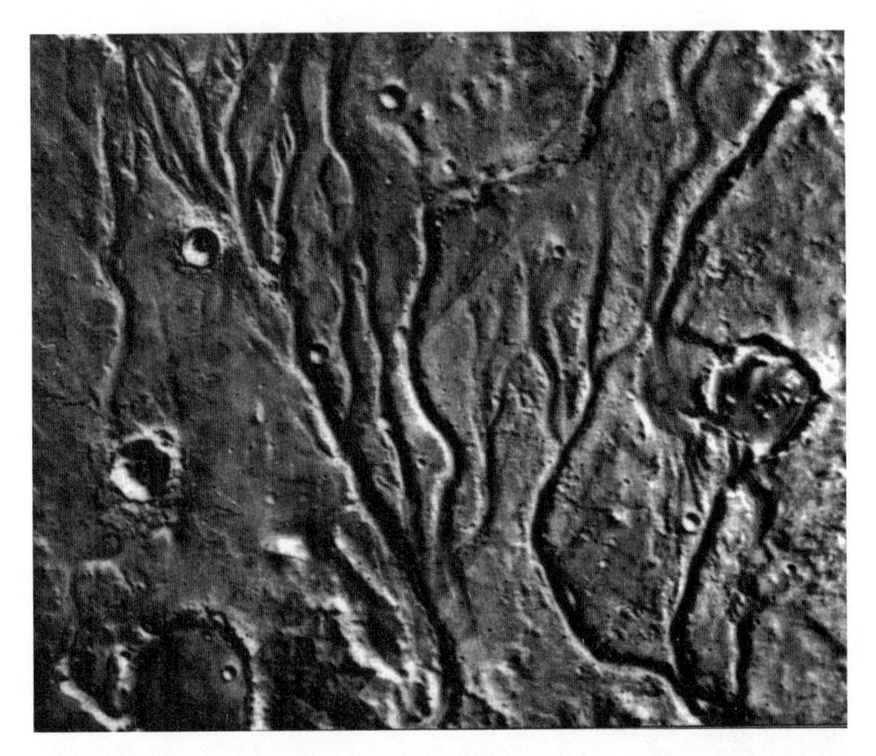

***Fig. 5.6 - Dry river beds on Mars.*** *The surface of Mars is braided with dry river beds that must once have held running water. The day is not far off when streams will once again splash through these dry courses.* NASA.

There have apparently been eras in Mars' geologic past when the planet has warmed enough to melt ground ice. Close examination of the many stream beds on Mars reveals that they formed in a way indicative of such melting. On Earth, rain falls and runs off forming streams and rivers. On Mars, many stream beds were formed when permafrost melted and seeped out of the soil.[472] Our terraforming efforts will raise the temperature of Mars permanently. Ground ice will again melt and flow through the ancient river beds, accumulating in the now dry seas.

**The Boreal Ocean**

Our terraforming efforts may just be reconstituting a frozen fossil of Mars' original ocean. It is theorized that huge "super plumes" of hot magma may have risen from the Martian mantle at times in the distant past. These bubbles of molten rock may have created hot spots in the crust, melting millions of cubic kilometers of underground ice. The water then poured out in channels 10 kilometers wide, and filled the basins in the northern hemisphere, creating an ocean. This body of water was on average 1700 meters deep, had a volume of 60 million cubic kilometers, and covered a third of the planet's surface.[473] This great ocean, lies frozen in eternal slumber, awaiting our magic touch to awaken, and once again fill Mars' thirsty basins.

On a rehydrated Mars, temperatures will be the highest along the equator. Air pressure will be greatest at the lowest elevations. Parallel to Mars' equator there runs a tremendous chasm called the Valles Marineris (Mariner Valley). In places, this canyon is as much as two kilometers below the Martian 'sea-level'. In other places, like the Isidis Planitia, there are vast depressions a kilometer and more below 'sea-level'. In the southern hemisphere there are great basins like Argyre Planitia and Hellas Planitia which in places are as much as four kilometers deep. We can expect liquid water to collect first in these canyons and basins.

As the atmosphere thickens and the planet becomes warmer, the rains will fall. Lakes and streams will push north and south from the points of their first appearance on the equator. The waters will eventually fill the larger depressions, forming small seas. Between these seas, at least at lower elevations, will be thousands of smaller lakes formed by water filling the many impact craters that dot the Martian landscape. In addition to the Boreal Ocean, there will be an Australian Sea covering Mars' South Pole up to a latitude of 70º. This circular body of water, 2400 km. across and on average less than a kilometer deep, will contain altogether two million cubic kilometers of water. This sea may freeze over during the long southern winter.

The Boreal Ocean, covering much of the Northern hemisphere will be larger, wider, longer, and deeper than any other body of water on Mars. The Boreal Ocean will girdle Mars' northern latitudes, filling the entire basin of the Vastitas Borealis. It will create an island some 1100 km. across at the North Pole. Arms of the Boreal Ocean will extend south in places as far as the equator. The Northern Ocean's waters will completely surround the Elysium Highlands, creating of them an island continent the size of Earth's own Australia. From shore to shore, the Boreal Ocean will be almost as wide as the Atlantic and nearly as deep. If filled to capacity, the Boreal Ocean could contain 48 million cubic km. of water. It will never freeze completely. Where it penetrates below the equator, along the Hephaestus Gulf, the temperature will seldom fall below 75° F. Along the endless red-sand beaches of Elysium Island's southern shore, groves of coconut palms will sway in the tropical breeze, and surfers will catch enormous slow-motion waves, riding the wine-dark seas of Mars.

During the summer of Mars' Northern Hemisphere, the Boreal Ocean's ice sheets will retreat, the water will warm and turn over. The nutrient load from sediments will reach the surface. Bathed in the light of lengthening days, the plankton will bloom in the still cold waters. Populations of krill will dance in their billions, feeding on clouds of jewel-like diatoms. And there, as the brilliant blue and white paired stars of the Earth and Moon rise above the eastern horizon, a great gray back will break the waves, and leviathan will blow his misty breath into the mellow Martian air. This small miraculous moment will happen in a place where now only dust blows over a frozen and sterile plain.

**Martians**

Like most great miracles, establishing terrestrial life on Mars will take some time. The ecology of Mars is going to be different from that on Earth for a long while, if not always. At the beginning of our terraforming process, the atmosphere of Mars will consist almost wholly of carbon dioxide and water vapor. People venturing outside an ecosphere will need oxygen masks. Martians will not have to wear space suits, however. Even at this stage in the terraforming process, humans will be able to walk around in the open in just shirt sleeves.

Plants will be able to thrive in the new Martian environment almost immediately. In a replay of the early processes of life on Earth, the plants will pioneer the planetary ecosphere. Ultimately, the plants will create conditions hospitable for animal life. At first, plant-life will be confined to aquatic and marine species—primarily algae. In the warm bodies of water along the equator, these plants will find ideal environments for growth. Eventually, we will introduce genetically engineered plant species, designed for a high tolerance to $CO_2$. These new land plants and the aquatic plants will proliferate. Grass and trees will follow the spreading temperate zones north and south from the equator. The classic progression of species will spread over the terrain, clothing the naked red soil in a verdant blanket. At low elevations near the equator, tropical rain forests will proliferate. In the mid-latitudes, temperate deciduous forests of elms, maples and mighty oaks will spread their leafy branches. Further north and higher in elevation, will grow boreal coniferous forests. Finally, at the upper tenable limits of altitude and latitude the mosses and lichen of Arctic tundra will cling to the rocky ground. (See Plate No. 11)

The burgeoning plant population will steadily increase the concentration of oxygen in the atmosphere. Eventually, there will be enough oxygen to support animal life. Like the plants, these animal species will have to be engineered to tolerate high concentrations of carbon dioxide. Humans, who are not genetically engineered, will probably always require oxygen masks to breathe comfortably on Mars. With a full complement of plants, animals, insects and microorganisms, Mars will have been brought fully to Life.

Large scale settlement of the Martian surface will, of course, have to wait until the cometary bombardment phase is over. It is unlikely that one of our man-made meteors would actually strike an ecosphere, but it is even less likely that anyone would care to gamble on it. Ice miners and other Martian pioneers will probably evacuate to Phobos prior to impact. Comets will be arriving only every six years on average, so interruptions to surface operations should be minimal. The ice barrage will last for a couple of decades. As soon as it is over, we can begin putting permanent ecospheres on the surface.

Ecospheres will work as effectively for establishing bubble ecologies on Mars as they do on the Moon. All the systems developed in Asgard and translated to the Moon will work again on the Elysian fields of Mars.

Like the Moon, Mars has kept most of its meteorite impact craters more or less intact. The craters on Mars have been weathered over time, and so have a shallower profile than lunar craters. Even so, they will still make suitable foundations for our bubble habitats. Lunar and Martian ecospheres will be similar in most respects. On the Moon, no external forces, other than gravity, act on the shield domes. On Mars, we must contend with the planetary atmosphere.

At first the Martian atmosphere will be tenuous, and its effect on the domes will be small. Later, the atmosphere will thicken and the forces on the domes generated by winds must be taken into account. Mars' gravity, while low by Earth standards, is twice that on the Moon. This will allow the thickness of water shields to be cut in half.[474]

Some forethought must be given to the location of ecospheres on Mars. Eventually, the ocean basins will fill. Therefore a lot of good crater sites at low elevations must be left vacant—unless their inhabitants want to end up as Martian sub-mariners. The first ecospheres will be located in craters that are near the equator, but just above the future 'sea level'. A good location will be near the large crater Crommelin. Crommelin is above sea level and is right next to one of the equatorial basins—The Meridian Sea—where open water will begin to collect first.

Mars is a much larger planet than Luna—145 vs. 38 million square kilometers. When terraformed, half of Mars will be under water or ice. Nevertheless, it will still provide almost twice the habitable area of Luna. Fortunately, most of the good crater sites occur outside those areas that will be inundated. Mars will therefore be able to accommodate about double the number of ecospheres that will ever be built on the Moon.

Development of the Martian civilization will parallel that on Luna. A multitude of ecospheres will spring up on the craters scattered over the planet. By the time the atmosphere has thickened and humidified enough to create a shirt-sleeve environment on the equator, we will have thousands of ecospheres established all over Mars. Just as on Luna, a web of electromagnetic mono-rails will join these communities in an interconnected network. On Mars though, they must run inside tubes and tunnels to maintain vacuum. The planetary population will burgeon through immigration and natural productivity, until some optimal number has been reached.

**West of Eden**

As the Martian civilization grows, terraforming will continue. The Martian atmosphere will thicken; the planet's surface will warm; plants will spread across the barren red plains; and oxygen will accumulate in the new atmosphere. Eventually, people and animals will walk in the open beneath the "hurtling moons of Barsoom".[475]

The time will come when lovers will sit together on a thick carpet of green grass, high on a wooded hillside, looking out over a broad plain of fields and blossoming orchards. A dozen golden domes, gleaming like shields of beaten bronze, will reflect the first saffron rays of the rising sun; the hearts of the young lovers will stir, as the first day dawns on a new world.

When we have completed the metamorphosis of Mars, the universe will be a different place. Around mighty Sol, father of all Life, will circle now two exquisite gem planets. The cosmos will be home to a pair of living worlds: one feminine in her azure beauty, pregnant with animate matter; the other masculine in his ruddy newness, bursting with vitality. The rules of the great contest between Cosmos and Chaos will have been changed unalterably. We puny humans will have grown mighty; we will have become—like the Olympian gods themselves—the shapers of worlds.

# CHAPTER 6

# SOLARIA

***Dear Friend, theory is all gray***
***And the golden tree of life is green.***
**Goethe**

Life is a force of nature, not unlike electromagnetism or gravity. This force, which acts to transform simple elements into animate matter, has heretofore been only a curiosity in the Cosmos—a freak footnote to creation. Now though, Life is ready to burst upon the universal stage and take its rightful place among the forces of nature as the prime mover of reality. Life will explode from the Earth and spread throughout this solar system, transmuting raw materials into living tissues. Wherever Life finds a congenial combination of energy, water, and nutrients it will take hold. Life is a form of miraculous alchemy; once initiated the reaction will continue as long as there are reagents to sustain it and energy to drive it. We humans are the catalysts of this chemistry. We will trigger a chain reaction of elemental transmutation that will engulf the whole solar system. Once this process has started it probably can't be stopped.[476]

The result will be a living solar system. Just as Gaia is a living planet, Solaria will be a living solar system. With the advent of Solaria, Life will become a force of truly astronomic scale. In Solaria, the entire energy of a star and virtually the whole substance of a solar system will be transformed into animate matter. Sol—the G class star we call the Sun—will, in a very real sense, be alive.

At present, all but the tiniest fraction of the sun's prodigious energy is spewed into space as an utter waste. Only two parts in a billion of the sun's energy come to us on Earth.[477] The rest, 99.999999998%, pours into the void. Of the light that does reach the Earth, only .05% goes into the living substance of the biosphere.[478] Altogether, the energy invested in Life is the merest scintilla of the sun's total output. On an astronomic scale, Life hardly exists at all.

When Solaria reaches maturity, however, the energy flowing through the multifarious veins of living things will amount to a large fraction of the sun's total output. As light streams out from the sun, it will be captured by quintillions of living cells. The myriad life forms will use the raw energy of sunlight to coil organic elements into the complex molecules that form algae, flowers, trees, and people. The light of the sun will pass through a vast filter of living stuff, its radiance absorbed and down shifted from white to green.

Life has heretofore been limited to this lone world, but we are going to change that. We humans are the sentient, tool-using, manifestation of the Life Force. Through us, the Force is capable of expanding its horizons beyond the confines of a single planetary surface. Like seeds from a ripened pod, we will burst out into the solar system. Bubbles of Life will fleck the face of the Moon with an effervescent green-gold froth. Mars will be girdled about with azure seas and mantled with virescent forests. Next, the asteroid belt will be transformed into a living halo, surrounding the sun with an iridescent cloud of billions of billions of golden ecospheres. This luminescent nimbus will form Solaria—a living, even sentient, solar system.

**Sparkle Still the Right Promethean Fire[479]**

Life is a species of fire. In a universe of cold cinders, the Earth may be the only glowing spark. Now, at last, the long smoldering ember is ready to ignite new blazes. Beacons of Life will burn brightly on the Moon, on Mars, and in the space around Earth. The sun is surrounded by windrows of combustibles, like chopped kindling heaped up by a pyromaniac. Like a bonfire built in a lumber yard, the Living flames will leap into conflagration and spread through this solar system like wildfire.

In contemplating the settlement of the solar system, we tend to focus on the planetary bodies. This is not unnatural; they stride across the sky like gods and dominate the imagination. Until now, the Project too has focused on the Moon and Mars, and on free space around Earth. By the time we reach the level of Solaria, however, we will need resources on a larger scale than even these environs can provide. By chance, or by design, this solar system has been provided with a resource base that is perfect for our expanding needs—the asteroid belt. This region of space holds a treasure trove of vital stores. Here, human expansion into the solar system will reach its peak.

The availability of abundant resources at minimal costs will inevitably attract large populations to the asteroid belt. Like other forces of nature, Life follows reality's ground rules. Life always concentrates itself in the places with the best combination of sunlight, water, and nutrients. In this solar system, the asteroid belt is unsurpassed as a region for the efficient expansion of Life. As our human population expands, we will need more resources and more elbow room. The asteroid belt provides both these commodities in astronomic abundance. It is just as well that we find ourselves in a solar system so abundantly endowed with riches—we're going to need them.

**Be Fruitflies and Multiply**

We are accustomed, in the late twentieth century, to thinking of population growth as an unmitigated disaster. The ecosystem is in crisis, our numbers are redoubling, and a catastrophe appears inevitable. Like the yeast in the bottle, we seem certain to choke on our own poisons. That will come true, however, only if we remain confined to the land mass of this single planet. But we are not trapped in this bottle. We can blow the lid off and escape! Once we do that, we will find ourselves awash in an ocean of space. Freed of our planetary confines, we will have at our disposal the entire resources of the solar system. At that point, our potential is unlimited. Once in space, we can go on to develop a celestial civilization of awesome grandeur.

Prophets of doom are currently in fashion. Some of these Cassandras strike me as being decidedly anti-human. A few of them seem to think the world—even the universe—might be a better place without us. This is so wrong. Humans are the source of all light: poetry, music, art, love, laughter, hope, dreams; none of these would exist without us. Without us, the universe itself might not even exist. Reality may depend on our consciousness to perceive it and give it tangible form. Without us, all might be without form, and void; and darkness would remain upon the face of the deep. I believe that humans are good, and that more humans are better. True, a population explosion, within the confines of a single ecosphere, is certainly suicidal. But we need not remain restricted to our present land mass. We can expand. First, into the unsettled frontiers of the world's oceans. Then, into space. Once we are out of the bottle, we need never turn back.

As we expand our presence in space, the importance of Earth as the tap-root and well-spring of all Life will become ever more compelling. Preserving and maximizing natural diversity and ecological complexity is sure to become one of mankind's top priorities. Within the next Millennium, we will come into an era when the Earth is actually benefited by the growing magnitude of man's powers. When we have entered such a phase, the continued growth of our species will become an unmitigated anti-disaster. Our maturing powers will allow us to repair the ravages of the past. We can restore our Mother planet to health and then protect her—forever.

Humanity is presently expanding at around 2% per year.[480] This is actually a very modest rate of growth compared to our absolute potential. As we progress through the next Millennium, two developments will rapidly accelerate: One, death rates will fall; and two, life spans will increase. Together, these two trends will lead to a dramatic increase in our rate of growth, even without any change in birth rates. The low cost of living and high quality of life in the space colonies will be conducive to large families. The rate of population growth in space could be very high. Historically, humanity has grown rapidly when pioneering new frontiers. For example, during settlement of the American West, annual growth rates of 8% were the norm.[481] Growth at frontier rates of seven or eight percent results in a doubling of population every decade. In just a century, a single pioneer couple could become the progenitors of 2000 descendants. Let's assume that during the Third Millennium our species will continue to grow at an average rate of about 2%. This assumption, modest as it is, leads, nevertheless, to some remarkable extrapolations.

Exponential growth is easy to understand when viewed as the time needed for a population to double. To determine doubling time, divide 70 by the rate of growth. At a 2% rate of growth, the population will double every 35 years. The Earth's population has doubled in the past 35 years and is likely to double again in the next 35. Over the course of a thousand years, a growth rate of just over 2% leads to an increase of one billion fold.[482] In the year 3000 A.D., the population of the solar system could easily exceed five billion billion! [483]

***Fig. 6.1 - The Overcrowded Earth.*** TOLES copyright 1990 *The Buffalo News*. Reprinted with permission of UNIVERSAL PRESS SYNDICATE. 

**The Brain Is Wider Than the Sky**[484]

For people of limited imagination, this phase of our existence may hold particular terrors. Until a few ticks of the cosmic clock ago, the Earth itself appeared a boundless sphere of infinite horizons. The wealth of the world must have seemed limitless to the fur sealers and buffalo hunters of the last century. So must the yolk of every egg seem an endless source of nourishment to its embryo. But all too soon, we have outgrown our amniotic environment.

Having just become accustomed to the idea and implications of limits, some people are hard pressed to stretch their brains around the concept of a living solar system. The only thing more frightening to some minds than the awesome grandeur of space is the idea that we might grow to match it.

Whether or not humanity should expand to such dimensions can, and should be, the never ending subject of debate. Do we have the moral right to impose ourselves on the planets? Does the prospect of billions of billions of people mean more art, poetry, and knowledge; or just more greed, misery, and corruption? These are interesting questions which will doubtless stimulate countless fascinating arguments. That such discussions will never, and can never, reach definitive conclusions is moot. While academics, philosophers, and bureaucrats debate the merits of growth; pioneers, entrepreneurs, and holy warriors will be blazing their way through the solar system.

For better or worse, there is really no stopping us. This thing is bigger than mere human intentions. Life is a motive force. That force will do whatever it can do. Life can expand into space, and it will. A green fire storm of Life is going to engulf this solar system. How bright the blaze depends only on the availability and flammability of the fuel. Regardless of its scope, this conflagration will not be halted by an argument. The momentum behind three billion years of evolution can't be quelled by a theory, any more than a forest fire can be extinguished with a squirt gun.

**The Wonder Years**

Long before the end of the Third Millennium, we will need to accommodate large populations in the solar system. The Solarian phase begins in earnest two and a half centuries into the Third Millennium and embraces a span of another 250 years. We are not talking about a time hidden in the dim mists of an unfathomable future. Solaria is no further from us in time than Benjamin Franklin and his kite. By the year 2250, the total population of the solar system is likely to be approaching the trillion mark. That is a nice round number—200 times our present population. If there are ten or twenty billion people on Earth—an upper desirable limit—then the other 980 billion must live somewhere else.

There is certainly room for many hundreds of billions of people in space colonies near Earth. These habitats will eventually surround the Earth with a shining ring. Many more will orbit the gravitational islands at $L_4$ and $L_5$, girding these spaces in loops of beaded gold. The numbers of people who ultimately make their homes in these locales are impossible to predict. What is certain, is that once the Apollo/Amor asteroids have been mined out, people living in the Earth/Moon system will face significant cost penalties for water and other resources.

The AA asteroids amount to only a millionth of the mass of all asteroids. Altogether, they can supply three trillion tons of ore for human uses. About half the AA asteroids are carbonaceous chondrites, so water supplies from them could amount to 300 billion tons. Carbonaceous asteroids can supply a wide variety of valuable substances, but water will be the limiting resource. Its availability will determine the potential for growth in the Earth/Moon network. If we assume that each person requires 60 tons of water—the same as in Asgard—then the AA asteroids can supply enough water for a population of just five billion.[485]

The Martian population is not limited by the water supply. Mars has oceans of it. Mars is, however, a planetary ecosphere, and, like the Earth, must not be suffocated by humanity. Mars is a smaller planet than the Earth, so limiting its population to ten billion will be none too few. Altogether, the four realms of Earth, Asgard, Moon, and Mars, will accommodate a human population of less than 50 billion.[486] The other 950 billion people of the year 2250 A.D. must make themselves at home elsewhere.

By the year 2250, ninety five percent of all humans will live in the asteroid belt—the realm of Solaria. Within the main belt, there are three billion billion tons of asteroids, containing over 450 million billion tons of water. That is a small ocean—16 times the liquid volume of all Earth's rivers, lakes, and streams combined.[487] There are enough water supplies and other resources in the asteroid belt to sustain 7500 trillion people.[488]

The asteroid belt is really another world, a gigantic world, lush with riches, unique and spectacular in the opportunities and abundance it promises us. The best thing about this world is that it is chopped up into bite-sized pieces in a banquet of easily digestible morsels. This moveable feast has been thoughtfully laid out for us in the perfect place for a cosmic picnic.

Within the context of the Solar System as a geographic realm for the settlement of mankind, the asteroid belt is the vast new frontier of the untrammeled West. Its exploration and ultimate development is apt to parallel the American frontier experience; except this time, no indigenous peoples will be shoved aside, no life forms massacred, and no ecologies ruined. These negative ramifications of pioneering will be absent. The positive synergies of freedom, stimulation, and growth that come with the new frontier will be correspondingly amplified.

**Asteroid Ho!**

The outposts on Deimos and Phobos will always attract a special brand of rugged individualist. Not even the Moon or Mars will satisfy the desire of these pioneers to settle in unpeopled places. Not content with the ease and comfort of life in civilization, pilgrims with itchy feet and travelin' bones will head for the high frontier; wherever they find it, they'll push it back. These trail blazers, on their way to conquer new lands, will come to the Martian moons in an ever-growing stream. Like the boomers going to Saint Joe, Mo., these space settlers will outfit themselves at the 'rail-head' on Deimos.[489] They will convert their hard-won grubstakes into tools and provisions, and strike out for the new territories. Most will head for the myriad micro-worlds of the asteroid belt. Other hardy souls will venture further afield, to the Trojan asteroids, and even the moons of Jupiter.

Many of these outbound immigrants will be moved by the simple desire to flee the cloying multitudes of the inner worlds. Others will be impelled by that unfailing motivator of mankind: greed. Settlement of the solar system will be greatly encouraged by the adoption of a Millennial Mining Law. The administrative code underpinning the Law can be complex in its particulars, but the essence of the Law itself should be utter simplicity: "First in time is first in right."

If you are the first person to reach a satellite, and stake and register your claim with the Solar Mining Office, it's yours! It is unlikely that an individual could mount a solo expedition to the asteroid belt; so you would probably have to sell shares in your venture to get the capital. At any rate, you and your partners are now the proud owners of a celestial mother lode.

Lets say you've staked your claim to an average carbonaceous chondrite in the outer asteroid belt. What's it worth? Much interesting speculation about asteroid mining has centered on those of the metallic variety.[490] When one visits asteroids at the museum, it certainly appears that the nickel-iron types would be worth the most. After all, metals form the back-bone of industrial infrastructure, and their extraction and refinement is at the root of civilization. That is true enough today, but in the New Millennium, metals will be a side show. The real money is in water, carbon, nitrogen, sulfur, and the other light elements that make up the substance of animate matter. There will be claims staked on metallic asteroids to be sure, but in the first great "land rush", the savvy entrepreneurs will be after carbonaceous chondrites.

If you managed to be the first claimant on a typical smallish asteroid—10 km. in diameter—you would possess a trillion tons of the most valuable resources in the solar system. If, after paying for extraction and transportation costs, consumers in the realms of Asgard and Avallon paid an average of only one dollar per kilogram for light elements, your asteroid would be worth over one hundred trillion dollars.

Every person who has ever dreamed of having his own little world (including me) will be headed pell-mell for the asteroid belt. In this special region of space, vacant planetesimals teem in uncounted numbers. Any pioneer brave enough to make the journey can claim for himself and his descendants forever, sole title to a world of his own—there to be king of all he surveys. I can see it now: the Little Prince of Pallas meeting with the Duke of Dembowska in the castle of the Queen of Kleopatra. (See Table 6.2)

There are millions of asteroids that are more than a few kilometers in diameter, and billions and billions of smaller ones. Eventually, they will all be claimed, mined, and settled. As the human race grows to maturity, filling up the solar system, the market for asteroid organics will expand continually. Any asteroid prospector can stake his claim with full confidence that eventually there will be a demand for his minerals. It may take time. But asteroid settlers will be some of the only people in the solar system who can live with near perfect self-sufficiency. They can afford to wait.

Like all pioneers, the first settlers in the asteroid belt must be hardy souls with iron wills and rawhide constitutions. They must be capable of facing isolation and deprivation; they must be able to fend for themselves; and most importantly they should be endowed with infinite faith in the future. Like Europeans headed for the California gold fields in 1849, new immigrants to the asteroid belt will flood in from the Old World. They will book their passage on the trans-oceanic vessels of their time—orbital habitats cycling between Earth and Mars. When the pioneers arrive on Deimos, they will equip themselves for their perilous expedition to the new frontier. The planetary civilization on Mars will be able to provide all the technology needed to successfully homestead an asteroid.

In all likelihood, these space pilgrims will join up in "wagon trains" for the trip out. Each band of hardy trail blazers will form their own society, agreeing among themselves on its structure and systems of self-governance.

Each family buys a launch pod and a cargo pod, both tanked up with enough fuel for rendezvous maneuvers. In addition to housekeeping basics, each adult carries the obligatory space suit—the only major piece of equipment brought from Earth. The group pools its resources to buy the essentials: a small ecosphere habitat, a semi-autonomous robotic miner, a pair of multi-purpose utility vehicles, a universal fabricator, and some power bubbles with thermionic generators.

The colonists arrange financing for their expedition long before leaving Earth. All the adults in the group have special skills, and like most people in the Third Millennium, they customarily work as telecommuters. In the group there are engineers, designers, writers, counselors, researchers, and consultants. All of these people work by logging on to interactive telecommunications networks. Most of their jobs are insensitive to short time delays, so they can work without regard to their location in the solar system. The data professionals in such groups have all been selected as much for their ability to work through telepresence as anything else. As telecommuters, they can continue to work in their careers despite being on their way to the asteroid belt.

This has important economic ramifications for the colonists. First, they do not face a cut off in earnings during the two years or so it takes them to reach their destination. Second, it means that they can continue to work and generate revenue after their arrival, even though otherwise isolated on a barren asteroid. This ability to participate in the general economy while physically isolated, is one of the most important factors enabling our rapid expansion into space.

**InVestament Bankers**

Financial institutions in the Earth-Moon system have very little interest in bankrolling colonists headed for the asteroid belt; too remote, too risky. There is, however, a ready source of funding. The first big colonial push out to the asteroids targeted the giant asteroid Vesta. Vesta orbits close to Mars and is a big asteroid with a unique and valuable composition. The original Vestans flourished, building their tiny world into a major economic force in the outer solar system. Very quickly, the Vestans realized that the highest returns could be made by financing other colonists who desired to settle the wide frontier of the asteroid belt. Psychological barriers and remoteness from the inner worlds, plus very real risks, kept home-world banks out of the game. This allowed the Vestans to charge very high interest rates for colonization loans.

The inVestament bankers understood that loans made to colonists were doubly lucrative. First, the colonists invariably consist of highly motivated groups of mature professionals. (This Darwinian selection is self-enforcing; groups not meeting the rigorous standards are simply not approved for loans.) These hard-core groups inevitably hit the ground running, and very quickly transform their new habitats into productive resource mines. The telecommuting professionals give the groups an economic underpinning which allows them to carry heavy interest burdens. New productivity from asteroid resources enables them to pay back their loans very quickly. The asteroid bankers are consequently able to turn their capital over rapidly, so the investment pool doubles in size every five to ten years. Second, each new group of colonists adds a jolt of synergy to the whole economy of the region, bringing with them demands for goods and services which can be supplied by yet other colonists.

This double-barreled effect—rapid capital turn-over plus exploding demand—creates a very powerful positive feedback loop in the economy. As the Vestans make more and more money on their loans, the investment pool grows, making it possible to make more loans, generating greater profits, in turn increasing the investment pool available for loans, etc. The result is runaway growth of real wealth.

The settlers find it fairly easy, therefore, to borrow substantial amounts of money with no more collateral than their résumés. Colonists typically head out to the edge of the new frontier with an ample stock of capital equipment and a generous line of credit at the FIBV (First Interplanetary Bank of Vesta).

From our 20th Century perspective, the amounts of money involved are flabbergasting. A typical group of colonists might borrow the equivalent of $200 million of today's dollars. This appears a crushing debt to us, especially at 15% interest with a ten year balloon. But we must adjust our perspective to that of 23rd Century economics. We are getting used to the idea of the exponential growth of human numbers, now we have to acclimate ourselves to the idea of a similar explosion in personal wealth.

We can look forward to a steady increase in the level of individual wealth over the next Millennium. The implications of this type of growth are even more staggering than the implications for population growth. In recent history, the global rate of growth in per capita income has averaged 3%.[491] At this rate of growth, personal income would be $250 million dollars a year in just a few centuries. That seems unlikely. More modest rates of sustained growth are certainly possible though. For data professionals like those colonizing the asteroid belt, personal incomes of $2 - $3 million a year—in real dollars—are quite likely.

It may at first appear incredible that an average white collar worker could possibly generate that kind of economic voltage. To grasp it, we need to look at trends evolving over the prior three centuries—from 1950 to 2250 A.D. The exponential explosion in the capacities of our computers and telecommunications equipment will have continued almost without abatement during that time. It is dangerous to make direct extrapolations of technologies that are presently growing at such terrific rates, but it really doesn't matter. Right now, computer power is doubling every two years, and telecommunications gear is on a similar near vertical track. Even if these rates of progress slow dramatically in centuries to come, human capabilities in these vital arenas are likely to be quite impressive by the time the colonists set out for the asteroid belt. The result of this progress in hardware and software development will be a dramatic leverage in the economic capacities of individual workers. It will be an easy and routine thing for an engineer of the mid 23rd Century to fly through virtual space and perform $100,000 worth of engineering in a week.

Telecommunications, coupled with other trends, will leverage the economics of the human race to very high planes. Twin engines will power this expansion: 1) growth in overall population, and 2) access to extraterrestrial resources. Try to imagine the markets available in the year 2250. The total population in the solar system will be approaching a thousand billion; a trillion consumers, and all of them packing quarter million dollar credit limits on their Iridium Visa cards. If you were planning to introduce a product for a tiny niche market, you might be projecting a billion customers a year. A two million dollar salary in such an economy will not be any more noteworthy than pulling down $30,000 is today. For the 15 or 20 adults in a pioneering group, the $200 million dollar expedition loan will not be much more onerous than a present day home mortgage.

**Space Family Robinson**

(Let's join a group of these pioneers, and go along to see what it's like to settle a virgin asteroid.)

*Long before their expedition even started out, the Robinson group had targeted their bit of celestial real estate. They selected a destination only after detailed consultations with the solar mining office. They chose an asteroid of moderate size—10 km. in diameter—orbiting in the middle of the belt. This particular asteroid is already owned by its original claimant who is in residence there. The leaders of the expedition have been in radio negotiations with the old prospector and have come to terms. The new colonists have agreed to pay the owner an up-front cash bonus and a royalty of 8% on all exports.*

*There are still many millions of unclaimed asteroids; but this is a group of families and they don't want to face the extra risks of an unknown planetoid. The Solar mining office has conducted extensive surveys and has detailed spectroscopic information on almost every asteroid larger than a few kilometers in diameter. Little else is known about the unclaimed asteroids. By contrast, this asteroid—Sykes 1011—has been pretty thoroughly evaluated. In order to hold his claim, Sykes has been "proving up" his asteroid ever since he first staked it fifteen years before. He has done extensive exploratory drilling, made detailed assays of his cores, taken seismic readings, and otherwise scrutinized his private little world.*

*The old prospector's diligence has paid off. The rich data base has served its purpose and has attracted a well-heeled group of settlers. Sykes, who has lived for a decade and a half in almost complete isolation, now anticipates a rich pay day. He is going to be amply recompensed for his years of unrequited labor. The prospector's risky investment will pay rich dividends for the rest of his life.*

*The colonists—named for the largest family, the Robinsons—have been willing to pay a premium price for Sykes' asteroid. His certified assays show that S1011 has an especially high water content of 18%, and unusually high concentrations of tantalum, and cobalt. For extra bonus money and a higher royalty, Sykes has been willing to assign all future development rights to the Robinson group. The Robinsons have done detailed modeling of the economics of their risky venture; they have good reason to believe that they too will ultimately enjoy rich rewards.*

*The Robinsons launch from Deimos in a replay of their flight from Avallon. Families are catapulted off in their individual pods, and the wave riders skim across the top of Mars' thickening atmosphere, picking up a slingshot boost from the red planet's gravity well. The pods shoot off at precisely calculated trajectories to rendezvous with an orbital habitat cycling between Mars and the main asteroid belt. Unmanned cargo pods, bearing precious equipment follow. After a few days in cramped discomfort on the wave riders, the colonists rendezvous with a cycling habitat. They dock their riders and move into temporary but homey quarters in the outbound ecosphere. There they live comfortably for the 15 to 20 months it takes to reach the asteroid belt. During that time, educations, and careers carry on, without much interruption.*

*After the cycling habitat reaches an optimal point in its orbit, the colonists will jump off to their destination asteroid. The entire journey has been carefully timed to minimize the duration and energy requirements of this final leg. Everything has gone according to the schedule—arranged years in advance. Delays at any stage of their odyssey could have stranded the travelers at some intermediate point for years. As it is, the colonists are in a perfect position to rendezvous with their chosen planetoid. The Robinsons again board their wave riders and make the short hop to Sykes 1011, arriving at the asteroid with only a few kilograms of propellant left.*

*Touching down on their new world is more like docking with a giant ship than landing on a planet. The gravity is less than a thousandth of that on Earth. The little convoy sets down in the midst of a scene of splendid desolation. The horizons of the little world crowd close, unbroken by anything but Syke's radio strobe and a hand-made banner with* "Wellcum Robinsuns" *spelled out in strips of gold foil.*

*Like the pioneers of another age—who on arrival in their new Canaans, unhitched the oxen from the wagon and immediately yoked them to the plow—this new breed of settlers also sets promptly to work. The Robinson's group leader, Dr. Zachariah Smith, meets with Sykes—a grizzled old varmint who looks four centuries out of place. Sykes has spent 15 years on the asteroid, living in an empty propellant tank buried under the regolith. He has been utterly alone except for the company of a holographic dog named Rimmer. From the musky smell inside the old prospector's digs, Dr. Smith judges that Sykes hasn't been out of his space suit in the past decade. True to form, Sykes conducts the entire closing over the upturned bottom of a supply canister, never removing anything but his helmet. 'How in the world,' Dr. Smith wonders, looking at the old character's stained beard, 'can any one chew tobacco gum inside a space helmet?'*

*With legal formalities out of the way, the first order of business is to launch power bubbles. The group has brought three thermionic generators, each with enough capacity to provide all of the colony's power needs. Fuel is pooled for one of the wave riders, and the generators are towed into close polar orbits around the asteroid. This takes just a small amount of propellant as orbital velocity is only a few meters per second. Once the generators are in stable orbits outside the asteroid's shadow, the bubble reflectors are inflated around them. Properly focused and adjusted, the generators begin to crank out power. The electricity is converted to a narrow beam of microwaves and is transmitted directly to receiving antennas strung along the asteroid's equator.*[492]

*With a good supply of power now at their disposal, the colonists begin to transform their Lilliputian planet. They set up and activate the mining station. The robotic augers immediately begin shunting loose regolith into the maw of the machine. Inside the miner, the soil is heated in a plasma arc furnace. As the temperature rises, steam hisses out of the hot ore, and the vapor is pulled off and condensed. When the first silver trickle of liquid water begins to dribble into the transparent collection tanks, a cheer bursts from the jubilant colonists. At this point they know they have succeeded. Their years of sacrifice and hard work have paid off. They have water. Where there is water there is life. Now they know they can survive on this orbital slag heap. They have everything they need to sustain life and build a new world.*

*The robotic miner is not a large machine; its throughput is only a couple of kilograms per minute. Watching powdered regolith pass through the holding bin is like watching sand trickle through a big hour glass. Even so, the machine produces over 500 kilograms of distilled water and half a ton of oxygen per day. In a highly efficient process, the crushed asteroid ore is melted and most of its usable elements extracted. The molten rock preheats the incoming ore stream and is then extruded from the miner in the form of cooled bricks, slabs, and other structural components.*

*The colonists garner everything they need from the asteroid. At this stage of development, the principal commodities required by the colony: power, oxygen, and water, are available in abundance.*

**Table 6.1**
**Useful Materials Produced by Robotic Miner Every Day**

| Element | Kg./Day | Element | Kg./Day |
|---|---|---|---|
| Oxygen | 1152 | Sodium | 23 |
| Silicon | 576 | Chromium | 12 |
| Water | 518 | Cobalt | 5 |
| Iron | 430 | Manganese | 5 |
| Magnesium | 403 | Nitrogen | 5 |
| Carbon | 57 | Titanium | 4 |
| Calcium | 52 | Phosphorus | 3 |
| Sulfur | 52 | Potassium | 2 |
| Nickel | 32 | Copper | .3 |
| Aluminum | 45 | | |

*The next pressing project is to erect a temporary habitat. The temporary structure is just a large fabric bubble with a cable net thrown over the top. The colonists hastily level the bottom of a small crater with their utility vehicles. The bubble is inflated and regolith is shoveled in to cover the whole thing three meters deep. Though palatial by Sykes' standards, it will be rough living for many months, until a proper ecosphere can be built. The colonists move into their temporary domicile, pitching their family tents inside the Spartan space. The temporary habitat is not roomy, and is harshly lit by yellow tritium bulbs. It provides a shirt-sleeve environment nonetheless. On the face of this hostile bit of real estate, adrift in the void, even this rough bubble is a welcoming haven.*

*Establishing a self-sustaining ecocycle to produce food and recycle wastes is now the colonist's top priority. Coils of algae tubing are deployed and an ecocycle centered on a super critical water oxidizer is organized. At this point, the colonists are dug in for the long haul.*

*Now they can turn their attention to raising the colony's standard of living. The first step in that direction is the construction of a proper ecosphere. The surface of the asteroid, closely resembles that of the Martian moon Deimos. It is pocked with craters of various sizes, up to a kilometer in diameter. A crater 200 meters in diameter is chosen, and work begins to transform it into a permanent ecosphere. The robotic miner forms a ring of fused regolith around the crater rim while the utility tractors terrace the inner slopes. While forming the foundation ring, the miner extrudes bulk materials and some simple finished goods like reinforcing cable and anchoring bolts.*

*Refined silicon and other elements are fed into the uni-fab (universal fabricator) which produces the silicone bubble membrane. The uni-fab is a remarkable piece of machinery and represents one of the colony's most expensive and valuable capital assets. The uni-fab is capable of producing virtually any material or machine component.*[493] *All it requires is a supply of the appropriate raw materials and detailed design instructions for its computer.*

*The fabricator uses MBE (Molecular Beam Epitaxy) technology to produce parts and materials. MBE is extremely simple in concept: a beam of charged atoms is sprayed onto a substrate, not unlike painting a car. Successive layers of atoms are beamed on until the desired thickness is built up. The composition of the molecular beam can be varied at will, as can the shape and thickness of the final products. With the right materials and instructions, the uni-fab can produce anything, from ball bearings to saran wrap.*

*The only materials the uni-fab can't produce are living tissues. You could feed in the appropriate instructions and materials needed to form a frog, but all you would get is frog soup.*

*The uni-fab can, however, easily manufacture just about anything else. It could readily produce all the parts of a camera, for example. You would still have to assemble the components, but all the lenses, fittings, and tiny screws could be manufactured by the uni-fab. Depending on its design, it could be a very fine camera. All the parts would be made with a finish and precision accurate to a couple of angstroms—the width of a single atom. The uni-fab enables the colonists to be self-sufficient in virtually all manufactured products, from toys to computer chips.*

*While the uni-fab produces the ecosphere's dome material, work progresses on crater preparation. When the inner terraces and anchoring ring are completed, the bubble membrane is installed, and inflation of the ecosphere begins.*

*An ecosphere 200 meters in diameter will require about half a million kilograms of oxygen. It will take the robotic miner a little over a year to produce this much air. As the ecosphere is inflated, the crater is terraformed to provide a life-rich habitat. Trees, grass, and flowers are planted in profusion. An open stream meanders across the terraces and slowly cascades down the slopes to a small pond. A fountain sprays water in dramatic slow-motion arcs. The soft splashing of water gently falling through the micro gravity will fill the interior with the unmistakable sounds of Earth. Slowly, the robotic miner will produce enough water to form a water shield for the crater dome. When complete, the shield will allow the colonists to pitch pavilions in the gardens and live in the open, among the eight acres of grass and trees under the dome.*

*The Robinsons will construct the first ecosphere for somewhat the same reasons that Sykes did his core drilling—to attract new settlers. With the ecosphere completed, the otherwise barren little asteroid now beckons with the welcoming green glow of a miniature Eden.*

*New colonists, looking for opportunities, but less willing than Sykes or the Robinsons to face risk and hardship, will be attracted to the new habitat. This first permanent ecosphere is easily large enough for dozens of families.*[494] *Several groups of immigrants join the growing colony. This third wave of settlers does not face the harsh wilderness that greeted old Sykes and the Robinsons.*

*These newcomers can move right in to a comfortable habitat, resuming their lives without much of an interruption. Since the hazards and discomforts are low, individual families, even single people, will be able to immigrate to the new colony.*

*Despite this lack of hardship, the new colonists will enjoy a large share of the asteroid's abundant wealth. The new comers will pay cash bonuses to the Robinson group. They will literally be buying a 'piece of the rock'. In return, they are supplied with dwelling space, food, water, energy, and amenities. The new arrivals also get a stake in the colony. They earn a position in the corporate identity of the colony and a share of future royalties and profits.*

*As more and more colonists arrive, the earlier waves grow wealthier. Sykes becomes rich as Midas, but he never does move out of his rickety fuel cylinder. The Robinson group took a big gamble and hit the jackpot. The Robinson's and the other founding families become enormously wealthy, building their own private compounds in some of the asteroid's choicest craters. For generations to come, the descendants of these pioneers will enjoy the status and bank accounts that go with "old money".*

**New Bern**

As the growing community accrues wealth, dramatic new projects can be undertaken. The first of these will be—as on most new worlds—the construction of a mass launcher. The launcher will be built almost entirely of local materials. For example, low-temperature superconducting electromagnets can be made of Alnico—an alloy of aluminum, nickel, iron, and cobalt. All these metals are available on the asteroid. Ultra-low temperatures can be maintained in the magnets by shielding them from radiant heat under a blanket of vacuum insulated regolith and by circulating liquid hydrogen at -252°C. Once the mass driver is complete, the asteroid colony can begin to export commodities profitably.

There will always be an insatiable demand for water and other light elements back in the Earth-Moon system, but importing hydrogen from the asteroids will not be cheap. It requires a change of velocity ($\Delta v$) of 11 kilometers per second (kps) to move payloads from the inner asteroid belt to the Earth's orbit. That is the same $\Delta v$ required to lift payloads off the Earth. This fact will put a premium on the cost of living in the vicinity of the Earth. Escaping this cost penalty will be one of the motivations fueling immigration to the asteroid belt. In the self-sufficient colonies of the asteroids, the cost of living will be attractively low. There will always be people willing to pay the premium to live near the Earth, however; so the thirst for hydrogen from the asteroid belt is apt to remain unquenchable.

The colonists on Sykes 1011—now renamed New Bern—produce elemental hydrogen which they liquefy and hurl into space in vacuum insulated canisters of chromium cobalt alloy. The canisters need not be propelled to 11 kps—a very energy intensive proposition. Instead, they are impelled at a few tenths of a kps onto an orbital path that rendezvouses with the large asteroid Ceres. On Ceres there is a large interplanetary mass driver with huge cargo capacity. Hydrogen, metals, and other commodities from all over the asteroid belt are consolidated into bulk shipments. Big cargo carriers are flung off on journeys to the inner planets that can take years. The colonists' shipments of hydrogen, vitallium and other commodities are automatically credited to their accounts on Vesta.

The economy of New Bern thrives. Underpinned by the work of telecommuters, and supplemented by commodity exports, the colony grows richer. The colonists invest heavily in semi-autonomous factories. Highly specialized products are manufactured for niche markets all over the Solar system. The colonies on New Bern specialize in precision medical instruments and implants. Their vitallium endoskeletons come to be highly prized by people going through trans-geriatric metamorphosis.[495]

New ecospheres spring up. All the best crater sites on the asteroid are domed over. As new immigrants arrive and children are born, the population burgeons. There is no shortage of room on the asteroid. It seems a tiny world, but it is big enough to accommodate a large city. The surface area of the asteroid is 314 million square meters. That is enough room for a population of three million people. It's hard to fathom, but even this many colonists wouldn't be crowded. Each person would have as much room as the marine colonists on Aquarius, where almost 40% of the area is dedicated to park land and open space. The same ratios would apply on the surface of the asteroid. Half the surface area could be dedicated to gardens, lakes, playing fields, and other open spaces, and there would still be ample room for a large population. All the industry, and much of the colony's supporting infrastructure, is put underground. The surface is left free for living.

Eventually, the whole surface of the asteroid is enclosed inside an ecosphere. With a membrane surrounding the asteroid, the whole surface can be terraformed and inhabited. Lush plant life will cover the asteroid's surface. People will live in their pavilions, set among the trees and flowers blanketing the once barren landscape.

A water shield for an ecosphere 12 kilometers in diameter would weigh 2.25 billion tons. This would require just over two percent of the asteroid's water supply. Inside the bubble membrane will be an oxygen atmosphere amounting to 86 million tons—requiring only a tiny fraction of the asteroid's oxygen.[496]

The ecosphere will provide a tremendous volume of livable space. At some places, the asteroid's 'sky' will be two or three kilometers high. Trees will be able to grow thousands of meters tall, dwarfing even the Never-trees in the craters of the Moon.[497] In the minuscule gravity, flight will be almost effortless. As in Lothlorien, many people will live in the branches and hollow trunks of the gigantic trees. Swiss Family Robinson will have come full circle.

As mining progresses, the asteroid's interior will become honeycombed with caverns and tunnels. Robotic miners will cut through the rich carbonaceous ore of the planetesimal like termites boring through fruit cake. Even after the asteroid is fully enclosed in an ecosphere, mining will continue deep in the interior. Exports will be spit out through rigid magnetic launch tubes that penetrate the outer shield membrane. Over time, the interior of the asteroid will be hollowed out and terraformed. (See Plate No. 14.)

A large fraction of the asteroid's bulk will be converted into the living substance of people, plants, and animals. Every kilogram of water or metal ever removed from New Bern will be found somewhere in the solar system: bound up in the radiation shield of some other colony, rustling in the leaves of a tree, or coursing through the veins of a child. Nothing will be wasted, but everything will be transformed.

Over time, New Bern will mature into a vibrant miniature world. It will possess a unique history and culture all its own. The process that creates New Bern will be played out all over the asteroid belt at various times. The transformation of the belt will begin in the inner zones closest to the Sun and will proceed apace until the remotest asteroids have been settled, encapsulated, and terraformed. As the human population grows, each planetesimal will become the center of an expanding community, nourished and sustained by its asteroid's resources. As Solaria blossoms and ripens, the entire asteroid belt will become thickly sprinkled with free-floating ecospheres and terraformed asteroids. (See Plate No. 15)

**The Myriad Micro-Worlds**

The asteroids present us with raw real estate of no small dimensions. The largest asteroid, Ceres, provides almost three million square kilometers (1 million sq. mi.) of new territory—an area larger than central Europe.[498] The 32 largest asteroids, each of which is over 200 km. in diameter, provide a combined surface area of nearly ten million square kilometers.

One of the by-products of mining resources from planetesimals will be the creation of an abundance of livable space inside them. Huge, fusion-powered mining machines will cut through the interior with gamma ray lasers, vaporizing the ore and instantly separating purified materials from the gas stream. The early result of these mining operations will be the creation of vast networks of underground tunnels and enormous chambers honeycombing every large asteroid.

In the low gravity, these underground chambers can be expansive. Caverns, kilometers high and tens of kilometers wide can be carved out. Hydrogen plasma tubes bolted along the roof will flood the interior with the mellow white light of natural sunshine. The interior spaces can be thickly planted with vegetation, and the chambers can become a mini-ecospheres. A person inside such a chamber might be hard put to tell that he was underground at all.

The potential for such habitats is almost unlimited. If half the material were mined out of a large asteroid like Vesta, it would provide forty million cubic kilometers of interior space. This space, divided into an endless labyrinth of tunnels, passages, and chambers would seem almost infinite. These underground caverns will provide a fantastic realm of unparalleled variety. In the micro-gravity environment, one will be able to explore these internal worlds on the wing. The Gordian grottos, honeycombing the asteroid like Swiss cheese, could provide a life-time of exploration and surprise.

The tunnels and caverns will expand as ore is mined out, until eventually, even the largest asteroids are completely hollow. The whole interior surface can then be terraformed, creating a world turned inside out. Like David Innes in Pellucidar, one can stand in such a world and look up at the horizon, curving away into the distance.[499] Overhead, spinning in the gravitational center of the hollow asteroid, is a ball of fusing plasma—the interior's own sun. A hemispheric reflector rotates slowly around the miniature sun, creating the darkness of night over one side of the habitat.

**Table 6.2**[500]
**The Largest Main-Belt Asteroids**[501]

| Asteroid | Radius-km. | Area-km$^2$ | Mass-tons |
|---|---|---|---|
| Ceres | 466 | 2,728,869 | 1.5E+18 |
| Vesta | 264 | 875,828 | 2.7E+17 |
| Pallas | 262 | 859,319 | 2.6E+17 |
| Hygiea | 207 | 538,458 | 1.3E+17 |
| Interamnia | 161 | 323,714 | 6.0E+16 |
| Davida | 159 | 317,691 | 5.9E+16 |
| Europa | 138 | 241,052 | 3.9E+16 |
| Juno | 136 | 230,722 | 3.6E+16 |
| Patientia | 134 | 223,962 | 3.5E+16 |
| Euphrosyne | 129 | 207,500 | 3.1E+16 |
| Eunomia | 124 | 193,221 | 2.8E+16 |
| Bamberga | 121 | 185,508 | 2.6E+16 |
| Camilla | 119 | 179,451 | 2.5E+16 |
| Sylvia | 119 | 177,953 | 2.5E+16 |
| Eugenia | 118 | 176,461 | 2.4E+16 |
| Psyche | 118 | 174,975 | 2.4E+16 |
| Themis | 118 | 174,975 | 2.4E+16 |
| Egeria | 117 | 170,554 | 2.3E+16 |
| Thisbe | 116 | 169,093 | 2.3E+16 |
| Cybele | 115 | 166,191 | 2.2E+16 |
| Kleopatra | 112 | 157,633 | 2.1E+16 |
| Loreley | 109 | 147,935 | 1.9E+16 |
| Fortuna | 108 | 145,220 | 1.8E+16 |
| Iris | 106 | 139,867 | 1.7E+16 |
| Herculina | 104 | 135,918 | 1.6E+16 |
| Alauda | 103 | 133,317 | 1.6E+16 |
| Hermione | 102 | 130,743 | 1.6E+16 |
| Bettina | 100 | 125,664 | 1.5E+16 |
| Hebe | 100 | 125,664 | 1.5E+16 |
| Siegena | 100 | 125,664 | 1.5E+16 |
| Ursula | 100 | 125,664 | 1.5E+16 |
| Winchester | 100 | 125,664 | 1.5E+16 |
| **TOTALS** | | **9,934,448** | **3E+18** |

With an outer membrane enclosing the entire asteroid, flyers can come and go between the outer and inner surfaces, soaring through the wide apertures of old craters. Hollow asteroids, terraformed both within and without, will form worlds of wonder where Edgar Rice Burroughs himself would be right at home.

The number of smaller asteroids is incredible. There are ten billion asteroids larger than 100 meters in diameter.[502] Virtually all of these—9,990,000,000—are between 100 meters and 25 kilometers across. These small asteroids don't contain much mass. Altogether they amount to only ten percent of the mass of Ceres. What they do have in abundance is surface area. The total area of the small asteroids amounts to billions of square kilometers.

Even a small asteroid—a hundred meters in diameter, barely larger than a big rock—weighs a million and a half tons. Little crumbs like these contain enough water, carbon, and other resources to sustain a population the size of a small city. An asteroid just a couple of kilometers in diameter weighs billions of tons and can support the population of whole continents. An asteroid 25 kilometers in diameter, weighs tens of trillions of tons and can support the population of whole planets. There are no fewer than ten million asteroids as large as 25 km. in diameter.[503]

In the fashion of New Bern, the smaller asteroids will make wonderful ecospheres—like lunar craters turned inside out. Asteroids up to 50 kilometers in diameter are small enough to be entirely enclosed inside a bubble membrane. Terraformed into miniature versions of living planets, with their interiors hollowed out, each of these cavernous micro-worlds could accommodate billions of people living in spacious comfort.[504]

Asteroid communities of all sizes are possible, from planets to villages, and even single-family homes. Homesteading a single-family asteroid would involve finding a carbonaceous chondrite about 20 meters in diameter. The homesteaders would mine out the interior of the asteroid, leaving a shell of rock intact. This crust could provide protection from cosmic rays and small meteoroids. The asteroid's hollow interior would provide enough room for a small family to live comfortably. Such an asteroid would weigh 14,000 tons, providing 1400 tons of water and hundreds of tons of valuable organic elements. A small fraction of these resources can be retained to sustain the family's closed-cycle ecosphere. The rest can be exported. There are 100 billion asteroids between 10 and 100 meters in diameter.[505] Most of these would make suitable homesteads for families or tribes.

As settlements spring up among the asteroids, trade and commerce among them will flourish. Trade always thrives when the costs of transportation are reduced. This effect was noted as a significant advantage on the Moon, where superconductive magnetic monorails whisked goods and people between domed craters at high speed and minimal cost. The civilization in the asteroid belt will enjoy even greater advantages. On the Moon, it was necessary to build the monorails. In the asteroid belt, ships will be able to fly between colonies with little or no infrastructure involved. Commerce between asteroids on the same plane and at the same distance from the Sun will be the easiest. For example, flight between asteroids like Ceres and Hygiea will involve very small changes in velocity.[506]

Asteroids tend to occur, not as isolated islands, but as archipelagos. Hundreds of small asteroids usually travel in a swarm around one big mother asteroid. Typical of such asteroid "families" is that of 8Flora. Flora is about 150 km. in diameter. This big asteroid is accompanied by five asteroids 30 km. in diameter, eight 15 km. in diameter, and hundreds of smaller ones.[507] The large asteroid will become the focus of development, with satellite communities springing up on the smaller companion asteroids. In this way, the process of development and diffusion will naturally progress from larger to smaller bodies.

As in the many craters of the Moon, the multiplicity of habitats will allow people to create for themselves a wide variety of communities, each characterized by its own cultural mix. Eventually, there will arise a wondrous medley of cultures, not unlike Bohemian and ethnic neighborhoods in today's cities. The freedom of unlimited space will allow diversification to proceed to its ultimate, and desirable, extreme. It will be a wonderful thing to travel among a multitude of habitats, sampling their diverse offerings. It will be like the experience of travel on Earth, where the real attractions are people and their cultural varieties.

**Pool of Radiance**

The asteroid belt orbits between Mars and Jupiter. This puts it just within the outer limits of the sun's warmth. Mars is only half again as far from the Sun as the Earth; an average of 228 million kilometers, compared to Earth's 150 million. By comparison, Jupiter is in the hinterland. Great Jove passes in stately procession around the Sun, attended by his own court of miniature planets; like a minor god, too proud to acknowledge the ascendancy of his shining superior. Jupiter is 780 million kilometers out—five times as far from the Sun as the Earth. At Jupiter's distance, the sun provides barely four percent of the light it lavishes on the Home Planet. Jupiter and the worlds beyond are really in the dark, exiled forever to roam outside the circle of the sun's friendly light.

The asteroids, by contrast, share space on the sun-warmed hearth with the other inner planets. The asteroid belt is fairly wide, but the innermost asteroids are only about 70 million kilometers further out than Mars.

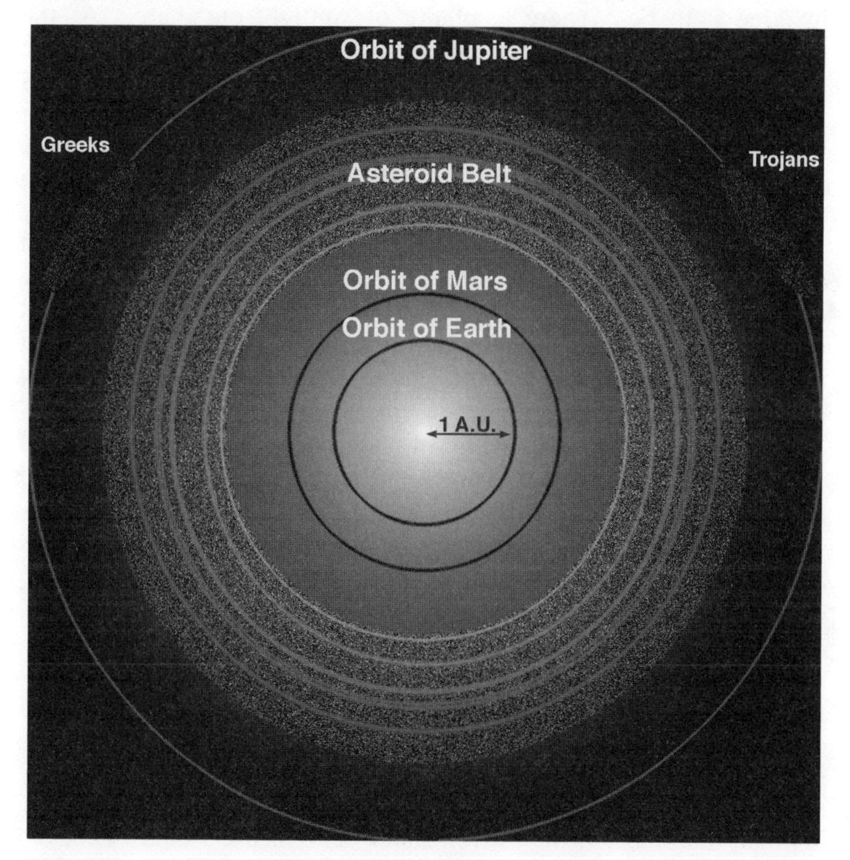


***Fig. 6.2 - Scale diagram of Inner Solar System.*** *Light diminishes rapidly at modest distances from the Sun. Compared to the Earth and Mars, Jupiter and the outer planets dwell in darkness. The asteroids fill most of the space on the plane of the ecliptic between Jupiter and Mars. The asteroids are grouped in concentric bands, separated by "Kirkwood Gaps", caused by gravitational resonance with Jupiter.*

At one astronomical unit—the distance of the Earth from the Sun—the solar constant is 1300 watts per square meter. This is a flood tide of radiant energy. For habitats in free space, like Asgard, it is really too much. One of the biggest problems in ecospheres is the elimination of waste heat. In Asgard, we had to institute countermeasures to cope with the overabundance of sunlight. In the asteroid belt these problems of over exposure are reduced.

The asteroid belt is not the dim realm of Orpheus one might imagine it to be. There is plenty of sunlight out there. At the middle of the belt, the solar constant is over 140 watts per square meter.[508] Light intensities within the belt vary from around 325 watts in the inner belt to around 80 watts in the outer belt. Incident light at these intensities is not much different from what we experience routinely on Earth. Of the 1300 watts striking the top of the Earth's atmosphere, a third is scattered and reflected by clouds and dust. Another 10 to 15% is absorbed by carbon dioxide, water vapor, and ozone.[509] Only about 675 watts actually impinge on each square meter of the Earth's surface. This amount is further attenuated by the latitude and season.[510] The light intensity in the inner asteroid belt is about the same as that experienced on a sunny winter day in Montreal.

Sunlight illuminates the space around the Earth at an intensity of 224,000 lumens per square meter.[511] In the middle of the asteroid belt, this falls to 25,000 $l/M^2$. By comparison, reading requires only 300 lumens per square meter; even fine needle work requires only 2000 lumens per $M^2$. In the outer fringes of the asteroid belt, the ambient sunlight will be brighter than in a well lit modern office.

One might think the asteroid belt would be too cold, but this is not the case. Bettina, one of the main belt asteroids, is about the same size as Sicily. While Bettina is three times further from the sun than Sicily, the asteroid is not three times colder. Sicily's cozy warmth is due almost entirely to the blanket of Earth's atmosphere. Without that blanket, the saffron sands of Syracuse would be locked in the iron grip of eternal icebergs. The average surface temperature on Earth is a balmy 15° C. (59° F.). Without an atmosphere, the Earth would be permanently frozen at -20° C. (-4° F.).[512] A planet's climate is determined more by the nature of its atmosphere than by its distance from the sun. We control the atmospheric properties of our ecospheres, so the climate inside our bubble habitats in the asteroid belt can be tailor-made for our own comfort. When we have finished terraforming the asteroids, compared with Sicily, there will be better beaches on Bettina.

In the realm of Asgard, in orbit around Earth, each person requires about five square meters of solar bubble to meet his total power needs.[513] In the inner asteroid belt, twenty square meters are needed. This increases to 45 $M^2$ in the middle of the belt, and 80 $M^2$ in the outer fringes. The most expensive item in a solar power bubble is the generator, which stays the same size, regardless of the size of the reflective collector which is relatively cheap. So, as solar intensity falls, the cost of electric power goes up only incrementally. Solar power will still be cheap and abundant, even in the outer asteroid belt

**Climb K2[514]**

Half way through the next Millennium, human numbers will have grown to 330 trillion—66,000 times that of the Earth today. In 2500 A.D. the human race will possess astonishing powers. Our increased wealth will invest us with a hundred million times the economic muscle of the present world economy. Our computational machines will be pushing the envelope of theoretical possibility. Technologies only glimmering today will be mature, and others, yet undreamed of will be burgeoning. With such powers, and the material resources of an entire solar system at our disposal, there will be few imaginable projects which still exceed our grasp: terraforming planets, building our own artificial worlds, restoring the Earth's natural splendor; all these shrink in perspective to become not only conceivable, but even easy.

A Soviet astrophysicist, N.S. Kardashev, proposed a classification of civilizations into three broad categories:

1—civilizations using all the power available on a single planet,
2—civilizations using the entire power output of a star, and
3—civilizations using all the power in a galaxy.

Our own civilization is on the way to Kardashev Level one (K1). Once we have begun to grow in the solar system, we will probably not stop until we have at least approached Kardashev Level two (K2).

The power output of the sun is 380,000 billion billion kilowatts. A true K2 civilization with an average per capita energy consumption of three kilowatts, would number over 100,000 billion billion people. The population in the solar system will probably never climb to such staggering proportions. But a population of five billion billion within a thousand years is a very real possibility.

It is hard to imagine such a population. For every man, woman, and child alive today, in a thousand years, there will be a population the size of China's. Just the descendants of the people in your car-pool will be enough to fill an entire world. While the vision of your neighbor Ed—a guy who wears plaid ties and tells 'knock-knock' jokes—as the patriarch of a billion descendants is indeed chilling, there is an upside. Albert Einstein is the kind of human genius who only comes to us once in every ten billion births. That's about the total number of human beings who ever existed, both living and dead. In the world of the future, at any given time, there will be 500 million Einsteins! There will be millions of Michelangelos, billions of Beethovens, trillions of Tennysons. Such a civilization will be awash in art, music, poetry, and science—all of immortal quality.

With just 5% of the labor force employed as scientists and engineers, there would be 250 million billion researchers at work.[515] It is impossible to conceive of any problem we could not solve. Take for example a particular tiny question that only one scientist in a million investigates. On Earth today, that means there are only 50 or 100 scientists focusing on this question. By the end of the Third Millennium, the number of scientists working on just one such problem would exceed 250 billion. To hold a minor symposium, they would have to rent an entire planet! Now imagine these billions and billions of scientists all interconnected in a single vast telecommunications network, able to contact any one of their myriad colleagues instantaneously. Envision all of them assisted in their work by artificial intelligences a billion times more powerful than all the computers on Earth today combined. Their aggregate powers will approach omniscience. There will be no discoverable fragment of knowledge left ungleaned from the ever growing field of the knowable universe.

**Vital Stores**

Can our little solar system really support five billion billion people? The surprising answer is yes, and easily at that. There are stores enough in the solar system to support even very large populations for billions of years.

In taking stock of the solar system, let's restrict our inventory to a tally of the available water supplies. We need a lot of different things, but water is the most fundamental of commodities. Life is, after all, mostly water—50 to 90%. By comparison, carbon, nitrogen, and all other elements amount to only fractions of the mass of living tissue. Water has many vital roles: it is a metabolite, a carrier, a diluter, a humidifier, a cleaner, and, at least early in the next Millennium, a radiation shield. So let's make the broad assumption that, if the solar system has enough water to support a large population, it will have enough of everything else too.

How much water does it take to make five billion billion people? The average person contains around 40 liters (10.5 gals.) of water.[516] Five billion billion people would require 200 million cubic kilometers of water—just for their own bodies. To provide such a population with the water needed for culturing algae, growing plants, cooling habitats, shielding from radiation, and other purposes, may require hundreds of times as much. For stock taking purposes, lets assume that the average water allotment will be the same throughout Solaria as that required in Asgard—60 tons per capita.[517] This would raise the total water demand to 300 billion cubic kilometers.

The oceans, glaciers, rivers, and springs of the Earth hold 1,326 million cubic kilometers of water.[518] If all the waters on Earth were collected into one gigantic reservoir, the pool would be 1300 kilometers across and 1000 kilometers deep. This amount of water forms a useful measure of one 'ocean mass'. Total water demand by the end of the Third Millennium could equal 226 ocean masses. Where can it all possibly come from?

As it happens, our solar system is richly endowed with this remarkable mineral—the stuff of life.[519] The oceans of Mother Earth justifiably impress us, but they contain only a fraction of the water available in the solar system. The moons of Jupiter alone contain many times as much water as there is on Earth. For example, Callisto, the size of the Planet Mercury, is about half ice, and contains forty times as much water as there is on Earth. Europa and Ganymede hold similar reservoirs. (See Plate No. 12.) Water can also be formed chemically from elemental hydrogen and oxygen, which are both abundant. Finally, the Oort cloud holds another huge supply of water and other useful materials.[520]

Not counting the Oort comets, the moons and other small bodies of the solar system contain just about exactly 300 billion cubic kilometers of water. It is an interesting coincidence that this is just the quantity the human population will require by the year 4000 A.D.

What is true of water is equally true of all the other elements and compounds needed to support the Solarian civilization. Jupiter alone weighs two and a half times as much as all the other planets combined. Even a very large civilization could not exhaust this store house in billions of years.

**Table 6.3**
**Water Supplies in the Solar System**

| Body | Radius-km | Mass-kg. | $H_2O$-tons |
|---|---|---|---|
| **Earth** | 6378 | 6E+24 | 1.3E+18 |
| **Martian-**Phobos | 11 | 9.6E+15 | 9.6E+11 |
| Deimos | 6 | 2E+15 | 2E+11 |
| **Asteroids-**Main Belt | | 2.88E+21 | 2.16E+17 |
| Trojans | | 1.2E+20 | 1.14E+16 |
| AA | | 3E+15 | 1.5E+11 |
| **Jovian-**Ganymede | 2638 | 1.50E+23 | 9E+19 |
| Callisto | 2410 | 1.06E+23 | 6.36E+19 |
| Io | 1816 | 8.92E+22 | * |
| Europa | 1563 | 4.87E+22 | 3E+18 |
| **Saturnian-**Titan | 2560 | 1.36E+23 | 8.16E+19 |
| Rhea | 765 | 2.28E+21 | 1.6E+18 |
| Iapetus | 720 | 1.9E+21 | 1.33E+18 |
| Dione | 560 | 7.36E+20 | 7.35E+17 |
| Tethys | 525 | 6.26E+20 | 4.40E+17 |
| Enceladus | 250 | 7.4E+19 | 5.18E+16 |
| Mimas | 195 | 3.76E+19 | 2.63E+16 |
| Hyperion | 145 | 1.1E+19 | 7.7E+15 |
| Phoebe | 70 | 1.43E+18 | 1.00E+15 |
| **Uranian** -Titania | 520 | 1.2E+21 | 6E+17 |
| Oberon | 460 | 8.2E+20 | 4.1E+17 |
| Umbriel | 450 | 7.6E+20 | 3.8E+17 |
| Ariel | 430 | 6.7E+20 | 3.35E+17 |
| Miranda | 160 | 3.4E+19 | 1.7E+16 |
| **Neptunian-**Triton | 1900 | 5.7E+22 | 2.85E+19 |
| Nereid | 470 | 1.13E+21 | 5.66E+17 |
| 1989 N1- Naiad | 100 | 8.38E+18 | 4.19E+15 |
| 1989 N2-Thalassa | 75 | 3.53E+18 | 1.77E+15 |
| 1989 N3-Despina | 50 | 1.05E+18 | 5.24E+14 |
| **Pluto-**Charon | 400 | 2.68E+20 | 2.68E+17 |
| **Comets-**Short Period | | 6.28E+14 | 1.57E+11 |
| Chiron** | 120 | 7.24E+18 | 1.81E+15 |
| **TOTAL** | | | **3E+20** |

**Io appears to be completely anhydrous. **Chiron, which orbits between Saturn and Uranus, was originally thought to be an asteroid, but has since revealed itself to be a giant comet.*

The bits and crumbs of the solar system—asteroids and moons—will be eaten up fairly early in the game. These are handy bite-sized chunks of material, like finely minced baby food, to be consumed during Solaria's infancy.

Large human settlements will probably not grow up as far out as Jupiter where sunlight is dim. Therefore, the huge material resources of the Jovian and Saturnian systems will need to be transported in to the sunnier climes of the asteroid belt. Jupiter orbits the sun at 13 kps.; the asteroids orbit at 16.5 kps; so payload can be sent from objects in the Jovian system to the asteroid belt for Δvs as low as 3.5 kps.[521] By comparison, sending material from the Jovian system to Earth's orbit would require Δvs of at least 17 kps. When multiplied by the gigatons involved, this difference in energy requirements becomes a compelling reason to concentrate our future growth in the asteroid belt.

After water supplies in the asteroid belt have been tapped out, Callisto is the next best source. Callisto is larger than Europa but has a shallower gravity well, due to its low density. Europa, which is mostly rock, has a density of 3.6 grams per cubic centimeter—a good deal denser than granite. Callisto, by contrast, is only about as dense as chalk, 2 g/cm$^3$. Gravity on Callisto is less than an eighth of that on Earth. A cubic meter of water, which would weigh 1000 kg. on Earth, will only weigh 125 kg. on Callisto.

In order to send cargo from Callisto to the Asteroid Belt, enough energy must be supplied to both escape the Moon's gravity and climb out of the awesome gravitational hole Jupiter creates in space. Escape velocity from Callisto is 2 kps, and the moon orbits Jupiter at 8.2 kps.[522] There is also a difference in orbital velocity of around 5 kps between Jupiter itself and the mid-belt asteroids. Happily, we do not have to overcome the additive total Δv of 15 kps. By carefully timing and aiming the launch of ice blocks from Callisto, we can send water to Ceres for a Δv of only 4.12 kps.[523] (See Plate No. 13.)[524]

Accessing water and other resources is a question of cost. The most accessible resources will be used first. If the cost is justified, there are abundant reserve supplies of all the vital elements needed for Life. Fully 92% of the mass of the solar system, other than the sun, is contained in Jupiter and Saturn. Jupiter alone weighs 318 times as much as the Earth.

Jupiter appears to be surprisingly dry. It contains only twice as much water as in the Earth's oceans. Jupiter should contain a thousand times more water than it seems to. Indeed it might, hidden beneath the methane and ammonia cloud tops of its upper atmosphere.[525]

Even without preformed water, however, the gas giants can supply almost limitless quantities of vital elements. The four largest planets are composed mostly of hydrogen. The terrestrial planets, like Venus are about half oxygen by weight. If oxygen from Venus were combined with hydrogen from Jupiter, a mass of water equal to seven times that found in the rest of the solar system could be produced.[526]

The cost of resources from the gas giants will be very high when compared with resources from the moons and asteroids. Jupiter's gravity well is 190,000 kilometers deep.[527] Hauling water up from the bottom of that well will be quite a chore. By comparison, Callisto's gravity well is only 480 kilometers deep.

Very little material can be harvested from the Sun itself. Mining the "surface" isn't feasible, and the solar wind is very tenuous. Material is continually blowing off the Sun, but at its present rate of evaporation, the Sun would take 100 trillion years to disappear. The mass of the solar wind, compared with the mass of the Sun is negligible. Total material flux in the solar wind amounts to 2 X $10^{16}$ kg./yr. That is barely more than twice the mass of Phobos. Water vapor amounts to about one part in a thousand of the sun. So, even if all the water in the solar wind could be collected, it would amount to only 20 billion tons a year. Even over the course of a millennium, this would add up to only a fraction of the water contained in Neptune's smallest moon.

**The Oort Cloud**

The resources of the central solar system need not limit our horizons. There is another resource pool that dwarfs all others—the Oort cloud. Surrounding our solar system is an egg-shaped cloud of comets. The dimensions of this cloud strain the ligaments of the imagination. Its inner margins are forty times as far from the sun as the Earth, and its outer limits extend to 100,000 A.U.—almost half the distance to the next star.[528] The cloud contains as many as 100 trillion comets.[529] The aggregate mass of the Oort cloud could be as high as one percent of the mass of the Sun—ten times more than all the planets in the solar system combined.[530]

Recent data indicate that the Oort cloud might be sprinkled with thousands of giant comets, some as large as Pluto.[531] Pluto may actually be the innermost of these big comets. Each of these giant comets could contain as much mass as 400 million comets of average size.[532]

If such estimates are correct, the Oort cloud will provide humanity with an almost limitless reservoir of vital materials.[533] Comets are mostly water mixed with a rich melange of valuable organic molecules, and sprinkled with precious metals.

**Table 6.4**
**Compounds in Typical Comets[534]**

| Compound | % of Mass |
|---|---|
| Water - $H_2O$ | 57.50 |
| Carbon dioxide - $CO_2$ | 14.00 |
| Carbon monoxide - CO | 6.00 |
| Formaldehyde - HCHO | 5.75 |
| Hydrogen cyanide - 2HCN | 5.75 |
| Carbon Disulfide - $CS_2$ | 4.60 |
| Acetylene - $C_2H_2$ | 4.00 |

Balance of 2.4% in elemental metals and other materials.

**Table 6.5**
**Materials Yielded from One Comet**
**10 km. in Diameter**

| Element | Yield - MM Tons |
|---|---|
| Oxygen | 313,000 |
| Iron | 44,000 |
| Carbon | 39,000 |
| Silicon | 24,000 |
| Magnesium | 23,000 |
| Hydrogen | 21,000 |
| Nitrogen | 19,000 |
| Sulfur | 15,500 |
| Nickel | 3000 |
| Chromium | 3000 |
| Aluminum | 2000 |
| Copper | 26 |
| Lithium | .3 |

Like asteroids, comets are served up in easily chewed bits. Most comets are less than 15 kilometers in diameter, with an average diameter of three or four km.[535] A good sized comet—10 km. in diameter—would yield a wealth of elemental treasures.[536]

Importing these valuable materials from the Oort cloud is feasible, but time consuming. A comet 64 A.U. from the sun has an orbital velocity of just 3.75 kilometers per second (kps). If this velocity is neutralized, the comet will fall towards the sun on a parabolic trajectory.[537] It will take 40 years for the comet to reach the inner solar system, but the energy cost to send it on its way is relatively small. Using a mass driver to impart the required change in velocity would consume only about 6% of the comet's mass as propellant.[538] Comets further out in the cloud require even smaller changes in velocity to be tipped over the lip of the Sun's gravity well. A comet at a distance of 1000 A.U. moves around the sun with an orbital velocity of less than one kilometer per second. Stopping such a comet in its orbital track is relatively easy, but it then takes 2400 years for the comet to fall into the inner solar system.[539] Once the first shipment has arrived, and a continuous stream of comets are on their way, it won't matter very much how long they take. The incoming comets will form a sort of bucket brigade, bringing in fresh supplies of water and elements to resupply Solaria. A steady rain of new resources will fall into the inner solar system for millions of years. Eventually, the mass of imported cometary material could equal and then exceed that of the native planets.

**Dyson Shells**

There is nothing to prevent us from attaining the pinnacle of K2 status. If the giant gas planets cannot meet our demands, the cometary mass of the Oort cloud surely can. There is no shortage of energy, and it is easy to get. The sun is going to produce its power anyway. Right now it just shines into the vacuum in a vain attempt to raise the temperature of the universe. Whether we collect this free bounty or not is immaterial to Father Sol. He will go on radiating for the next five billion years in any case.

The sun's energy treasures are gratuitous. The sun is the central bank, the mint of the solar system. Beams of golden sunshine are the coin of the realm. This inexhaustible fountain of wealth gushes with riches beyond the dreams of men. The creative Force has endowed us with a true cornucopia; this brilliant horn of Amalthea has flowed for eons past and will flow still for thousands of millions of years to come. We need but dip our silver cups in this unending river of gold.

The average American uses around 100 thousand kilowatt hours of power per year from all sources.[540] This is wanton profligacy, even for a planet-bound Earthling of the late 20th Century. Space dwelling Millennialists will not be such wastrels. An advanced human civilization will be energy efficient. For a space colonist living in an ecosphere it will be virtually impossible to use as much energy as a modern American. Inside ecospheres, the climate is controlled by natural heat balance; the sun provides most needed light; transportation is virtually frictionless and for the most part over very short distances.

Even so, let us assume that each inhabitant of Solaria requires the same amount of energy as a typical American. At this rate of consumption, the total power demands of the five billion billion inhabitants of Solaria would be around $5 \times 10^{23}$ kwh/yr.—one seven-thousandth of the sun's supply.[541]

Getting energy from the sun is one of the easiest things we will do. In addition to the light falling on our habitats, it will at some point become necessary to tap the considerable fraction of solar energy that is not beamed in our direction. This would at first seem to be a problem. Solaria's cloud of ecospheres is concentrated along the plane of the ecliptic, and extends only 20° above and below the plane. If we erect solar collectors in close orbit around the sun, they must intercept some of the light that would otherwise illuminate the cloud.[542] To achieve K2 status, we must eventually harvest all available sunlight, without obstructing light in the plane of the ecliptic. Therefore, light has to be collected from the areas of the sun above and below the equatorial band.

Fortunately, this problem is amenable to a simple solution. Countering the gravitational attraction of most celestial bodies, requires maintaining an orbit around them. This is not the case with the Sun, however. The Sun is uniquely a source of energy, and that energy can be used to counteract its gravitational force. Sunlight actually exerts a tangible pressure on objects it shines on. (This effect should not be confused with the solar wind, which is thousands of times weaker than the pressure of the sunlight itself.) The pressure of sunlight can keep solar collectors aloft, without having to orbit them around the Sun. A thin dark sheet of photovoltaic material can float motionless in space, held up by beams of light.[543]

Some fraction of the solar energy impinging on the collector surface can be harvested as electricity. Present technologies have achieved conversion efficiencies of 25%. By the time large-scale collectors are actually deployed, perhaps half of the energy in sunlight could be harvested.

Electricity can be concentrated and converted into microwaves or laser beams for broadcast to regions where the power is needed. Depending on the density of the collector membrane, some given surface area will be needed to support the weight of energy conversion and transmission facilities. Each of these units could form a self-contained hexagonal power station capable of hovering in space indefinitely. Any number of these power stations could be suspended in space inside the orbit of Mercury.[544] By multiplying the number of stations, any desired amount of solar energy can be captured, up to 95% of the total output of the sun. Eventually, the interlinked collectors would form gigantic hemispheric shells hovering over the sun.

***Fig. 6.3 - Solar collectors.*** *Huge solar power collectors will float above the sun, supported by the pressure of sunlight. The energy they collect will be converted to microwaves or lasers and beamed to outlying parts of the solar system.*

The only limitation to the extent of the collectors is the necessity to leave light in the plane of the ecliptic unobscured.[545] Any obstruction of light illuminating the ecliptic could have serious adverse effects on the ecospheres of Earth and Mars. Therefore, the collector shells won't extend below the solar poles. The collectors will form a spherical shell around the sun with a gap along the equator equal to one solar diameter. Only 5% of total solar energy will pass out through this gap. The balance of the sun's energy will be collected on the inner surfaces of the shells.

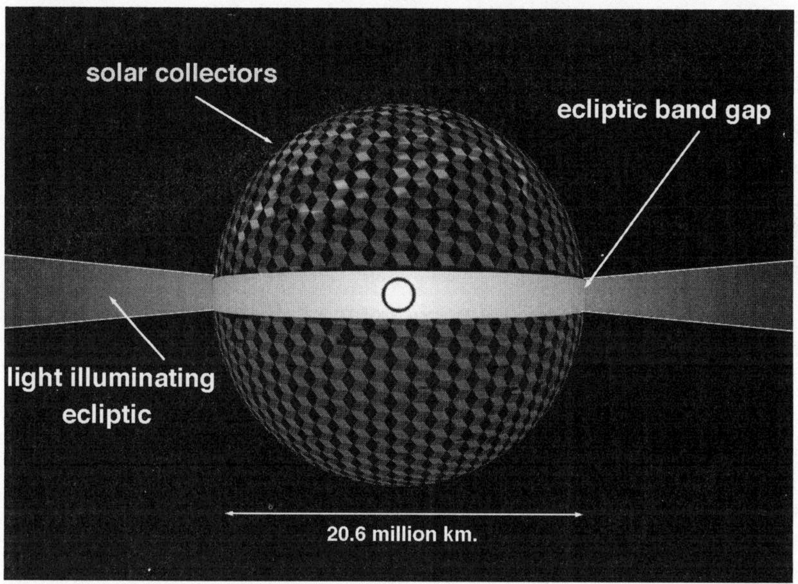


***Fig. 6.4 - Solar collector shells*** *will enclose most of the space around the sun. A gap along the equator will allow the sun to illuminate the planets as it always has. The shells will be invisible from planetary surfaces.*

There is enough mass in the planet Mercury to satisfy all our needs for solar collector and power plant material.[546] When the solar system was first formed, elements were conveniently sorted out of the primordial accretion disk. Lighter compounds, like water, condensed in the outer regions, while heavier elements, like metals, condensed in the inner regions. As a result of this beneficiation, Mercury is remarkably dense, consisting mostly of iron and other heavy metals. These materials have been conveniently concentrated near the sun, making construction of the collector shells much more economical.

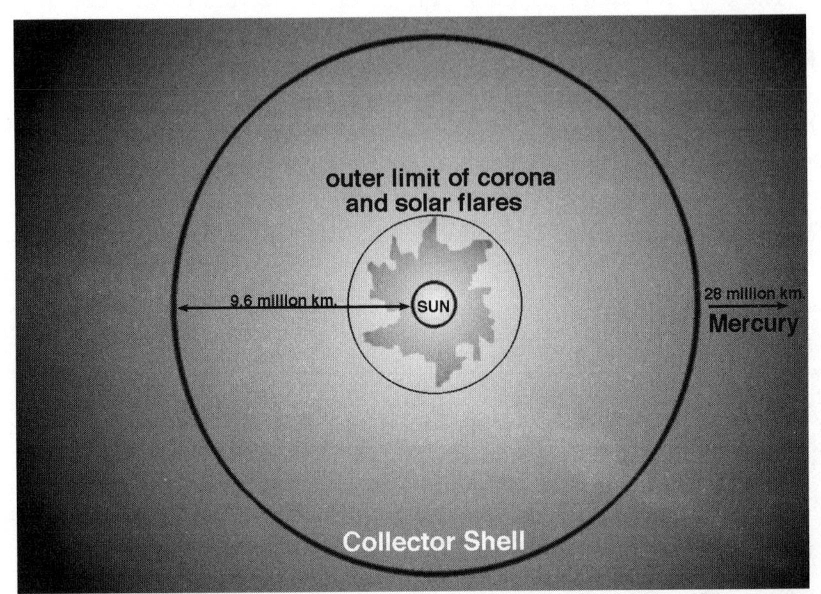


***Fig. 6.5 - Plan view of Solar collector shell to scale.*** *The shell's diameter is twelve times that of the sun, putting the collectors well outside the range of even the largest solar flares.*

The solar collector shell is a modified version of the 'Dyson Sphere'. This is a term usually associated with Freeman Dyson's concept for the ultimate form of a stellar civilization. Dyson reasoned that as a highly advanced civilization grew to require the entire power output of its home star, it would eventually surround its sun with orbiting structures. These structures would absorb the star's energy, downgrading it to the infra-red part of the spectrum. Dyson theorized that this shift in a star's spectral signature might reveal extra-terrestrial civilizations.[547] Dyson's concept is one of the most dramatic ideas about the evolution of civilization ever to emerge. Not surprisingly, it has caught the interest of thinkers ranging from science-fiction writers to serious scientists.

The solar collector shell will not be inhabited. The collectors are gossamer affairs of hot thin membranes, ill suited to habitation, though it would theoretically be possible. Asgard style habitats will be concentrated in a band along the ecliptic, probably not extending much beyond ten or twenty degrees north and south of the solar equator. The multitude of free-floating habitats will form a sparkling smoke-ring shaped cloud around the solar system. (See Plate No. 16) Such a swarm of habitats is what Dyson had in mind when he postulated that stars might be red-shifted by advanced extra-terrestrial civilizations. Dyson was not, however, specific about the numbers, types, or locations of the habitats, leaving the aliens to work that out for themselves. Rather than a "Dyson sphere" then, Dyson's original concept might more aptly be described as a "Dyson cloud".

**Kings of Infinite Space**[548]

In tonight's sky, if it is very clear and very dark, you can see about ten thousand stars. Imagine seeing billions and billions of new stars, many brighter than Venus. In Solaria, the night sky will glisten with countless points of light. A golden arc will span the sky in a frosted glowing band, like a new Milky Way.[549]

If we continued to expand our population at the historic 2% rate, we could reach a maximum K2 population of 100,000 billion billion people half way through the Fourth Millennium—1500 years from now.[550] That may happen, but it is far more likely that at some point we will feel we have reached a comfortable maximum population. What that level is we can't really say. It's like a Kalahari Bushman trying to speculate about how many people could live on Manhattan Island. But we can suppose that at some level, well within the theoretical limit, our species will content itself with some odd number.

Such extrapolations are (thankfully!) outside the scope of this book. We have only to concern ourselves here with what the next thousand years hold in store for us. The bottom line is that the Solar system has ample resources to support our continued growth through the next Millennium. What happens after that is not my department.

When thinking about populations numbering in the billions of billions, it is important to make a quantum leap of the imagination. Contemplation of such hordes can bring to mind images of steaming crowds, packed cheek-by-jowl like human cattle. But the Solarian world of the future could not be more different. Solaria is a world of wide open spaces; more like Montana than Manhattan.

If you can't imagine the size of space, then go out tonight and just look at it. There it is, spread before you like an infinite black velvet blanket. Ignore the stars for now, they are far far away, both in space and time. Look only at those few lonely objects which populate the void inside our own solar system. There is the Moon, the only object other than the sun which appears to possess real size; shining Venus is just going down on the western horizon; there is Mars, faintly glowing red; just coming up on the opposite horizon is mighty Jupiter, barely a speck at this distance; all in all, just a few glints of light to alleviate the emptiness. Jupiter is so tiny you can't even distinguish a disk of any size; it is just a geometric pinpoint. Yet, into that point you could dump a thousand Earths and more.[551] How many more Earths could be piled into the volumes surrounding that dot? We will never crowd such vistas, no matter what our numbers.

A thousand years from now, our population will be colossal, but the spaces we occupy will be nothing less than astronomical. Billions of billions are hard numbers to imagine, and can be scary. Just remember that these multitudes will be scattered through the reaches of space, from one side of the solar system to the other. The asteroid belt occupies a toroidal volume equal to the spherical volume of space inside the orbit of Mars—$5 \times 10^{25}$ cubic kilometers.[552]

If the average population of each ecosphere in Solaria is 100,000, the number of habitats in the year 2500 will total fifty trillion. Viewed from outside the asteroid belt, the swarm of habitats will look like a thick golden mist. Inside, the cloud is mostly empty space. The reality is startling in its expansiveness. With fifty trillion ecospheres spread throughout the area of the asteroid belt, there will be no more than one in every trillion cubic kilometers. On average, ecospheres will be separated from each other by twelve thousand kilometers. Our teeming trillions of ecospheres will barely amount to a sprinkling of powdered sugar on a cake the size of the Solar system.

That the Solar system as a whole is home to billions of billions of other people will be of only academic interest to individual Solarians. The numbers form aggregate statistics of megalithic proportions, but the people live in individual communities of various sizes. Some people will opt to live on metropolitan asteroids with large populations; others will prefer small ecospheres with closely knit communities. The gross population of the Solar system will be no more relevant to the residents of a small asteroid than the population of the United States is to the people living in Hicksville, Ohio.[553] There are advantages in being a member of a larger society, but you don't have to rub shoulders with the masses unless you want to.

Even after the solar system is heavily populated, there will still be limitless horizons for restless individualists. Outer space really is "big sky country". No one living in the Solar system ever need feel overcrowded or fenced in. Beyond the asteroid belt are many more worlds. Millions of Trojan asteroids share Jupiter's orbit, numerous moons circle the remote gas giants, Saturn, Uranus, and Neptune. And beyond these frontiers are the nearly infinite spaces of the Oort comet cloud.

**Archangel Gabriel and the Host of Heaven**

In the coming Millennium, we will catapult to a new level of physical existence. Our very nature as a species and as a universal phenomenon will be transfigured. In the process of transforming the inanimate slag of the solar system into living matter, we will also be metamorphosing ourselves.

To form a collective entity of a higher order, a minimum threshold must be exceeded. For individual sub-units, the critical threshold appears to be around 100 billion:[554]

• A hundred billion atoms can organize to form an individual cell.
• A hundred billion cells can organize to form an individual brain.
• A hundred billion human minds can organize to form...

Something Wonderful!

When we succeed in creating organized structures of human beings numbering more than 100 billion, we are likely to experience a quantum evolutionary leap. A hundred billion people joined together in a sophisticated telecommunications network are apt to become something more than the mere sum of their parts. You and I are something more than mere aggregations of brain cells. So too will networks of a hundred billion people be something more than mere mobs. What exactly they will be is difficult to say—like a neuron trying to define Einstein.

These transcendent beings of aggregate consciousness, must be formed of people within close proximity. Despite the best efforts of our feverish imaginations, the universe does impose certain limits. The emperor of all limits is the speed of light. If the entire solar system were to act as a common mind, it would be a singularly ponderous one. How nimble a brain could it be with an average distance between neurons of ten light minutes?

By the end of the next Millennium, there will be 100 billion people within each million trillion cubic kilometers of the Dyson cloud. At this density, the neurons in each macro being will be at most four light seconds apart.[555] (See Plate No. 15.) As the population in the solar system increases, the density of neurons will rise, and the "intelligence" of the macro beings will grow. By the end of the Third Millennium, there may be fifty million such transcendent entities. Of course, they would not be discreet individuals in the sense we are used to; there would be no telling where one such being ended and another began.

To ascend to the next level will presumably require 100 billion of these macro-beings, which would require ten thousand billion billion individual persons. At full-blown K2 status, there will be one hundred thousand billion billion of us—enough people to form a thousand billion macro-beings, and presumably ten mega-macro entities. I try to imagine what consciousness on that level might be like, but it gives me a headache.

As discreet individuals, we can not directly share the consciousness of such entities, but we will be indirectly aware of their existence. Waves of unexplained synchrony will sweep rhythmically through the networks, washing over us like an ocean of harmony. The 'unseen hand' will become an active force, evident in the everyday workings of Solaria. The sense of being a part of something greater than oneself will be compellingly real. An abiding sense of identity will permeate the whole of human consciousness. Though we won't be able to invite them to dinner or engage them in a game of chess, the macro beings will nevertheless exist, as real, if intangible, parts of our universe. They will possess a species of higher consciousness; they will, in a sense, be demigods.

**Nomad's Land**

Human expansion into space will follow a natural progression, from our roots on Earth, out to our spreading branches among the stars. The cometary Oort cloud marks the furthest frontier of the sun's empire. Before the end of the New Millennium we will have begun to settle even this remote hinterland. At first, autonomous robotic ships will be sent out to capture and return comets. Eventually, human immigrants will be drawn to the splendid isolation of deep space. A small group of people could fly out on a long transfer orbit from one of the outposts in the outer solar system—a mining colony on Pluto for example. When they reached their destination comet they could set up a self-sustaining colony, very much in the mold of settlers in the asteroid belt. This cometary community would exist in almost complete isolation from the home star system. The comet colonists would be virtually self sufficient, engaging in limited trade with other cometary colonies. These people would be true nomads of deep space.[556] (See Plate No. 17.)

Production of energy is a different proposition for the cometary colonists. Sunlight is very dim in the outer regions of the solar system. At a distance of 55 A.U., there is only one three thousandth as much sunlight as there is in Earth orbit. The comet colony will get its power directly from the comet itself. A comet 10 kilometers in diameter will contain over six million tons of Deuterium.[557] With this fuel supply, the colonists need never want for energy. Every kilogram of Deuterium can produce a hundred million kilowatt hours of power. The Deuterium supplies on the comet could sustain a community of one thousand colonists for 70 million years.[558]

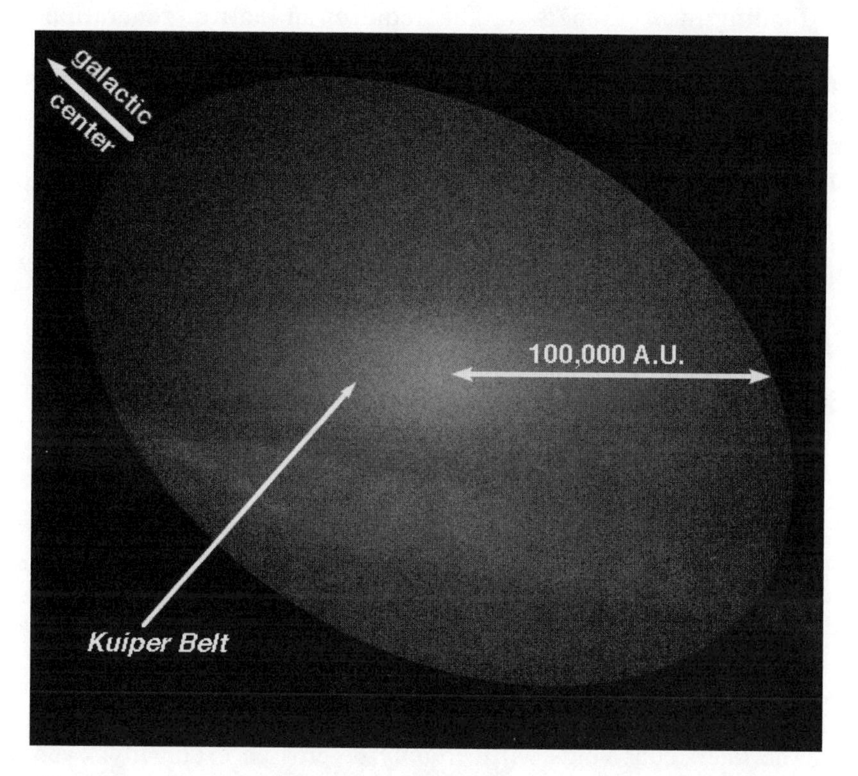


***Fig. 6.6 - Oort Cloud.*** *Our solar system is surrounded by a vast swarm of comets. Because the plane of the ecliptic is tilted at 60º to the plane of the galaxy, gravity pulls the cloud into an ovoid shape. At this scale, the smallest dot you can perceive is bigger than our entire solar system.*

The cometary colonists could gradually populate the whole of the Oort cloud. Comets in the cloud are separated from each other by an average distance of 36 A.U.[559] Each colony will therefore occupy a space equivalent to that of the Solar system inside the orbit of Uranus. The limited resource base available to a comet colony necessitates that its population remain relatively stable.

Since life spans are apt to be very long, emigration will be essential. Each generation, young colonists will set out for a new comet located further out in the cloud. The colonists would spread through the cloud at the sedate pace of one and a half A.U. per year, migrating in jumps of 40 or 50 A.U. per generation. At this rate, the expanding wave of colonists would reach the outer fringes of the Oort cloud in 66,000 years.

Finding new comets might require an entire generation. Locating objects the size of small islands in gulfs of space as wide as whole solar systems will not be easy. Optical searches are a possibility, but looking for tiny objects illuminated only by star light will require expansive telescopic reflectors. The colonists might prefer to operate powerful radars to probe through deep space in search of comets. The flickering beams of billions of these bright radars will be one of the characteristic signatures of a living galaxy.

Beyond the limits of the Oort cloud is a sprinkling of interstellar comets, which have been thrown completely out of our own and other solar systems. These rootless rogues wander the wide gulfs between the stars. Such drifters are widely dispersed, separated from each other by 6000 A.U.—100 times the diameter of the solar system.

These galactic comets can serve as stepping stones between stars. By settling these intermediate comets, colonists could migrate across interstellar space. Eventually, succeeding generations would reach the outer limits of the Oort cloud surrounding the next star. Crossing the gap between stars would mean accelerating the speed of the migration to 240 A.U. per year. At that rate, it would take the cometary colonists only 300 years to cross the gap between clouds.[560]

The interstellar voids will certainly be crossed eventually; even by cometary colonists who stay home. Comets are routinely expelled from the Oort cloud by gravitational perturbations.[561] Some cometary colonies would inevitably be flung into interstellar space. Permanent settlements on these interstellar comets will carry their cargoes of life with them as they wander the spaces between the stars. It might take millions of years for a wandering colony to enter the domain of another star system, but when it did, it would carry humans and other life forms with it. The eventual settlement of the galaxy is therefore assured, if through no other means than cometary osmosis.[562]

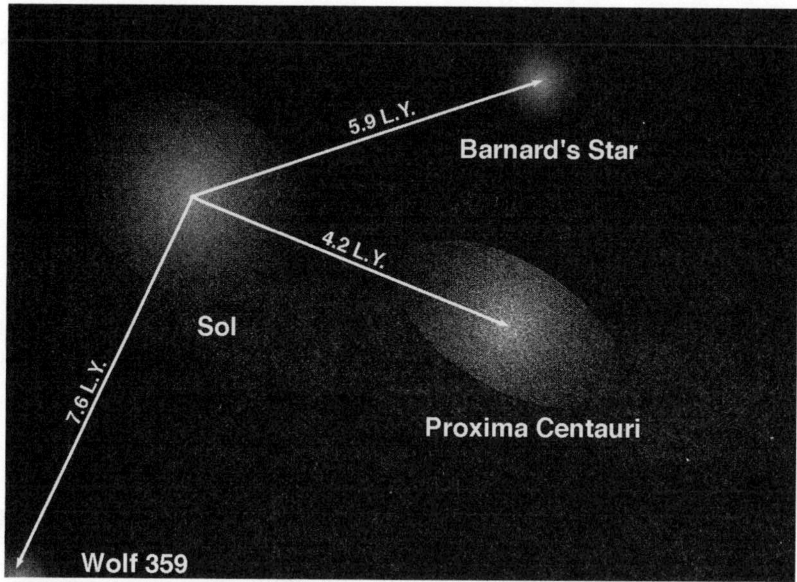


***Fig. 6.7 - Interstellar comets.*** *It is believed that every star system in the galaxy is surrounded by its own cometary cloud. Between these clouds is a smattering of interstellar comets. These comets could provide stepping-stones to the stars.*

If Life were to slowly percolate through space in this fashion, it might take six billion years to permeate the galaxy. It is unlikely to take so long, however. When colonists begin to reach comets held in the Oort cloud of the next star system, they might not be very surprised to find people already there. These will not be aliens; they will be the descendants of star colonists who traveled to the nearby stars tens of thousands of years before. The migration through the cometary cloud in the new solar system will follow the same progression as that in the original solar system—just 500 or a thousand years behind. When this meeting occurs—"Doctor Livingstone I presume?"—it will mark a major milestone for Life. The void between two stars will be filled with a cloud of blue-green sparks. Even the depths of space will have been brought to Life.

**Live From Solaria**

The living solar system will present an awesome sight. It will fill most of the sky like a miniature galaxy, a vast band of light encircling the sun. The close side of the band will glow with the unmistakable phosphorescent green signature of Life. The inner arc of the band, far away on the other side of the sun, will glisten like a cloud of shimmering gold dust. (See Plate 16) If one tunes to the right frequency, one will hear the pulsing harmonic chorus of a billion billion voices—music of the spheres.

Our solar system will be transformed. It will be as different from the sterile void as the Earth is from the Moon. Just as Life transformed a molten ball of rock into Gaia, the living planet; it will soon transform a frozen disk of orbiting rubble into Solaria, the living solar system. Having performed this act of transcendent alchemy, Life will be ready, at last, to leave its home star and carry the sacred process out into the galaxy.

# CHAPTER 7

# GALACTIA

***Man is but the place where I stand,***
***and the prospect hence is infinite.***
**Thoreau**

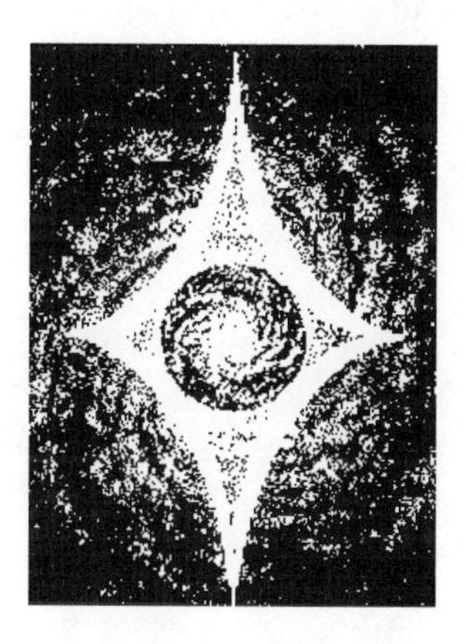

As far as we know or can tell, the universe—with the sole exception of our own magical planet—is entirely devoid of life. Life has utterly transformed the Earth. The atmosphere is an endoplasmic membrane of a unicellular photosynthetic organism we call Gaia. The Earth could not be mistaken for a dead planet any more than a live oak could be mistaken for a pillar of salt. A thousand years from now, Solaria will glitter and pulse with Life on a cosmic scale. Our solar system will be as obviously alive as the Earth. A million years from now the same transformation will have swept through the entire galaxy. The Milky Way will be completely transmogrified.

This is sure to happen, for it is the nature of Life to expand her empire if she can—and she can. When we look out on the heavens, we see not the slightest inkling that another living thing inhabits those frozen spaces. If Life is a common phenomenon in this universe, then the star clouds of our galaxy should shout its presence. But the sky is silent. There is no trace of Life making its metamorphic presence felt on a galactic scale. The roaring silence of deep space only confirms our worst fears—we are alone.

Until we learn otherwise, we must assume that we are the solitary guardians of Life's only spark. As far as we know, the fate of the cosmos is in our hands. Ours is an awesome and frightening responsibility.

We exist as a species for one reason: to bring Life to the universe. A living universe is the ultimate fulfillment of our Cosmic destiny. Gaia is a living planet, the seed of Solaria—a living solar system; Solaria will be the seed of Galactia—a living galaxy; Galactia will, in turn, be the seed of Cosmia—a living universe.

It is our destiny to ignite the chain reaction that enlivens the cosmos. Because of us, Life will erupt into the galaxy. Like windblown firebrands, we will carry the spark of Life from star to star. Wherever we go, the green conflagration of animate matter will set the skies ablaze!

**A Resonance of Emerald**[563]

A million years from now, sure as God made little green apples, the human race will have spread across this galaxy. Dyson clouds will envelop every stable star between the Orion Arm and the galactic nucleus. A hundred billion living solar systems will be linked together, forming a galactic culture of fantastic power and complexity. Our once puny race will have attained Kardashev Level Three. Human numbers, ten quadrillion quintillion ($10^{34}$), will exceed the limits of comprehension. These people, in their zillions, will have utterly transformed the substance of the galaxy. A million years from now, the Milky Way will be as different from its present incarnation as a rain-forest is from an ice-cap. Our unstoppable expansion into the galaxy will transform this lifeless congeries of fusing hydrogen into a vibrant metabolic organism.

Our species will survive its present gestation. And once we succeed in hatching out of our gravitational shell, we will go on to populate this entire galaxy. We are going to burst up out of this planet like a bean-stalk from one of Jack's magic beans. Once our planetary seed has sprouted, and our first tendrils have twined themselves about the trellis of the Moon, nothing can stop us. We—not just We as humans, but We as all Life forms—will assimilate the raw materials of our solar system and convert these base elements into living matter. Once we have grown a mighty forest around our own Sun, we will cast forth our seeds upon the celestial winds. Our spores will cross the gulfs of space and come to rest in the warm, well-lighted places around neighboring stars. There, they will take root in the rich soil of comets and planetesimals. Warmed by the rays of these alien suns and nourished by abundant carbonaceous fertilizers, these seeds will grow into new groves of Life. These celestial orchards will, in their turn, bear fruit and scatter seeds of their own. Spreading thus from star to star, the pollen of Life will permeate the galaxy.

The galactic metamorphosis will be unmistakable. For one thing, the Milky Way will turn green. Dyson clouds at high densities will red-shift the color of the stars they surround. This happens because the light is absorbed and reemitted at longer frequencies as a consequence of the Second Law of thermodynamics. The Second Law demands that energy used to create higher levels of order must be degraded in the process. In Dyson clouds composed of ecospheres, that degradation will carry an unmistakable signature. The transparent ecospheres, with their algae coils and lush gardens of floating foliage, will act as selective light filters. The very high chlorophyll content of the ecospheres in Dyson clouds will give them a greenish tinge when viewed en masse. Light passing through such a cloud will come out green. Seen from a distance, a star surrounded by a Dyson cloud will have a distinct aquamarine hue.[564]

There are yellow stars, and orange stars, red, white, and blue stars, even brown dwarfs, and black holes; but there are no green stars.[565] A green star in the night sky, can mean only one thing—a living solar system. It will be as obvious to the eye as the blue -green glow of the living Earth. As we suffuse through the galaxy, a growing patch of green stars will expand with us, like a widening oasis, pushing back the white sands of the galactic desert. (See Plate No. 19.) This will be, quite literally, what Freeman Dyson has described as "The Greening of the Galaxy."[566] A million years from now, our descendants will look up at the night sky and see a diffuse band of celadon green arcing the vault of heaven—the 'Mossy Way', no doubt.

**Teen Titans**

To go among the stars, our grasp must exceed the reach of interstellar space. To cross the gulfs between stars requires of us the stature of Titans. It is an undertaking on the scale of gods. Kindergartners do not climb K2, nor should infant civilizations attempt expeditions to the stars. Interstellar flight is the province of solar civilizations. Our own solar system offers us tremendous opportunities to expand into space and convert inanimate matter into living substance. We must carry that process to fruition, if not completion, before we will be ready or able to set out for the distant stars. By the middle of the next millennium, our civilization will have attained the necessary maturity.

Conquering the galaxy is going to require powers on an incredible scale. The source of all power is not really energy. The universe is awash in energy, but powerless in the grip of chaos. The real power in the universe lies in the minds and hands of human beings. To accomplish the enormous undertakings before us will require a great multitude.

Therefore, breaking out of Earth's planetary shell, is not only possible, it is essential. Only a civilization on the scale of Solaria can contemplate travel to the stars. Restricting humanity to our home planet would be like trying to keep a brontosaurus in its egg. Even if it could survive such stultification, the magnificent beast would never achieve a fraction of its potential. To fulfill our stellar destiny, we must hatch out and leave the Earth. Only then can we mature into the colossi we are meant to be, and stride across the star fields with the stature of the Sons of Jove.

**Star Flight**

The challenge of traveling between the stars is titanic. The distances are so immense as to be almost incomprehensible. The nearest star is over four light years away. The light year is a measure of distance that everyone uses flippantly enough, but few stop to think about what it means. Light travels 300,000 kilometers every second. If a loop of optical fiber was strung around the Earth and a pulse of light fired through it, one could see the light beam flash by seven times in a single second. That is faster than you can blink. Light can cross the plains of Kansas, bridge the wide Pacific, hurtle across Eurasia, skip across the Atlantic, and be back in Topeka in the flick of an eyelash. That is fast. Yet, light takes four long years to cross to the nearest star. How many blinks of an eye are there in four years? How many gulfs of space as wide as Kansas are there in such a distance? Have you ever driven across Kansas? It's endless.

Trying to derive a way to understand interstellar distances is almost hopeless. People wonder about infinity and say they just can't imagine it. Forget infinity; I defy anyone to imagine just the distance to Alpha Centauri. It may be helpful to imagine the distances in terms of relative sizes: If we were the size of bacteria, the Earth, on the same scale, would be about the size of a house; the Sun, a ball 1.4 kilometers across, would be 150 kilometers away; the next star would be over half the distance to Mars. Still incomprehensible. If we were the size of atoms, the Earth would be the size of a grain of sand; the sun, 15 meters away, would be the size of a cantaloupe; and the nearest star would be 4,000 kilometers away. So to get some conception of the distances involved, imagine the Earth as a grain of sand in New York City, and the next star system as a melon in Los Angeles.

Spanning such voids will require either prodigious quantities of energy or very long stretches of time. To fly to Alpha Centauri in a spacecraft like the one that carried the Apollo astronauts to the Moon would take almost a million years—better take along plenty of Tang. Compared to our interplanetary fast fusion shuttles, though, Apollo is a horse and buggy. Our fusion powered clippers will be capable of speeds in excess of 300 kps. Within the Solar system, a few hundred kps is blisteringly fast. At such speeds, one can travel between Earth and Mars over the weekend. What passes for high speed within the Solar system, however, is an imperceptible crawl in interstellar space. At 300 kps, traveling to the nearest star system would take 4,200 years.

Many proposals for star flight accept such protracted travel times. Usually, such schemes involve sending out gigantic self-contained "world ships", equipped to support whole colonies in interstellar space for many millennia. Supposedly, generation after generation would live in these micro worlds as they crept along toward the next star.

Even if the problems attendant to such lengthy voyages were solved, travelers on world-ships would suffer certain frustration. The sure fate of the inhabitants of a world-ship would be to watch by telescope as their destination star was enveloped in a Dyson cloud. Travelers who started centuries later would nonetheless arrive at the stars thousands of years ahead of the world ships. World-ship travelers would finally arrive at their destination, only to find a mature K2 civilization already in place. The destination star system would have been long since colonized by people who had set out centuries later with better technology in faster transports. The world-ship travelers could then only wonder at the monumental folly of their forbearers.

Star flight at slow speed is no answer. Rather than sending out primitive world-ships, which take thousands of years to get anywhere, it is better to spend a few extra centuries developing technology, and then build fast ships which can cross the void in at most a few decades. Never set out for the stars if there is a reasonable chance that the people you leave behind are going to beat you there. If your maximum speed is only 1% of the speed of light—3000 kps—you may as well stay home. At that rate it will take over 400 years to get anywhere. During that time, the huge and thriving civilization you left behind is bound to come up with something better. You will find yourself being greeted at your destination by the descendants of pioneers who departed centuries behind you, but arrived centuries ahead of you.

The practical lower limit on speeds for interstellar travel is 30,000 kps, 10% of the speed of light—.1*c*. If you can't travel at least this fast, don't bother going. At .1*c*, a ship will reach the next star system in 43 years. During four decades it is unlikely that technology will take an unexpected quantum leap. Even if the scientists back home develop warp drive while you're gone, they can only beat you by a few decades.

By Solar system standards, .1*c* is an outrageous speed. At 30,000 kps, one could travel from the Earth to Mars in under an hour. Nevertheless, it is a velocity that can be readily, if not easily, attained. At a constant acceleration of one *g*, a speed of 30,000 kps can be reached in just over a month.

Accelerating to this velocity with conventional rockets, however, is quite impossible. Even super fuels like monatomic hydrogen and triplet helium have specific impulses of only 15 - 30 kps. A step-rocket capable of achieving a thousand times its own exhaust velocity would outweigh the entire universe.[567]

Fusion rockets, although theoretically capable of taking us to the stars, are probably impractical in this application. In a fusion reaction, atomic nuclei are expelled at up to 10,000 kps.[568] If a system can be developed which directs all of the reacting nuclei in the same direction, then a rocket exhaust with that velocity can be produced.[569] To reach 10% of the speed of light, a ship would need to achieve three times its own exhaust velocity. A rocket carrying a 10,000 ton payload in a 40,000 ton structure would require about a million tons of fuel.

Such a ship would be of immense proportions, but it would not be outside the scope of even today's engineering capabilities. Unfortunately, this fusion rocket would have one very severe shortcoming: it couldn't stop. If the rocket must stop at the end of its journey—a decided advantage to star colonists—then it must also carry enough fuel to decelerate. This requires that the mass ratio be, not doubled, but squared. To deliver ten thousand tons to the next star system would require twenty million tons of fuel. A fusion rocket would have to leave our solar system with two thousand tons of deuterium for every ton of payload it carried.

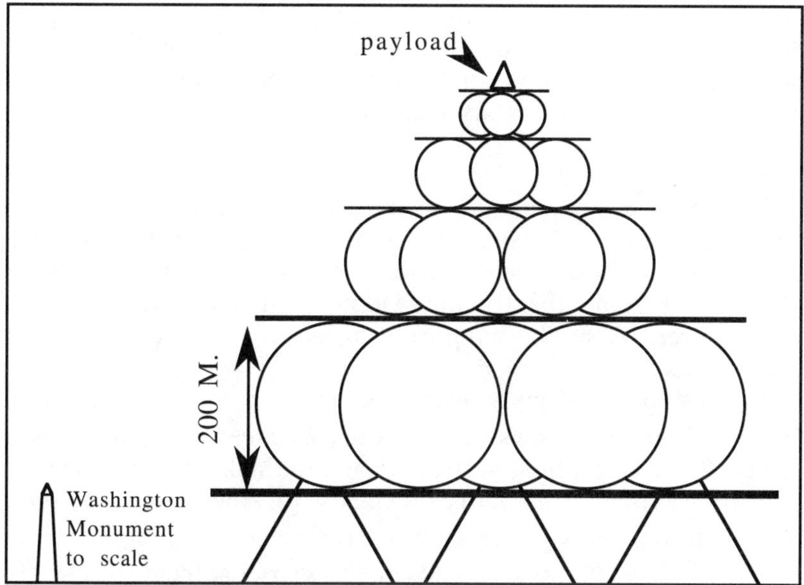


***Fig. 7.1 - Fusion rocket.*** *Interstellar rockets fueled by deuterium would be enormous. Such ungainly monsters will probably never be built. By the time we are ready for star travel, power sources more intense than fusion will be available.*

There is no shortage of innovative proposals designed to get around the colossal problems of interstellar flight: nuclear pulse propulsion, beamed energy drives, solar sail starships, fusion ramjets, interstellar ion scoops, and even gravitational wormholes have all been invoked as means of star travel.[570] All of these ideas are fascinating and make for interesting speculations, but they are for the most part only exercises. None of these elaborate and ingenious schemes will ever be put into effect any more than 40 million ton world-ships will ever be embarked on thousand year journeys.[571]

It may be fruitless to try to anticipate the technology of the year 2500 A.D. There are, however, two things we can be pretty sure of: First, we can safely bet that when such journeys begin, the people making them will not depend on 20th Century technology. To presume so is akin to extrapolating an SST from Columbus' caravels.[572] Second, we can be reasonably certain, that whatever force drives our star ships, it will operate within the limits of physical laws. (We may not have discovered all the statutes yet, but the laws have long since been written.) Therefore, we should try to anticipate a technology which, though unreachable with today's tools, is nevertheless allowed by our present understanding of physics. If such a method is conceptually available at this stage of the game, then we can be confident that the Solarian engineers will be able to build the necessary machinery.

**Anti-Matter**

We need to develop a power source as awesome as the chasms we must cross. Interstellar space is the ultimate frontier; to conquer it we need the ultimate fuel—anti-matter. We opened up the solar system by making the technical jump from chemical power to fusion power. A similar jump from fusion to anti-matter will open up the galaxy.

As Einstein made clear, matter is simply a form of condensed energy. Everyone knows the equation, $E=mc^2$; it is practically a cliché. But, as with light years, few stop to think what the equation is really saying. The letter $c$ stands for the speed of light, 299,792.8 kilometers per second, $3 \times 10^8$ meters per second. If we square $c$, that is, multiply it by itself, we derive the fantastic quantity of $9 \times 10^{16}$ meters per second. This huge number is then multiplied by the mass, $m$, in kilograms. The result is an expression of energy in joules. Every kilogram of matter contains 90 million billion joules of energy.[573] That is enough power to launch 9000 Saturn V rockets to the Moon.[574] The mass in a penny contains more energy than four atom bombs.[575] If we can unleash the titanic forces locked inside every substance from plutonium to banana peels, we can go to the stars.

The total conversion of matter to energy requires no breakthrough in our understanding of the universe, it just requires some anti-matter. Anti-matter, as its name implies, is an opposite form of matter. In normal matter, the proton is positively charged, and the electron is negative; in anti-matter this is reversed. When particles of matter and anti-matter meet they instantly annihilate each other. The combined masses are converted into pure energy—mostly in the form of gamma rays.

Supplied with anti-matter, we can fuel star ships capable of approaching the speed of light. Unfortunately, we can't go prospecting, hoping to find an anti-matter deposit. Any anti-matter that ever existed was annihilated long ago.[576] If we want anti-matter, we will have to manufacture it ourselves.

Anti-matter is not a theoretical dream material. It is routinely produced using available technology, albeit in minuscule quantities. If a proton is accelerated to a velocity so high that its kinetic energy matches its rest energy—.93 GeV (Giga-electron Volts)—then it is capable of producing an anti-proton when it collides with a target.[577] If a high-energy beam of protons is blasted into a block of dense material, then about one in a million to one in a billion proton collisions will form an anti-proton. Gigavolt electron beams have already been produced. Beams with energies on the order of 10 Gev are on the drawing boards for Star Wars applications.

The low efficiency of the random collision process makes anti-matter unbelievably expensive. To produce a single kilogram of anti-matter by bombarding a target with electron beams would require the entire electrical output of the United States for a million years. With current production methods, anti-matter costs 300 billion dollars per milligram. (It's too bad we can't go panning for anti matter; the mother-lode could fit on the head of a pin.)

If anti-matter is ever to become a viable starship fuel, we must develop a production process that is much more efficient and a lot cheaper. Einstein's equation, $E=mc^2$, means matter can be converted into energy; it also means energy can be converted into matter. Both matter and anti-matter can be forged out of raw space, if enough energy is focused on one spot. Producing electron-positron pairs requires an electrical field strength of $2.4 \times 10^{18}$ volts per square meter.[578] By concentrating many photons in a small area, this field strength can be created with lasers. Producing a field of this intensity requires laser power on the order of $1.5 \times 10^{34}$ watts per square meter.[579]

Creating such a concentration of power will not be easy. An energy level of $10^{34}$ watts is equivalent to 26 million times the power of the Sun. Achieving such energy densities is not, however, impossible. The energy field must be maintained for only billionths of a second, and the power can be focused into areas not much larger than atomic nuclei.[580]

Twenty five years ago, lasers operating in the visible portion of the spectrum had already attained peak power densities of $10^{15}$ watts/cm$^2$. Laser power rises as a function of frequency, which is itself determined by the laser's wavelength. The shorter the wavelength, the higher the frequency, and the greater the corresponding power of the laser. Today's most powerful lasers operate in the ultraviolet portion of the spectrum, at wavelengths of 1000 ångstroms. Soon, lasers operating in the X-ray region of the spectrum will make their advent, with wavelengths as short as one ångstrom. Each time the wavelength is reduced by a factor of ten, the power output of a laser makes a quantum leap. X-ray lasers, for example, will be able to pack a thousand times more power into a square centimeter than ultraviolet lasers.

When Solaria's engineers undertake to build anti-matter production machinery, they will be working with gamma ray and even cosmic ray lasers. At wavelengths of .0001 ångstroms and less, cosmic ray lasers will pack ten thousand to a million times the energy of x-ray lasers. The maximum theoretical energy density of an x-ray laser is around 4 $X10^{23}$ volts per square meter.[581] When cosmic ray lasers are developed, they will create power densities of 4 $X10^{29}$ volts per square meter.

By building large arrays of such lasers, Solarian engineers will be able to create the field densities necessary for the production of anti-matter. These laser arrays will have to be enormous and will suck up prodigious amounts of energy. With the entire power output of a star at their disposal, however, the engineers of Solaria should be able to meet the demand. Whole solar flares can be tapped, and their gigantic energies funneled into anti-matter production lasers.

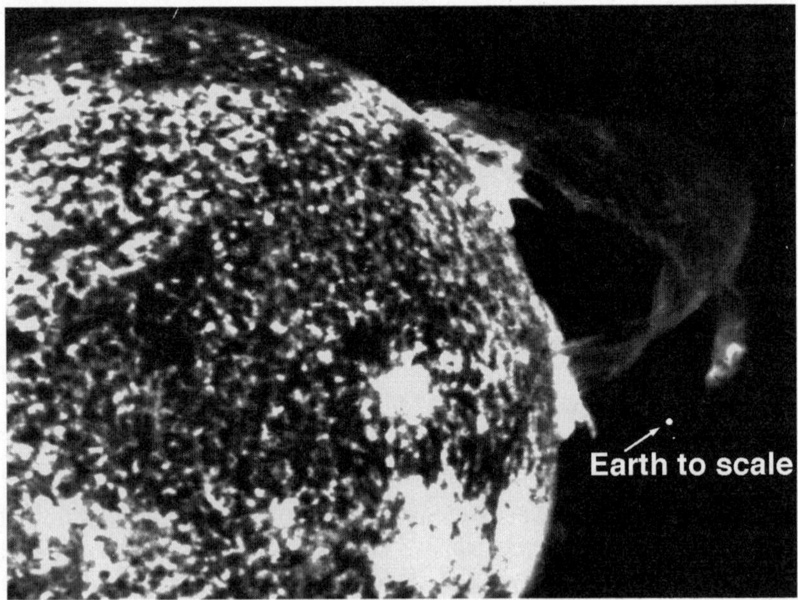


***Fig. 7.2 - Solar Flare.*** *Note the Earth, drawn to scale. With powers of such magnitude at their disposal, the engineers of the Third Millennium will be able to undertake virtually any enterprise they care to imagine.* NASA photo.

For the engineers of Solaria, producing bulk anti-matter should be no more challenging than creating the first atomic bomb was for the engineers of Los Alamos. Once perfected, the laser arrays could convert over 95% of the laser energy into electron/positron pairs.[582] At such high efficiencies, anti-matter should not be much more expensive than the energy required to create it. Anti-matter will simply be an intensely condensed form of energy—the perfect fuel for star ships.

With an abundance of anti-matter available, the stars will be within reach. Anti-matter reactions produce mostly pions (pi-mesons, a form of charged particles) moving at 97% of the speed of light.[583] Confined in a magnetic bottle, controlled with magnetic mirrors, and exhausted through a magnetic nozzle, these fast-moving particles can form an incredibly powerful rocket plume. An anti-matter rocket with an exhaust velocity of 150,000 kps or higher, is a very real possibility.[584] A single-stage rocket with this exhaust velocity could readily attain half the speed of light.

It will probably be most efficient to use anti-matter in a dilute form, mixing it with a volatile propellant, like water. For velocities up to 30% of the speed of light, the most efficient recipe is to use four tons of water per ton of ship, with varying amounts of anti-matter.[585] For short hops, inside the solar system, these amounts are very small. A one ton ship could fly to the Moon in just four hours, using four tons of water and 30 milligrams of anti-matter.[586] A ten ton fast clipper could make the passage to Mars in a week, propelled by 40 tons of water and 10 grams of anti matter. Sending a thousand ton payload four light years, to the next star system at 10% of the speed of light, would require 4000 tons of water and ten tons of anti-matter.

**A Fiery Steed at the Speed of Light**[587]

There is no way to exceed the speed of light and remain part of this universe, but the speed limit can be approached. Attempting to reach the speed of light involves scaling an asymptotic mountain without a summit. One can climb that curve forever, but never reach the top.

As the speed of light is approached, some real physical effects set in which make it impossible to ever break through the barrier. Relativistic velocities (those approaching the speed of light) generate a detectable effect called "gamma", designated as ¥. Gamma effects include an increase in mass, slowing of the passage of time, and a contraction in the direction of motion—all 'relative' to the outside observer, of course. At low speeds, like those we experience every day, gamma effects are undetectably small. Up to about half the speed of light these effects can be ignored. At 50% of light speed, for example, gamma increases the mass of a moving object by just 15%. As speeds increase, gamma quickly becomes more important. A starship moving at ninety percent of the speed of light will be two and a third times more massive than when it is at rest. As velocities increase above 90% of light speed, .9*c*, gamma rises more and more rapidly. At .99*c,* the ship is seven times more massive than when at rest; at .9999*c* the mass has increased by a factor of 70; at .999999*c* the mass has multiplied by 700 times. Every time two more nines are added to the decimal string, the mass of the ship will increase by a factor of ten. Therefore, smaller and smaller increases in velocity can only be achieved by expending greater and greater quantities of energy.

Imagine that you are trying to buy one of the pyramids, from Honest Zeno, purveyor of fine used camels.[588] Zeno seems a dim-witted fellow when he sells you the bottom half of the pyramid for $115.[589] He doesn't seem much smarter when he agrees to sell you the next 40% for $230.[590] A whole pyramid for $345 is a steal, but you don't have the whole thing yet. Zeno agrees to sell another 9% of the pyramid, but now he charges you $710.[591] This 9% has cost you more than twice as much as the first 90%, but you still figure $1055 for a pyramid is a bargain—just think of the souvenir concession. Zeno is getting tougher, but you're sure he's a true maroon and can be finagled out of the measly 1% he has left. Zeno agrees to sell you 99% of his remaining 1%, but he charges you $70,000. Now the price is getting steep, and you still don't own the pyramid. After hard wrangling, Zeno agrees to sell 99% of what he has left for $700,000. At this point, less than one percent of one percent of the pyramid is costing you close to a million dollars. Zeno's game is becoming clearer. The honest camel merchant will now continue to sell you 99% of what he has left at a price equal to ten times what you paid for the previous piece. It quickly becomes obvious that Zeno will always retain an interest in the pyramid. No matter how much money you spend, you will never own the whole thing.

Trying to attain the speed of light is the same proposition. To go just a billionth of a millimeter per second faster can cost millions of times more energy. There isn't enough power in the whole universe to propel even the tiniest starship over the infinite picket of nines separating it from the speed of light.

The increase in mass that comes with gamma, is mirrored by a decrease in time. For example: The nearest star is 4.3 light years away; reaching it in a starship traveling at ninety percent of the speed of light should take 57 months. On Earth, that amount of time—4 years and nine months—will pass normally, but aboard the speeding space ship, the time will contract to only 25 months. "Relative" to the people back on Earth, the space travelers experience time at a much slower rate. At higher speeds, the time dilation effect, like the increase in mass, becomes very dramatic. At 99.99% of the speed of light, the journey to a star 70 light years away would appear to take only one year. If we could somehow accelerate a star ship to 99.99999999% of the speed of light, we could make the two million light year journey to the Andromeda galaxy in 30 years. If we made the round trip to Andromeda and back, we would be in our eighties or nineties when we returned. While we were gone though, the Earth would have aged more than four million years! (In the movie *Planet of the Apes*, Charlton Heston and his crew flew off on a space expedition at the speed of light. When they returned, so much time had passed on Earth that chimpanzees had evolved into the dominant species.)[592]

Relativistic effects like time dilation are mind boggling, but they can be understood a little better by looking at thought models. Explanations of things like the increase in mass with velocity, or the foreshortening effects of the FitzGerald contraction are too mind bending to include here. We can get a grasp on time dilation though.

Imagine that you are floating in space and every second a flash-bulb is going off over your head. A sphere of light is spreading out away from you—obviously at the speed of light. Now, think of time as a measurement of the radial distance between you and the sphere of expanding light. In one year, the radius of the bubble of light will be one light year. While you are sitting in one place the distance between you and the light from the flash bulb increases at a uniform rate of $3X10^8$ meters per second. Now you take off on a magic silver surf board, accelerating to high velocity. The direction you choose is irrelevant, since the light is in a sphere surrounding you. Now, as you move away from the flash bulb, the sphere of light is moving away from you more slowly than it was when you were stationary. Since time is a measure of the rate at which the distance between you and the sphere increases, and that rate is decreased by your motion, time is slowed.[593]

To get a better understanding, look back at the flash bulb, and imagine that as it flashes it displays the time, like the LED on a

VCR. Say the first flash was at 12:00.00. The next flash is at 12:00.01. The 12:00.00 flash is ahead of you, and looking back you see the 12:00.01 flash go by. But, because you are moving away, the 12:00.01 flash took longer to overtake you. The faster you go, the longer it will take each flash to catch up with you. As your velocity approaches that of light, the flashes take longer and longer to reach you. If you actually attained the speed of light, the next flash would never overtake you and time would stop. If you exceeded the speed of light, you would begin overtaking the flashes of light that had been ahead of you, and time would flow backwards.

To achieve velocities approaching the speed of light requires incredible amounts of energy. At the speed needed to make a round trip to Andromeda in one life time, each kilogram will weigh 70 tons. (Remember that to the travelers aboard the ship, none of these relativistic effects are apparent. To them, everything—the passage of time, their weight, the shape of objects—remains unchanged.) If one is planning to hurtle through space at close to the speed of light one had better be in a big hurry or have an awfully long way to go. In the near term, mankind has enough challenges in the sky immediately over head. For at least the duration of the Third Millennium, we can content ourselves with travel at some modest fraction of light speed.

**Arrows of Outrageous Fortune**

Traveling at high velocities creates some unexpected problems. Every school boy knows that space is a vacuum and that we can travel through a vacuum without resistance. This is true only up to a point. If we speed along fast enough, we have to deal with the gas and dust in interstellar space.

The interstellar medium is highly rarefied, but as a vacuum it is far from perfect. The matter in interstellar space constitutes a rather tenuous gas with a density on the order of one nuclear particle per cubic centimeter.[594] The matter in 30,000 cubic kilometers of interstellar space would barely fill a thimble.[595] Passing through this thin gas at high speed does, however, create aerodynamic resistance. At close to the speed of light, the pressure of interstellar gas on the front of a space ship will amount to 37 milligrams per square centimeter.[596] This is a gas pressure equal to that of the Earth's atmosphere at an altitude of 90 kilometers. To fly through the interstellar medium, therefore, starships should be streamlined. Even ships traveling at relatively low fractions of the speed of light will benefit from smooth aerodynamic designs.[597]

Interstellar gas also creates a radiation hazard. At high speeds, gas molecules will strike the ship with energies comparable to cosmic radiation. As speeds increase above half the speed of light, induced radiation intensities go up dramatically. At two-thirds of light speed—200,000 kps—induced radiation is 100 billion times more intense than background radiation at the top of Earth's atmosphere.[598] Ship design must address this problem. Passive shielding can reduce some of the radiation hazard. The reactive mass for anti-matter drives will probably be water, carried in the form of ice. This ice could form the outer layers of a ship and help protect the passengers and cargo from induced radiation. Active electrostatic or electromagnetic shields will probably also be required.

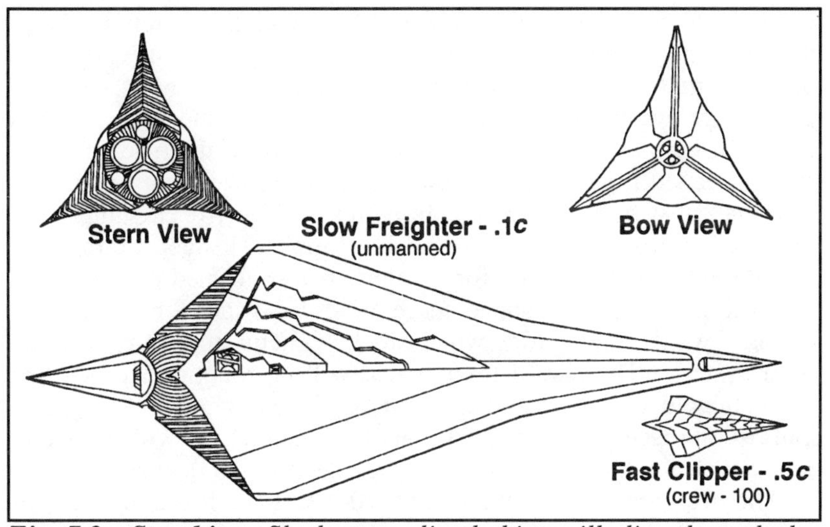


***Fig. 7.3 - Starships.*** *Sleek streamlined ships will slice through the interstellar medium at high speeds, carrying cargo and colonists to the stars.*

In addition to being gassy, interstellar space is also dusty. In every cubic kilometer, there are fifty or sixty specks of galactic flotsam, each about the size of a bacterium.[599] When a star ship, traveling at relativistic velocities, collides with one of these micro-meteoroids, little craters are gouged in the hull. Over the course of a long journey at high speeds, this sand blasting could erode away passive shields several meters thick.[600] Countering these effects will require active defense systems. One countermeasure would be a high intensity beam of short wave radiation projected in front of the ship to vaporize the dust particles. The ionized plasma could then be shunted out of the way by active radiation shields.

Together, the gas and dust in interstellar space will combine to force designers to minimize the frontal cross section of star ships. Ironically, Flash Gordon's sexy, streamlined rockets turn out to be good designs. When we travel to the stars, we will slice through space aboard sleek ships, honed and pointed like the weapons of destiny.

**Interstellar Freighters**

Expeditions to the stars can profitably be broken into two phases. In the first phase, the Solarians will dispatch heavy freighters, traveling at ten percent of the speed of light. These cargo ships will be entirely robotic and will carry everything needed to colonize and terraform a new star system—except people. The most important and least massive part of the cargo will be information. Megalithic computer banks will carry a comprehensive library of human knowledge. Included in the data pool will be information on how to produce all of the machines and robots needed to construct a new K2 civilization. Another type of information carried aboard the freighter will be frozen samples of DNA, taken from a wide variety of life forms. This genetic material will provide the seed stock for whole new terraformed planets. The cargo ships will be unmanned space arks. Like spores blown on the celestial winds, these ships will bear the genetic imprint of a living solar system. The rest of the payload will be robotic factories like sophisticated uni-fabs, and hard to produce high-tech equipment like fusion reactors. With this K2 starter kit, colonists will be able to produce all the tools and materials they need in the new star system.

The freighters will take about 50 years to make the transit between stars. Interstellar travel takes a lot of energy, and the higher the speed, the greater the energy required. By sending the cargo ships at relatively slow speed, huge quantities of energy can be saved.[601] Accelerating a kilogram of mass to ten percent of the speed of light requires a minimum of 125 million kilowatt hours of power.[602] Decelerating requires at least the same amount. At an energy cost of 5¢/kwh, this is a twelve million dollar proposition.[603] At $12 million per kilogram ($5.5 million/lb.), any available star flight economies are apt to receive serious consideration.

Accelerating a kilogram of mass to 50% of the speed of light instead of 10% requires not five times more energy, but 25 times more.[604] Shipping costs at half the speed of light will add up to more than 300 million dollars a kilogram![605] Sending the cargo third class and the people first class will save a lot on postage.

**Interstellar Clippers**

Forty-five years after the cargo ships have departed, smaller, faster ships will leave, carrying the human colonists. These fast transports will travel at half the speed of light, and will make the journey in under ten years. Because they carry only passengers, these ships can be relatively small and light.

At a constant acceleration of 1*g*, it will take a fast clipper six months to accelerate to half the speed of light. At this velocity, the colonists could cross to the next star system in only nine years—Earth time. The star ship crew will experience some time dilation at this speed, cutting apparent travel time by about a year.[606] While an eight year trip is no overnight jaunt, it is a brief interlude compared to the centuries or even millennia it might take to make the voyage at slower speeds.

Even with an economical means for producing anti-matter, and even on a very small ship, the costs of an interstellar voyage are going to be outrageous. Each person may require 50 tons of anti-matter fuel to make this trip.[607] Producing the fuel allotment for each passenger could require 2,500 trillion kwh of power.[608] Each person riding the Jefferson Starship will need an amount of energy equivalent to three decades of the present total power output of the world.[609] In round terms, a one way ticket to the stars might cost 125 trillion dollars.

Such expenditures of energy and money are intimidating, but star travel is the province of solar civilizations approaching Kardashev Level Two. When we begin making these journeys five centuries hence, we will have at our disposal a significant fraction of the Sun's energy. The sun generates enough power to fuel hundreds of such star ships every second. The energy needed for an interstellar flight will represent only a tiny part of Solaria's resources.

A mature K2 civilization will have access to power on the order of $10^{27}$ kilowatt hours per year. Our own planetary civilization has a total power output of around $10^{13}$ kwh each year. The energy requirements for an interstellar flight are approximately $10^{17}$ kwh, a mere ten-billionth of a K2 civilization's annual output. On the scale of our present planetary economy, a proportionate effort would involve driving a car across the United States.

Interstellar expeditions will be major undertakings, but they will be well within our capacities 500 years from now. The costs of an expedition to colonize a star system might be on the order of $15,000 trillion.[610] In the year 2500 A.D, the hundred odd trillion inhabitants of Solaria will enjoy average annual incomes of $millions per capita. If we were sending out one complete expedition every year, each citizen of the Solar System would have to chip in a couple of hundred bucks—the equivalent to us of the cost of a Sunday newspaper.[611]

As gigantic as the task of colonizing the stars might seem, it will require only a trivial expenditure of Solaria's powers and riches. Launching one seed ship expedition every year will require less than .0002% of the gross solar system product.[612] An equivalent expenditure to the economy of the United States would be one billion dollars—an amount spent annually on TV advertising for cosmetics and other fine toiletries.[613]

**Seed People**

Expeditions to the stars probably won't carry more than a few hundred people, a very tiny handful of seeds from which to grow a K2 civilization.[614] It may seem implausible to send such trivial expeditionary forces out to do battle with whole inanimate stars. Incredibly though, one or two hundred people are more than enough to found a thriving solar system.

Life holds the ultimate trump card in the struggle against Chaos. In the cosmic game, pitting the emerald forces of light against the sable minions of darkness, the deck is stacked in favor of Chaos: the expanses of frozen sterile space are effectively infinite; the hostility and violence of alien worlds and stars are virtually unlimited; even on Earth—the lone bastion of Life—everything ultimately dies. Against these overwhelming forces, Cosmos can play but a single card—exponential growth. Fortunately, in this universe, geometric growth is the trump suit.

The power of exponential growth is of such transcendent magnitude that it will allow us to annihilate annihilation itself. Life is a force of unlimited potency. The capacity of Life to animate the inanimate can overwhelm even the forces of entropy. A single mushroom cap can release a cloud of spores numbering in the hundreds of billions. If all the spores in the world grew at once, they could bury the surface of the earth meters deep in mushrooms. A pair of butterflies could, in just a few hundred generations, produce a mass of descendants outweighing the universe.

Exponential explosions are virtually the sole province of Life. The only other exponential reactions are fleeting events like stellar super-novae. Life is not going to be a transitory event in this universe. We are here to stay. We are going to overrun this galaxy in the blink of a Cosmic eye. After that, the rest of the chaotic universe will fall. We are going to spread through the universe from the epicenter of our explosion at a tenth the speed of light. We are the "Genesis Effect" incarnate.[615]

Humanity is capable of almost any rate of growth. We carry within us titanic forces of Life with unlimited potential. Each human female carries thousands of viable egg cells; adult males produce hundreds of millions of reproductive cells every day. Each of the trillions of other cells in the human body carries the full complement of DNA, and could—at least theoretically—generate a complete organism. If necessary, we could reproduce ourselves at a dizzying rate. All of the 'brave new world' weapons are in our potential arsenal: genetic engineering, cloning, in vitro gestation, and bizarre techniques we haven't even imagined yet. None, I'm happy to say, are necessary.

Techniques like cloning amount to nuclear warfare in the battle against Chaos. In some unforeseen terrible emergency, they might actually be needed. In the routine business of peopling space though, such methods won't be called for. We come, factory equipped, with everything we need to conquer the stars. And we can do it without running the risk of dehumanizing ourselves.

At a rate of growth not exceeding 8%, 150 colonists can produce a population the size of the present Earth's in just over two centuries. While 8% is certainly rapid in terms of population growth, it is in line with past human experience. Such a growth rate doesn't require anything more exotic of the colonists than having large families. The extended life spans, and minimal infant mortality rates of people endowed with technology five centuries in advance of our own, make such growth readily attainable. Longer life spans will both allow women to extend their child bearing years, and provide a much greater overlap between generations. Advanced medical techniques will certainly reduce the risks and discomfort attendant to pregnancy, lightening the burdens of motherhood. Within 500 years of their arrival in the new solar system, the colonist's descendants could begin sending out colonial expeditions of their own. In less than a thousand years, star colonists could build a full-blown K2 civilization. All of this can be accomplished without tinkering with our human reproductive machinery.

A surprisingly small founding population can reproduce indefinitely without risk of inbreeding. Sufficient genetic diversity can be derived from even a very few bloodlines. The minimum number for a viable colony could be as few as ten persons.[616] In a population with 100 child-bearing adults, depression of the gene pool due to inbreeding can be totally avoided.[617]

This is especially true if careful consideration is given to the genetic diversity of the starting population.[618] Colonists will be selected for the expedition from the very broad spectrum of humanity available in the Solar System. Couples of widely varying genetic backgrounds will be chosen, thereby insuring a deep gene pool to draw from. Careful genetic screening can insure wide diversity among genotypes, as well as eliminate potentially fatal genetic diseases from the founding population.

Adopting a cosmic perspective can have a profound effect on our present prejudices. Once we begin to see ourselves as the progenitors of an inhabited galaxy, the meaning of certain things changes dramatically. Too often, we look on others different from ourselves with jaundiced eyes. There is a xenophobic streak in us that makes us distrust and dislike the person with skin of a different color from our own. But when we come to see ourselves as the seed stock of a living universe, our views change. Now, those black and white and red and brown people of all flavors appear like seeds of various flowers. No star colony will long survive without a rich cross section of human races in its seed pack. Genetic variability is one of humanity's greatest treasures. 'Diversity' may seem a nettlesome symptom of our turbulent infancy, but from a cosmic perspective it is really essential to our destiny.

**Second Star to the Right**

As in all things, the Cosmos has provided magnificently for our needs, while at the same time setting up hurdles we must surmount in order to reach each new threshold. In this case, we have been kindly provided with an ideal stepping star to the galaxy—Rigil Kent.

Rigil Kent is better known as Alpha Centauri. This name, however, reflects a certain Northern bias. Rigil Kent can't be seen from the Northern Hemisphere, but it is the brightest star in the Southern sky, and the third brightest star overall. We do not normally call Sirius—the brightest star in the Northern sky—"Alpha Canis Majoris". If Alpha Centauri were in the Northern sky, outshining everything but Sirius and Canopus, it would surely be referred to by name and not by position. The problem is, Rigil is easy to confuse with Rigel—the brightest star in the constellation Orion. Therefore, Alpha Centauri is called Rigil Kent, meaning Rigil in Kentaurus (the constellation of the Centaur).

By choosing Rigil Kent as our first astral port-of-call we gain several advantages. First, we set our sights on the nearest star. Second, we secure, as our initial foot-hold in the galaxy, a star of the same type as our own sun. And third, we establish our first galactic outpost in a place where the resources for colonization are likely to be incredibly rich.

Rigil Kent is our nearest neighbor by a good third. The Rigilian system is 4.3 light years away. The next closest star of any kind is Barnard's Star, a red-dwarf, almost six light years off. Having a star as close as 4.3 light years is great good luck. The average distance between stars in this neck of the galactic woods is seven light years. Considering the magnitude of the distances involved, having a close neighbor is quite an advantage. The next nearest sun-like star is Epsilon Eridani, over ten light years away. Rigil Kent also happens to be heading this way. It is approaching us at the not inconsiderable velocity of 25 kps. During the cargo ship's 50 year voyage, 40 billion kilometers will be trimmed off the end of the journey.[619]

Rigil Kent is a triple star system. The two main stars are similar to each other and about the same size as our own Sun—Sol. These stars, designated A and B, are separated from each other by a distance equal to about the radius of our own solar system.[620] Rigil B, orbits Rigil A every 80 years at a distance of 24 A.U., 3.6 billion km.[621] If Rigil B were in our solar system, it would orbit between Uranus and Neptune. Rigil B is a K0 type star, only slightly smaller and cooler than Sol. If there were an Earth-like planet in orbit around Rigil A, Rigil B would appear in the night sky as a very bright star. The third star in the Rigil Kent system, Proxima Centauri, is a red dwarf, which orbits the central pair.

Rigil A is a G2 class star just like Sol.[622] Both Sol and Rigil A shine with the same mellow white light; both stars are extremely stable; and both stars will persist for another six billion years. Rigil A will provide us with a warm and familiar place to construct our first interstellar colony. The concentrations and frequencies of electromagnetic radiations emitted by stars are by no means uniform. Some stars produce x-rays, while others predominate in radio. Rigil A will shine with the familiar rainbow spectrum of our home star. This will have great advantages for all the immigrant life-forms, especially plants.

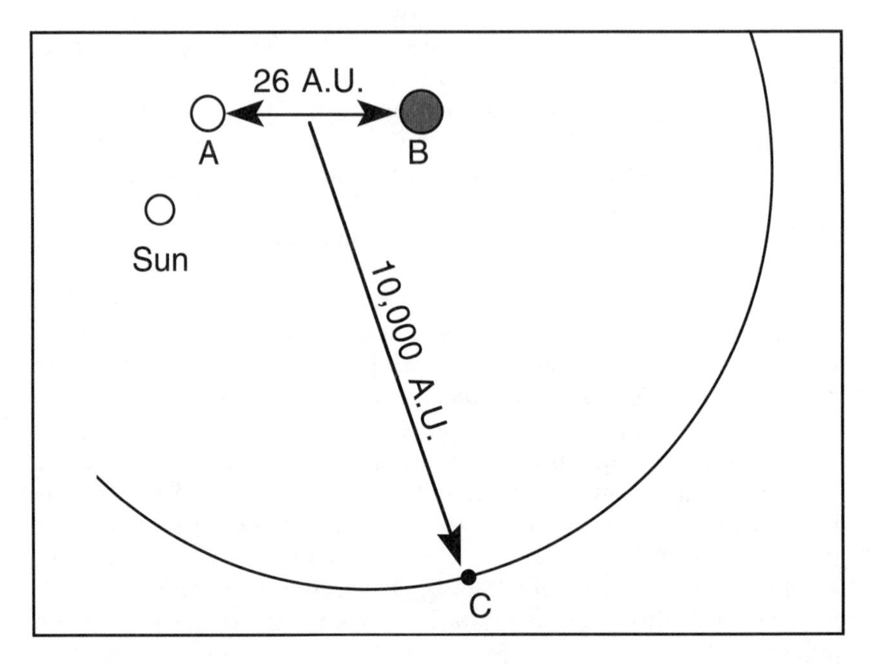


***Fig. 7.4 - Rigilian star system-not to scale.*** *The central pair of stars orbit each other in an area roughly the size of our Solar system just outside the orbit of Uranus. The dwarf companion, Rigil Kent C, orbits at a great distance.*

It is generally assumed by those who bother to speculate on these matters, that because Rigil Kent is a multiple star system, it is a poor candidate for space colonization. The presumption is that stable orbits are not possible in multiple star systems and so planets cannot form in them. The Rigilian system is therefore likely to be devoid of whole planets.

The absence of planets is irrelevant to building a space-based civilization in the Rigilian system. Virtually any star system will have orbiting debris in the form of planetoids, asteroids, meteoroids, and comets. As space dwellers, we are more interested in raw materials than in whole planets. In multiple star systems, most of the raw material is likely to have remained unaggregated, much like the asteroid belt in our own solar system.[623] Star systems with no planets, but possessed of thick swarms of asteroids will make ideal candidates for star colonies.

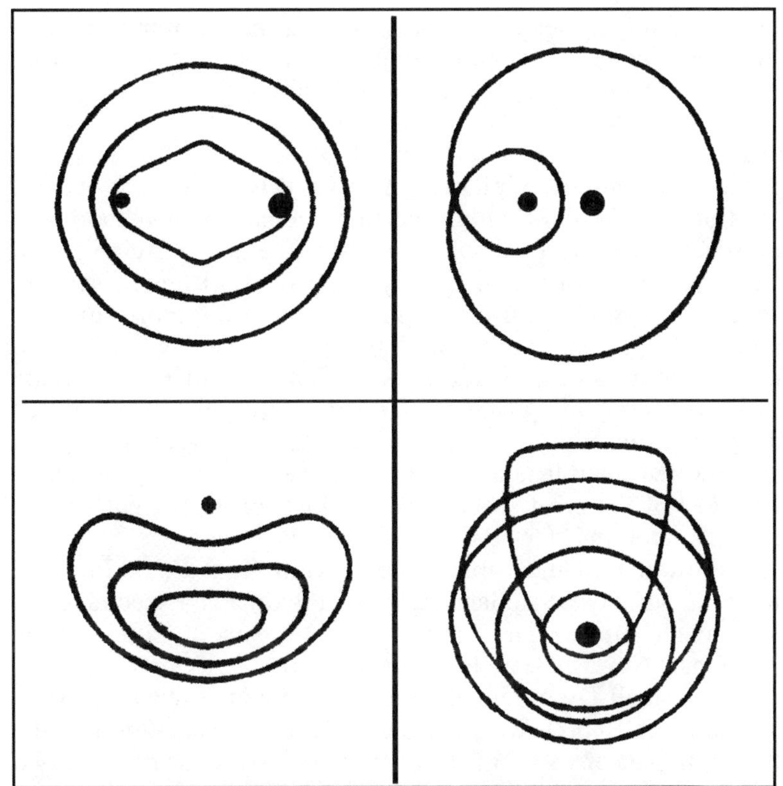

***Fig. 7.5 - Rigilian orbits.***[624] *There are numerous possible stable orbits in the Rigilian star system.*[625] *Planetary materials in this system may have remained unaggregated. Such a state of affairs could be enormously to the advantage of star colonists.*

The gravitational interactions of the stars in the Rigilian system will be inconsequential to the formation of Dyson clouds around them. On an astronomic scale, the stars of Rigil Kent are near neighbors. In terms of real distance, however, they are far apart. The Dyson cloud around Rigil A will be about as far away from Rigil B as we are from Uranus. Gravity, like light, decreases rapidly with distance. The gravitational attraction of Rigil Kent B will be barely a hundred thousandth of one *g*. The tidal forces generated by Rigil B will be virtually undetectable, and the gravitational effects of Proxima Centauri will be effectively nonexistent. Even if planets cannot or have not formed in the Rigilian system, it is certainly stable enough to accommodate space colonies in orbit around all three stars. Fully 85% of the stars in the Milky Way occur in multiple systems.[626] By mastering the techniques needed to orbit space colonies in the Rigilian system, we will be preparing for the task awaiting us in the rest of the galaxy.

**Rigilia**

The relative proximity of Rigil A to Rigil B means that two K2 civilizations can be built within the same immediate neighborhood. The distance from our own solar system to the next K2 civilization will always be measured in light years. In the Rigilian system, however, two such civilizations can grow up more or less simultaneously, separated only by interplanetary distances. Reaching another K2 civilization from Solaria will require years aboard an anti-matter powered star ship. Reaching another K2 civilization from Rigilia A or B will take only weeks aboard a fusion powered interplanetary clipper. Rather than being cast alone in the midst of the great spaces, the colonists of Rigil Kent will be able to call upon the star next door for aid and comfort.

The third star in the triple system, Rigil Kent C, or Proxima Centauri, is our closest stellar neighbor. Proxima is a spectral class M star much smaller and cooler than our sun or either of its companions. Nevertheless, Proxima will provide a good locus for growth of a small Dyson cloud. Proxima orbits the other two stars at a distance of 10,000 A.U., putting it .16 light years closer to the Earth during part of its orbit. Proxima is relatively far from its two bigger companions in terms of interplanetary distances, but it is very close by interstellar standards. The separation from the central star system is less than two light months. With anti-matter engines, the passage from Proxima to Rigil A will amount to hardly more than an inter-island hop.

Rigil Kent's many advantages will endow that system with an inherent scale and dynamism. My guess is that Rigil Kent will become somewhat equivalent to the Americas in the exploration of the Earth. At first it will be a distant outpost, but over time it will grow into a civilization on a scale commensurate with its resources. Ultimately, it will come to dwarf the Old World. Our original Solar system will always be revered as the ancient home of Life and mankind, but the Rigilian system will eventually become the epicenter of galactic colonization.

**Empire of the Sun**

By the end of the Third Millennium, our seed ships will have penetrated deep space to a distance of 25 light years in all directions.[627] Within this volume of space—roughly 1900 cubic parsecs—there are 190 stars.[628] Within a thousand years, human colonies will be established in each of these star systems. A sphere of inhabited space will be rapidly expanding around Solaria. This burgeoning interstellar civilization will be the gestating embryo of a pan-galactic human empire.

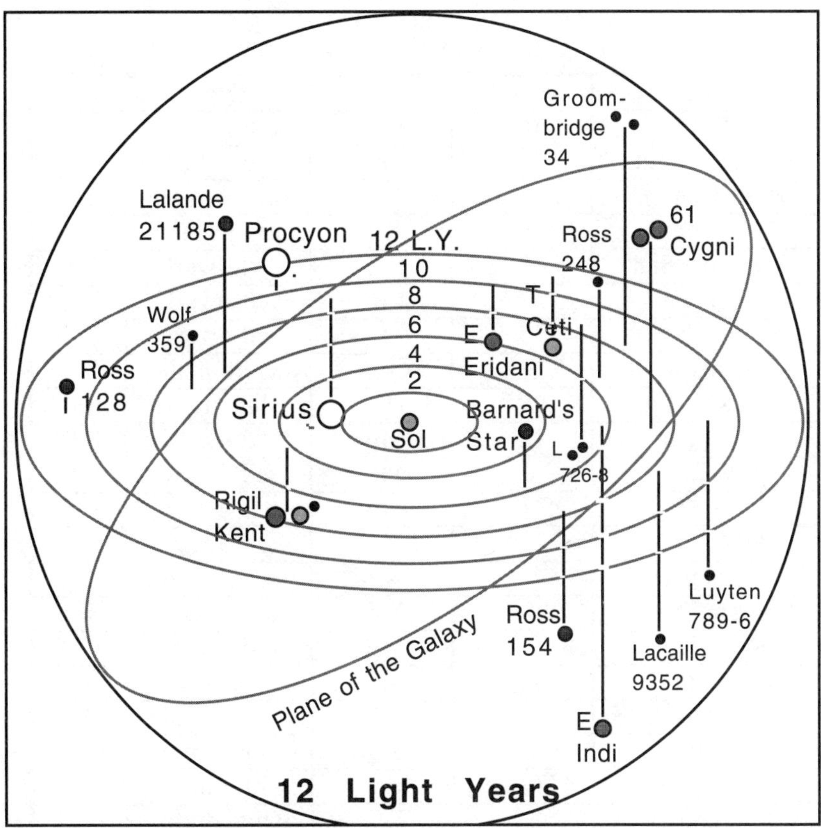


***Fig. 7.6 - Stellar neighborhood.*** *All stars out to a distance of 12 light years are shown.*

The stars within 50 light years of Solaria will become the core of our growing presence in the galaxy. During the last half of the Third Millennium, we will send out seed ships to the most distant stars within this central realm. Capella A/B will be one of the last big star systems to be targeted for colonization before the end of the Millennium. Our seed ships will depart Solaria, en route to Capella around the year 3000 A.D., and will finally arrive there midway through the Fourth Millennium. Within the boundaries of the 50 light year frontier, every stellar object will be targeted as the locus of a rapidly growing colony. These 1500 stars will eventually comprise the Solarian Foundation. As our species expands, the Foundation will become the hub of a nascent Kardashev Level Three civilization.

**Table 7.1**
**Stars to 12 Light Years**[629]

| Star | Mass* | Spec.** | Type | Dist. |
|---|---|---|---|---|
| Sol | 1.0 | G2 | yellow-main | - |
| Proxima Cent. | .1 | M5 | red dwarf | 4.3 |
| Rigil Kent A | 1.1 | G2 | yellow-main | 4.4 |
| Rigil Kent B | .89 | K6 | orange-main | 4.4 |
| Barnard's Star | .15 | M5 | red dwarf | 5.9 |
| Wolf 359 | .2 | M8 | red dwarf | 7.6 |
| Lalande 21185 | .35 | M2 | red dwarf | 8.1 |
| Sirius A | 2.3 | A1 | white-main | 6.4 |
| Sirius B | .98 | DA | white dwarf | 6.4 |
| Luyten 726-8 | .12 | M6 | red dwarf | 8.9 |
| UV Ceti | .1 | M6 | red dwarf | 8.9 |
| Ross 154 | .31 | M5 | red dwarf | 9.5 |
| Ross 248 | .25 | M6 | red dwarf | 10.3 |
| E Eridani | .8 | K2 | orange-main | 10.7 |
| Luyten 789-6 | .25 | M6 | red-dwarf | 10.8 |
| Ross 128 | .31 | M5 | red-dwarf | 10.8 |
| 61 Cygni A | .59 | K5 | orange-main | 11.2 |
| 61 Cygni B | .5 | K7 | orange-main | 11.2 |
| E Indi | .71 | K5 | orange-main | 11.2 |
| Procyon A | 1.77 | F5 | white-main | 11.4 |
| Procyon B | .63 | DF | white dwarf | 11.4 |
| 59 1915 A | .4 | M4 | red dwarf | 11.5 |
| 59 1915 B | .4 | M5 | red dwarf | 11.5 |
| Groombridge A | .38 | M2 | red dwarf | 11.6 |
| Groombridge B | -- | M4 | red dwarf | 11.6 |
| Lacaille 9352 | .47 | M2 | red dwarf | 11.7 |
| T Ceti | .82 | G8 | yellow main | 11.9 |

*Solar masses, **Spectral Type,

The stars within the Solar neighborhood will offer a wide variety of celestial habitats. Virtually all of these stars will provide fertile territory for space-borne Life. Sirius, the Dog Star, embodies the range of habitats to which we can adapt ourselves. One can easily spot Sirius in the night sky; it is the brightest star, and can be found just below and to the east of the constellation Orion. Sirius is a double star system with a large bright central star and a small dim white dwarf companion.

Sirius A is twice the size of the sun and 23 times as bright. Despite this, it will readily accommodate a Dyson cloud. Ecospheres orbiting Sirius A at a range of three A.U. would be fried, so the habitats around Sirius A must keep a respectful distance. The Dyson cloud around Sirius A will need to orbit at a distance of at least five A.U. In our own solar system, that is almost the distance to Jupiter. At that distance from Sirius A, the flux of radiant sunlight, would be the same as that in our own solar system at the distance of the Earth from the Sun. Dyson clouds will typically expand to sizes corresponding to the power outputs of the stars they orbit. The Dyson cloud around Sirius A could eventually be many times larger than Solaria.

Sirius B, a dim little white dwarf, is at the opposite extreme of stellar dimensions. This midget weighs as much as our own sun, but is only three times the diameter of the Earth.[630] Sirius B is the remnant cinder of a burned-out star. It puts out only a couple of thousandths of the energy produced by Sol. A K2 civilization will have to huddle close around this star's smoldering embers. The Dyson cloud around Sirius B will orbit just ten million kilometers from the surface of the tiny sun. That is just a tenth of the distance between our own Sun and the planet Mercury. Only a small solar civilization could grow up around little Sirius B—the K2 equivalent of a village. It might be a nice change of pace to leave the hustle and bustle of mighty Sirius A, teeming with its billions of trillions, and pay a visit to the pastoral back-water of Sirius B—home to a scant few trillion lonely hicks.

Only the blue super-giants are apt to prove inhospitable to space-borne Life. A Dyson cloud around Rigel (not to be confused with Rigil Kent) would have a radius on the order of 100 A.U., more than twice the distance to Pluto. It strains the limits of even the widest minds, to visualize a fully mature Kardashev Level Two civilization on this scale. Whether or not we would want to cultivate Dyson clouds around blue super-giants like Rigel is an open question. These stars pour out an impressive bounty of power, but they are short-lived by galactic standards. Rigel may only last a few million years. Blue super-giants burn out their fuel in an orgiastic bonfire, and then explode as super novae. The instantaneous destruction of a K2 civilization does not bear imagining.

Like cities on Earth, there will be just a few colossal metroplexes around big stars like Sirius, Procyon, and Altair; and a multitude of smaller burgs around the many stellar dwarves like Epsilon Eridani, Wolf 359, and Proxima. Most of the stars in the galaxy are not blazing super-stars like Regulus, Deneb, and Antares; just as most cities are not metropolises like London, Tokyo, and Rio.

Most of the stars in the galaxy—80% of those in our immediate neighborhood—are red dwarves. These modest stars, like Proxima Centauri, are half the size of our Sun and glow gently with barely 1% of Sol's energy. The great advantage of the red dwarves, aside from their abundance, is their extreme longevity. A K2 civilization around a red dwarf can rest easy in the knowledge that their star will never explode or expand into a red giant to destroy them.

Our own beloved Sol will someday betray us, puffing up to a livid grotesquerie. Five or six billion years from now Sol's girth will encompass the orbit of Mercury.[631] Our distant descendants will surely protect the sacred Earth, and can, if necessary, move their Dyson cloud further out, but it will be an epic inconvenience. Even so, after five or ten billion years as a red giant, Sol will cool. Eventually all his energy will radiate away. Fifteen or twenty billion years from now, our mighty sun will be nothing but a burnt out cinder. By contrast, the red dwarves will shine on, slow and steady, for hundreds or even thousands of billions of years—as close to eternity as we need imagine.[632]

**Genesis Effect**

By the end of the Third Millennium, the Solarians, here in the original Solar system, will have ended their seeding efforts. The seed ships sent out from Solaria in the year 3000 will have to travel 500 years before reaching their ultimate destinations, 50 light years out. It is senseless for Solaria to send ships beyond this limit. During the 500 years it takes the seed ships to make the journey, new K2 civilizations will have matured, dozens of light years closer to these distant stars. It will be better to let more remote members of the Foundation pick up the torch and carry it further into the void. They will be closer to the frontier than Solaria, and fresh and vigorous in the youth of their vibrant new civilizations.

The first star colonists will begin arriving in the Rigilian system around the year 2555 A.D. Within 500 years of their arrival, the Rigilian colonists will have developed large interplanetary civilizations. In five centuries, there can be populations of billions of billions around each of the stars in the Rigilian system. By the end of the Third Millennium, the Rigilians will have begun to send out seed ships of their own.

The Rigilian seed ships will have to be long distance transports, since Rigil Kent will be surrounded by a large sphere of stars already seeded by the Solarians.[633] The sphere of seeded stars will extend roughly 20 light years in one direction and 25 light years in the other. The Rigilians would therefore do best to limit their own efforts of dissemination in the direction away from Sol. They will send their ships out to the unseeded stars beyond the Solarian frontier.

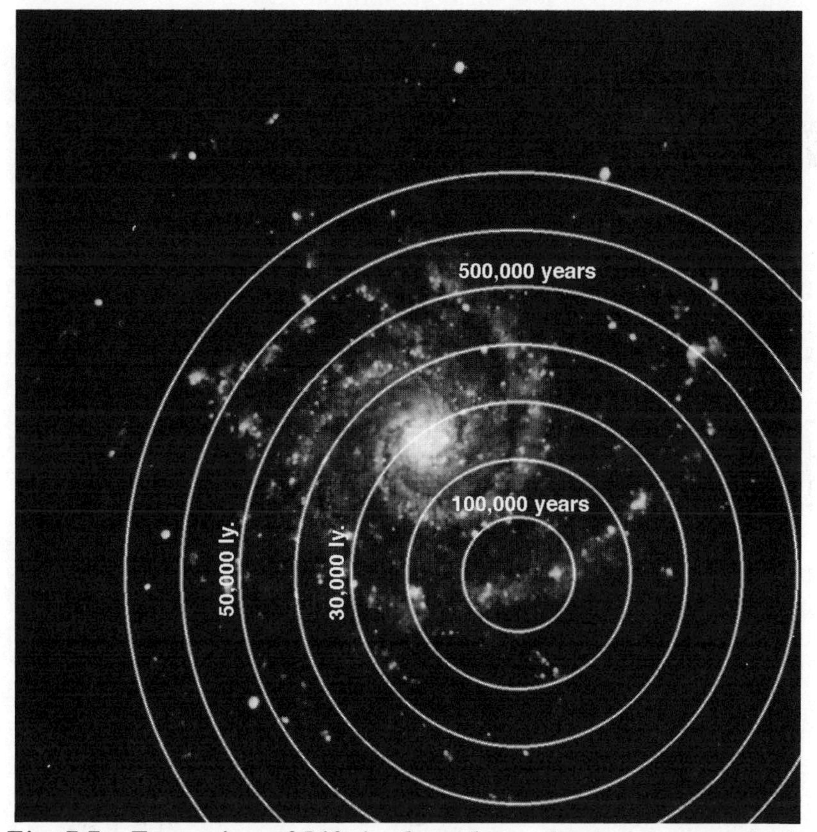


***Fig. 7.7 - Expansion of Life in the galaxy.*** *At an average velocity of 10% of the speed of light, the wave front of the human* Genesis Effect *will expand across the entire galaxy in about 700,000 years.* Photo by Hale Observatories.

As other star systems mature, they too will begin to send out seed ships. The unmanned seed ships will continue to travel at the sedate but economical speed of 10% of the speed of light. Each star system will project a cone of seeded stars out from itself, enlivening stars to distances of 25 or 30 light years. Each star system, after the Solarian's initial epic effort, will account for the insemination of only a small slice of intergalactic space. Younger civilizations, at a further remove from the home systems, will inject Life into the spaces beyond the frontier.

There will always be seed ships pushing out the envelope of Life. The radius of the galactic biosphere will grow ten light years per century. At this rate of expansion, the entire galaxy will be saturated with Life in less than one million years.

As the sphere of Life expands, the number of stars it encompasses will increase geometrically. The number of star systems attaining K2 status in the core of the sphere will be increasing at the same rate. Life will ripple through the star clouds like a shock wave, leaving a luminous green fireball in its wake.

**Gabriel's Horn**

Star systems like Solaria will reach the end of their seed bearing phase within a thousand years of their first sprouting. This is but a blink of the eye in the life span of these civilizations. Once established, a mature K2 culture should persist for the life of its home star—billions of years.

After their seed bearing phases, the ripened K2 societies can turn their attentions and energies to the massive task of forging links between the stars. Interstellar spaces are formidable barriers, but they can be bridged.

Building a true galactic civilization will necessitate condensing travel time between stars. Even with anti-matter star ships, voyages between K2 starplexes can take decades. We must eventually link the stars in a matrix which allows easy and relatively cheap travel throughout the galaxy.

Accomplishing this seemingly impossible task requires nothing more conceptually complex than an expansion of the Bifrost Bridge to interstellar scales. An electromagnetic launcher capable of accelerating a star pod to 99% of the speed of light would need to be about 450 billion kilometers long—roughly 100 times the radius of the Solar system.[634] The size of the components needed to bridge interstellar space can be intimidating, but we are dealing with mature K2 civilizations. (Even Leonardo would have thought the Golden Gate Bridge an impossibility.) By the time we undertake to construct these celestial viaducts, such acts of civil engineering will be well within our means.

With a 3000 A.U. launch tube, star pods can be pushed along at accelerations tolerable to the human body. At modest accelerations of around ten *g*s, it will take a month or so to attain terminal velocity. Any one making the trip, would almost certainly prefer to remain in a state of light hibernation for that time. A month at ten *g*s would be no picnic.

Once a high fraction of the speed of light had been attained, the pod would exit the launch tube and coast to the next star system. At 99.99% of the speed of light, apparent flight time would be only one seventieth as long as "real" time. Thanks to Einstein's time contraction, the four light year trip to Rigil Kent would appear to take only three weeks.

The star bridges will act like time machines. You can't travel into the past, of course, but you can travel into the future at a highly accelerated rate. Matter is really just energy in another form ($E=mc^2$); as it turns out, time is just energy in another form too. With the star bridges, we can trade energy for time. Anti-matter rockets can accomplish the same thing, but it is extremely expensive. The star bridges will make it practical to invest huge amounts of energy in the launch pods for one important reason—we can get our money back.

All bridges have two ends, and star bridges are no exception. They are built in pairs, connecting mature K2 civilizations. The first pair will be constructed to link Solaria with Rigilia.

A star pod is launched from the tube in Solaria and glides between the stars. At the end of its 'three week' glide phase, the pod enters the outer end of the Rigil Kent bridge. The pod then spends another month being decelerated inside the Rigilian launch tube. At the end of its deceleration phase, the pod emerges from the inner end of the Rigilian bridge. Now the pod is deep inside the Rigilian system, just a few dozen A.U. from the main body of the Dyson cloud around Rigil A or B. The pod leaves the bridge tube moving at only 30 or 40 kps., and makes its final rendezvous maneuvers using anti-matter engines.

As the pods decelerate, they will give up virtually all of the energy invested in them. The deceleration tube will generate power which can be stored in superconductive capacitors for use in the acceleration tube. This will make the bridge system very efficient. It won't matter that it takes huge amounts of extra energy to accelerate payloads to high fractions of the speed of light, since most of the energy in the system is recovered. The only energy lost is that wasted by friction with interstellar gas and dust. This too can be minimized by using wave riders. Stellar wave riders will actually surf the ultra-sonic shock wave created by their passage through the diffuse interstellar medium. This will reduce drag, and the lift thus created can be harnessed for maneuvers allowing flight to star systems not necessarily in line with the original launch tube.

Each of the launch tubes will be built like a double barreled shotgun, with one barrel for outgoing pods and a parallel barrel for incoming pods. As long as the two-way traffic is fairly balanced, the overall momentum of the system will remain near zero.[635] Each tube will originate in its home system at a distance of 40 or 50 A.U. from the central star.[636] The bridge will extend out into the inner edge of the Oort cloud. Constructing a bridge might require something on the order of ten to a hundred trillion tons of mass, depending on the design.[637] That is a bare crumb out of a solar system's basket of resources—equal to the mass from a single asteroid twenty kilometers in diameter.

If desired, travel times between portals of the time/space gateways could be made to seem instantaneous. All that is required is to increase the energy density of the launch tubes at their nether ends, to accommodate the relativistic increase in the mass of the launch pods. At 99.9999999999999999% of the speed of light, the pod—to those aboard—would appear to pass between the two star bridges in the blink of an eye.[638] To the traveler, it would seem that he left one bridge and an instant later entered another bridge, light years away.

As Honest Zeno taught us, energy densities go up exponentially as nines are tacked on to the final velocity. The 'instantaneous' journey would require nine million times more energy than the 'three week' trip. If the energy recovery system were efficient enough, it might not matter. Millions of times more energy is pumped into the star pod, but millions of times more energy is recovered.[639]

Such a bridge system could reduce the time and cost factors of interstellar flight to relative insignificance. Eventually a network of bridges will connect all the stars within humanity's growing empire. It may some day be possible to play tourist on a galactic scale, whizzing from one star system to another, dazzled by the wonders of a million different worlds.

### The Fermi Paradox

As we prepare to break our planetary bonds and emerge into the galaxy, taming stars and populating space, we have to wonder why it hasn't happened already. Why hasn't any other life form in the galaxy already spread from star to star? Why don't we look up in the night sky and see a living Mossy Way, instead of the sterile Milky Way? To all appearances, the galaxy is entirely vacant. There is not a shred of evidence that the galaxy has ever been colonized in the past, or that a galactic civilization now exists. As far as we can tell, we are utterly alone.

Enrico Fermi crystallized the issue of cosmic solitude when he posed his famous question: "Where is everybody?"[640] This rhetorical question, and its apparently inescapable answer, have come to be called the 'Fermi Paradox'. What Fermi means is 'Where are all the aliens?' The question implies there should be lots of extraterrestrials around, and there obviously aren't—hence the paradox.

A scant million years from now, we will have filled this galaxy. From the central core to the globular clusters, almost every star system in the Milky Way will pulsate with Life. If we presume that we are a typical species, then we have to ask why other intelligent life forms haven't colonized the galaxy already.

The Milky Way is at least ten billion years old. The diameter of the galaxy is only 100,000 light years. Even if extra terrestrial space arks crept along at the glacial pace of one light year per century, the aliens could inhabit the entire galaxy in just 10 million years.[641]

Assume that ten million years is a conservative estimate of the time needed for an intelligent species to colonize the galaxy. Further, assume there is never more than a single colonizing species in the galaxy at any given time. Even given these very conservative assumptions, the galaxy, and our little piece of it, should have been overrun by alien colonists five hundred times by now.[642]

Some very optimistic scientists estimate that at any given time millions of technical civilizations should be spread throughout the galaxy.[643] If this is true, why would our own jewel of a planet remain untouched? Perhaps we have just been overlooked. The galaxy is a big and crowded place. Stars teem in their billions, and ours is a mere dwarf among giants. Perhaps our little world has just remained hidden among the myriad stars roiling in the clouds of the Milky Way (see Plate 19).

Once we begin spreading through the galaxy ourselves, however, we will quickly discover that Earth-like worlds are rarer than sapphires in sand piles. No water-based life form engaged in colonizing this galaxy would ever by-pass a world like ours. The blue glow of our oceans, and the spectral signature of free oxygen in our atmosphere would attract any sentient beings within dozens of light years. Like bees swarming to the nectar of an opulent blossom they would quickly zero in on our sweet world. (Even if the ETs breathed methane and had liquid ammonia for blood, they would come to our solar system—drawn by the vacation paradise on Titan.)

Any life form will, by definition, be dependent on energy. Like us, they will be drawn to live closely huddled around stars. ETs should be drawn to long-lived stable stars like moths to candles. In a galaxy saturated with life, there is no good reason our congenial star should remain so neglected. Our own species is going to pervade the galaxy. Very few stars are going to escape a serious infestation of flowers. Our expanding waves of seed ships will inseminate virtually every solar system they come across. Dangerous exploding super-giants and extinct brown dwarfs are about the only celestial objects that will be bypassed. Certainly no juicy peach of a main-sequence star like our Sol will ever be ignored. Star systems like ours are prime galactic real estate. Yet, there is a manifest absence of alien condos here. Apparently, no wave of colonizing ETs has ever swept through our arm of the galaxy.

**Unidentified Flying Objections**

Many people—some of them otherwise sober-minded scientists—believe that the galaxy is not only populated, but that the residents of other star systems routinely visit the Earth. They point to countless eyewitness reports of UFO sightings and even close encounters of the third kind—abductions. Speculating on UFOs is one of the great American pastimes. Proponents are passionate in their insistence that the aliens are already here. They contend this would be obvious if it weren't for a governmental conspiracy covering up the truth. (Rumor has it that Oliver Stone is on the case.) No aliens have yet appeared on CNN, but everything from crop circles to the disappearance of Amelia Earhart is chalked up to them.

There is no one on Earth who wants to believe in UFOs more than I do. My keenest wish is to be whisked away in a luminous space ship; to be carried off to the heart of an advanced civilization where they already have fat-free chocolate and HDTV. If I could just rocket off to join someone else's pan-galactic empire... It makes me weep to think of all the trouble it would save me. But as fervently as I might hope for such deliverance, it seems unlikely that any sweet chariot is going to swing low for to carry me home any time soon. As much as I yearn to meet the little green men, there is no discernible reason to believe in them.

There is not a single shred of hard UFO evidence. Nothing I have heard of would even stand up in a court of law, let alone convince a hardened skeptic. The arrival of ETs on Earth would be the single greatest event in human history. By comparison, the discovery of fire, the fall of the Roman Empire, detonation of the atomic bomb, and landing on the Moon would all be reduced to trivialities. How could such an epoch-shaking affair transpire without producing any more evidence than a handful of blurry Polaroids? Belief in alien visitors requires hard evidence; at least a scrap, a smidgen, a particle, one iota, something. Anything! For my part, I would settle for a splinter of alien alloy, a corpuscle of alien blood, a fleck of alien dandruff. I will settle for anything you can actually put under an electron microscope and say of it, definitively: "It is not of this world." Is that too much to ask as evidence of the greatest thing since Moses? Of course, there is no such scintilla of evidence. And without it, no number of "eyewitness reports", duly chronicled by the *National Enquirer,* will ever make any difference.

**Chariots of the Frauds**

Not only are there not any aliens visiting the Earth now—Whitley Strieber not withstanding—there never have been.[644] There is no good evidence anywhere on Earth that aliens have ever been here. In space, that lack of evidence is more obvious and may be taken as conclusive.

In the 1970s, a popular charlatan named Eric Von Daniken proposed that Earth had been visited by "ancient astronauts". Von Daniken used every conceivable archaeological artifact, from the Pyramids of Egypt to the monoliths of Easter Island, to support his hypothesis. That his various points of proof were separated from each other by thousands of years of history didn't faze him.

One of Von Daniken's more intriguing bits of evidence was the presence of the so called Nazca Lines. The Nazca Plains are on the southern coast of Peru, in one of the Earth's most arid regions. The Nazca people flourished between 200 B.C. and 600 A.D. For reasons known only to themselves, they etched a variety of long lines and huge designs on the desert. The lines were inscribed simply by sweeping aside the dark gravel that covered the surface, revealing the lighter clay underneath. Due to the lack of rainfall in the region, these designs have persisted for ten or twenty centuries.

Von Daniken postulated that the lines represented guide markers pointing the way to a "space port", where alien ships supposedly landed. This is a pretty ludicrous notion. It's a little hard to believe that the aliens could find their way across trillions of miles of interstellar space without difficulty, but needed giant arrows scratched in the sand to find their space port.

Nevertheless, it does bring up an interesting point. If extraterrestrials had ever visited the Earth in the distant past, it is quite likely that the evidence of their sojourn here would have long since been rubbed out. On the arid Nazca Plains, such evidence might persist for a few thousand years. The span of a few millennia is, however, only a brief sliver of the Earth's history. In most other places, rain, wind, glaciers, and the plow would quickly blot out any evidence that aliens were ever here.

There is, however, a place where evidence of an alien visitation would persist for millions if not billions of years. Not only would the evidence persist there, but we can be fairly sure that the aliens would land on this particular place—the Moon. Sealed forever in the time-machine of hard vacuum, the ageless face of the Moon provides us with a permanent record. The tracks Neil Armstrong left on the Moon will persist longer than any engineering works on Earth. Long after Hoover Dam has been eroded away and washed into the sea, thousands of years after the Pyramids have been worn down to sand dunes, the boot-prints at Tranquility Base will still be visible.[645] Little bits of gold foil will still glitter in the sun, and the American flag will still wave in the vacuum's eternal breeze.[646] Even a billion years from now, the spindly legged descent stages of the lunar modules will still stand as shining monuments of that shining moment in history.

If any alien ship had landed on the Moon during the past umpteen million years, it would have left its indelible mark for us to see. The Moon is a natural destination for ETs interested in the Earth. It provides a very convenient base of operations with numerous advantages: In astronomical terms, the Moon is right next door, and the same side always faces the Earth; the Moon is free of an obscuring atmosphere, and has relatively low gravity. If aliens had ever really visited our solar system, we should find the Moon covered with three-toed boot prints and littered with galactic gum wrappers. Of course, we find nothing of the kind. The only artificial marks on the Moon's otherwise pristine face are the foot prints of a dozen men and a few wheeled vehicles.[647] Extra terrestrials should have colonized this solar system hundreds of times by now. You shouldn't be able to swing a dead cat on the

Moon without knocking over some alien artifact. But the truth is, there's nothing there. No one but us has ever been there; and no one but us has ever been here. Sadly, we must conclude that no ETs have ever called on our lonely planet.

This conclusion has cosmic ramifications. No visitation means no aliens. If they were out there, they would have been here by now. They've had five billion years to get here. If the universe is really conducive to the formation of living planets like the Earth and intelligent tool users like us, then tens of thousands of alien cultures should have matured during the life time of the galaxy. Some proportion of those should have risen to Kardashev Level Three. We're going to do it. Nothing can stop us but ourselves. If we can even conceive of doing it, then alien cultures, with millions of years of technical history at their disposal, must have done it. So, where are they?

**Radio Free Universe**

Life alters its environment in dramatic and unmistakable ways. The most cursory glance at the Earth identifies it as a living world. The atmosphere of the Earth is in chemical disequilibrium. Without life, the oxygen in our atmosphere would be gone in the blink of a geologic eyelash.

Just a thousand years from now, the appearance of our solar system will be totally different—transformed by life. As Freeman Dyson points out, a highly advanced technical civilization will alter the appearance of its home star. Once a K2 civilization has surrounded its mother star with solar collectors and habitats, that star will look very bizarre to a distant observer. When we have filled the space around our own star with ecospheres, it will no longer shine with unfiltered harsh-white light. The spectrum of the sun will be shifted. Its light will pass through the filter of a living green foam. Compared with other stars, it will look decidedly strange.

Just so, other living star systems should exhibit a characteristic "green" signature.[648] A growing interstellar civilization should show up as a fuzzy green patch on the star-fields of the Milky Way. Whatever its signature, a galactic civilization should change the appearance of the galaxy in some unmistakable way. We see no characteristic evidence of galactic civilizations anywhere in the sky.

If there were a galactic civilization, we wouldn't have to look very hard to find it. We need only cock a radio ear to the sky. The cacophony should be deafening. If the galaxy is indeed home to extra terrestrial civilizations, then we should be awash in reruns of alien TV shows. One shouldn't be able to point a radio telescope at the sky without being bombarded by images of the Arcturian Milton Berle. Unlike the spread of alien civilizations, which may creep along at only a light year per century, TV and radio cross space at the speed of light.

Our own radio signature is even now broadcasting our presence to the rest of the galaxy. Surrounding the Sun is a shell of intense radio noise, now about 180 light years in diameter.[649] Our radio presence has already expanded far enough to encompass 9000 stars. There is no way any radio astronomer on a planet within range of our transmissions could mistake the incoming signals. Even if he couldn't interpret the messages encoded within the radio and TV waves, he would nonetheless recognize the unmistakable signature of its artificial origin.

A century ago, our star was just a quiet yellow dwarf, with nothing to distinguish it. To outside observers, it would appear to have very suddenly erupted, like a radio volcano. Our little planet already outshines the sun in the radio portion of the spectrum.[650] Some of our most powerful planetary exploration radars are already ten billion times brighter than the sun in radio. An unexplained exponential doubling of radio emissions from an otherwise sedate main-sequence star would convene more than a few alien symposiums on radio astronomy.

Our civilization—just now at the crackling dawn of the telecommunications age—already raises a radio din of astronomic proportions. Try to imagine the electrostatic bedlam of Solaria: millions of massive deep-space radars probing for comets in the Oort cloud, billions of radar beacons sending homing signals to shuttle craft zipping through the solar system, trillions of TV channels broadcasting reruns of the Brady Bunch, and sextillion cellular phones—all of them on hold. Solaria's electromagnetic signature will be a continuous atomic blast of radio noise. Solaria will outshine any other radio source in the sky, and so should any alien civilization of like magnitude.

An alien K2 culture should glare with the radiance of a quasar.[651] (Quasars are point sources of radiation that shine with the luminosity of an entire galaxy.) But radio quasars are tens of billions of light years away. A K2 civilization would look like a radio quasar inside the galaxy. Having such a radio source in the galactic neighborhood would be like sitting next to Radio Raheem on the subway.[652] You really couldn't fail to notice. There is no way to hide a K2 civilization, even if you wanted to. Its radio signal would glare out at us from the star clouds of the Milky Way like a searchlight from a darkened shore. If the alien civilization is interstellar, the sky should shine in radio like the lights of L.A.

There is a program to actively search for signals from other civilizations in the galaxy: SETI (Search for Extra Terrestrial Intelligence). This is a noble cause, but it seems slightly absurd. Scientists huddle around radio telescopes listening intently to one star at a time for the sound of dripping water, when what they are seeking would sound like Niagara Falls. The most cursory radio snapshot of the sky should reveal K2 civilizations as clearly as the lights of great cities seen from orbit at night. That we don't see any such radio beacons in the skies probably means there are no Kardashev Level Two civilizations in this galaxy.

Perhaps advanced civilizations don't use radio, or radar, or microwaves. Advanced technology can be invoked as an explanation for the absence of extra terrestrial radio signals. But it seems unlikely that their technology would leave no imprint anywhere in the electromagnetic spectrum.[653] We have been compared to the aborigine who remains blissfully unaware of the storm of radio and TV saturating the airwaves around him. Presumably, the aliens use advanced means of communications which we cannot detect. What these means might be is, by definition, unknown, but they must be extremely exotic. We don't detect K2 signals in the form of laser pulses, gamma rays, cosmic rays, or even neutrinos. Therefore, the aliens must use some system we haven't even imagined.

This argument, appealing though it is, cannot survive contact with Occam's razor—in this case Occam's machete.[654] The evidence in hand is simply nothing—no signals. To explain the absence of signals in the presence of aliens, demands recourse to what is essentially magic. Unfortunately, the iron laws of logic demand that we reject such wishful thinking in favor of the simplest explanation which fits the data: No signals; no aliens.[655]

The skies are thunderous in their silence; the Moon eloquent in its blankness; the aliens are conclusive by their absence. The extraterrestrials aren't here. They've never been here. They're never coming here. They aren't coming because they don't exist. We are alone.

### The Empire Strikes Out

It may be just as well for us that the galaxy is as yet uninhabited. We like to envision potential aliens as benign entities with warm fuzzy smiles and glowing skins, ala *E.T.*, *Close Encounters*, *Cocoon*, etc., but there is no reason for them to be so cuddly. They are more apt to be like The Borg than the Ewoks.[656]

Fortunately, even if there are hostile and aggressive races in the galaxy, it is very unlikely that they can establish an Empire by conquest. Once a species has attained K2 status, there is virtually no way for an outside invader to conquer them. Barring some unknown process which can short circuit physical laws, no invader can bring enough fire power with him to compete with a mature K2 culture. It is simply an extension of Napoleon's time-honored maxim: that battle is incidental to the decisive question of supply. A K2 culture—able to harness whole solar flares at will—can easily fry any unwelcome raiders.

The only way for an aggressive species to expand its horizons at the expense of other life forms would be to invade star systems which have not yet tapped the powers of their suns. This consideration ought to add some impetus to our own race for K2 status. Until we have grown into at least solar adolescence, we will be vulnerable to a hostile takeover. In as little as five centuries though, we can attain the stature of a celestial teenager. At that point, we will be robust enough to give a warm reception to any nasty Vogons who might show up.[657]

**Wizard of Odds**

There are 200 billion stars in the Milky Way Galaxy. How could it be possible that ours is the only one harboring intelligent life? Actually, it goes far beyond that. Not only is our solar system the only source of intelligent life, it is probably the only source of any kind of life. Not only is our planet the only source of life in this galaxy, it is probably the only source of life in any galaxy. Hard as it may be to believe or accept, it is likely that our little world is the only speck of Living matter in the entire universe.

Those who tend to reflect on these issues, especially those who believe that life must be a common phenomenon, derive long elaborate formulae to prove their case. They point out there are hundreds of billions of stars in the Milky Way; of these, some 200 million are similar to the sun; around these other suns orbit 10 million earth-like worlds; life must have evolved on millions of these worlds; intelligent tool-users must then have developed hundreds of thousands of times; so there must be thousands of civilizations capable of star travel. Carl Sagan, the leading proponent of this viewpoint, calculates that the Milky Way has been home to no fewer than a <u>billion</u> technical civilizations![658] When this argument is extrapolated to the universe at large, the existence of ETs, at least somewhere, seems a virtual certainty. The odds of the Earth being the only living world in the universe are on the order of one in $10^{18}$.

With such an overwhelming number of chances, a billion billion Earth-like worlds, Life must have sprung up innumerable times—mustn't it? This argument is reasonable enough on its face, but as soon as speculators leave the realm of astronomy they enter *terra incognita*, where dwells an inscrutable mystery. No one knows what the odds are that life will evolve given an earth-like planet around a sun-like star. Sagan rates the chances at one in three. A close examination of the issue indicates that he may be off in his estimate by billions and billions.

The evolution of life is overwhelmingly improbable. The odds against life are so extreme that it is virtually impossible for it to occur twice in the same universe. That life ever evolved anywhere at all is a miracle of Biblical proportions. If it wasn't for our manifest presence, the creation of life could be dismissed as a wild fantasy. Generating animate matter through random chemistry is so unlikely as to be indistinguishable from impossible. Yet, here we are. Obviously, miracles do happen. But the question is: do they happen twice?

Proponents of the view that life is commonplace suggest that it is a simple process arising out of organic chemistry. Harlow Shapely, for example, concludes that because organic molecules are common: "Life must exist in nearly all star systems that have planets."[659] Carl Sagan writes of the origin of life:

> *In those early [primordial] days, lightning and ultraviolet light from the Sun were breaking apart the simple hydrogen-rich molecules of the primitive atmosphere, the fragments spontaneously recombining into more and more complex molecules. The products of this early chemistry were dissolved in the oceans, forming a kind of organic soup of gradually increasing complexity, until <u>one day, quite by accident, a molecule arose</u> that was able to make crude copies of itself, using as building blocks other molecules in the soup.*[660] (emphasis added)

There is no disputing Dr. Sagan's scenario for the origin of life. The only question is the likelihood of the "accident". If it is a probable sort of accident, given the right circumstances, then Harlow Shapely may be right; if it is not so probable, then it has serious implications for our place in the Cosmos.

Let's presume that all that is required for the evolution of life is the formation of a single self-replicating chain of DNA. (A great deal more than just this chemical accident is of course required to produce single celled organisms, and then a complete biosphere, and finally intelligent beings. But for the sake of argument, let's assume that a minimal chemical precondition is all that is required to set the chain of causality in motion that will eventually evolve you and me out of the mud.) As it turns out, the minimum chain length for self-replicating DNA is around 600 nucleotides.[661] (Nucleotides are the building blocks of DNA, consisting of the base pairs of adenine-thymine or guanine-cytosine that form the rungs, and the phosphates which form the backbone of the ladder in the double helix.) Six hundred links is an exceedingly short DNA chain. Consider that a very simple virus contains 170,000 links, and a bacterium seven million; your own DNA chain is six billion links long.[662]

How likely is it that the primordial soup, given enough time, will cook up a strand of "Genesis DNA"?[663] To calculate the odds of such an event occurring at random, we need to turn to 'information-theory'. This is an arcane branch of statistics developed to aid in the design of computers and telecommunications networks. Essentially, information-theory reduces the nebulous concept of 'information' to exact mathematical quantities relating to message length and content. According to information-theory, a message with meaning can be interpreted as a level of probability. In other words: how likely is it that the message will be generated at random? This probability is dependent on the number of bits of information required to encode the message. The number of bits is then the exponent (base 2) of the number of random trials it would take to generate that message. In plain English, this means that generating even a relatively short message by random trial and error takes an enormous number of tries.

Words, like those you're reading now, contain meaning—at least that's the intent. In theory, the same message content could be generated randomly (perhaps it would make more sense if it was). Using information-theory, we can find out what the odds are of a given message being generated by chance.

Let's use a very simple message, one I'm sure we're all familiar with from our earliest attempts to decode these alphabetic hieroglyphics: "See Spot run." This minimal message contains just thirteen elements: ten letters, two spaces, and a punctuation mark. Written English requires only about 50 symbols to convey any message: 26 letters, 10 figures, 13 punctuation marks, and blank spaces. The first position in our message has one chance in fifty of being an "S". The odds of generating a particular message one symbol long are 50 to 1. The second position has the same odds, so the chances of a message two symbols long turning up as "Se", are 50 X 50, or $50^2$ to one. Every time a symbol is added to the sequence, the odds against that sequence go up by one multiple: three symbols—50 X 50 X 50, or $50^3$ to one; four symbols—$50^4$ to one, etc. It is very easy to calculate the odds of any message being generated at random: the number of possible symbols is the base, and the base number is raised to an exponential power equal to the number of symbols in the message. The odds of generating "See Spot run." at random are $50^{13}$ to 1. To create this rudimentary message by accident would require 610,000,000,000,000,000,000,000 (six hundred billion trillion) trials. If a computer were programmed to generate a 13 character string at random, and created 10 million new strings every second, it would take the computer two billion years to come up with "See Spot run."

Information theory shows why generating a 600 nucleotide chain through random chemistry is—to put it mildly—unlikely. The genetic alphabet is much shorter, containing only four symbols: A-G, G-A, C-T, T-C; but this doesn't help matters very much. The same rules of chance apply. The odds of generating a particular string of nucleotides 600 base pairs long are $4^{600}$, or $10^{360}$ to 1. If these are the odds against the bob-tail nag, you'd better bet on the bay.

To generate a strand of "Genesis DNA" would take $10^{360}$ chemical reactions. That is a completely ridiculous number. Writing out such a number is an exercise in futility; it requires hundreds of zeroes. Describing it with words is just about as hopeless; a million billion trillion quadrillion quintillion sextillion septillion octillion nonillion decillion doesn't even touch it. The only way to describe it is as ten nonillion nonillion googol googol googol.[664] You can't even talk about such numbers without sounding like your brain has been fused into molten goo. If you persist in thinking about them it certainly will be.

Surely, there must be numbers of equal magnitude available to rescue us from such overwhelming odds. After all, DNA is just a large molecule. So we must be dealing with atomic numbers, and those are always mind boggling—right?

When Life arose, the Earth's ocean's were, as Carl Sagan suggests, one giant bowl of primordial soup. The number of chemical reactions going on in that stew must have been incredible. Over billions of years, any possible combination of DNA could have been cooked up—couldn't it? Well, let's take a look; the bottom line is always in the numbers.

The oceans of the early Earth contained, at most, $10^{44}$ carbon atoms.[665] This sets the upper limit on the possible number of nucleic acid molecules at $10^{43}$. (Assuming every atom of carbon in the ocean was locked up in a nucleic acid molecule—an unlikely state of affairs.) The oceans could therefor contain no more than about $10^{42}$ nucleotide chains, with an average length of ten base pairs. If all these nucleotides interacted with each other 100 times per second for ten billion years, they would undergo $3 \times 10^{61}$ reactions. This would still leave them woefully short of the sample needed to generate a strand of Genesis DNA. To get a self-replicating strand of DNA out of the global ocean, even if it was thick with a broth of nucleotides, would take ten billion googol googol googol years. Makes yours eyes spin counter-clockwise doesn't it?

But there are billions of stars in the galaxy and billions of galaxies in the universe. Over time, the right combination would come up somewhere—wouldn't it? Assume every star in every galaxy in the entire universe has an Earth-like planet in orbit around it; and assume every one of those planets is endowed with a global ocean thick with organic gumbo. This would give us 40,000 billion billion oceanic cauldrons in which to brew up the elixir of life. Now we're getting somewhere—aren't we? In such a universe, where the conditions for the creation of life are absolutely ideal, it will still take a hundred quadrillion nonillion nonillion googol googol years for the magic strand to appear. Sheesh!

Assuming some radically different form of life, independent of DNA, doesn't really help. By definition, life forms will always be complex arrangements of matter and/or energy. This complexity has to arise out of chaos. Therefore, some initial degree of order must first just happen. Whatever the form of life, its creation is dependent on the same sort of chance event that created our first strand of Genesis DNA. It doesn't matter what sort of coincidence is involved: the matching of base pairs, alignment of liquid crystals, or nesting of ammonia vortices; whatever the form of order, it will be subject to the same laws of probability. Consequently, any form of highly complex, self-replicating material is just as unlikely to occur as our form. Simply put, living is an unlikely state of affairs.

When all of the fundamental constants underlying the bare existence of the universe are also taken into account, it becomes all too obvious that life is a sheer impossibility.[666] How can a glop of mud like me possibly be walking around wondering why it exists?

**Gods R Us**

Our existence seems to defy the laws of nature. Our presence screams for an explanation. Could our existence be just a coincidence? According to the revealed wisdom of the divine bumper sticker: "Shit Happens". It is possible that life arose as the result of a spontaneous miracle against overwhelming odds. Possible, but unlikely.

The universe is a probability field. The future is an infinite continuum of possibilities. When an actuality is condensed out of that continuum, the present moment is defined. Definite moments are what we call reality. So, reality is created one moment at a time by condensing an infinite number of possibilities into a single actuality. Through this process the future becomes the present.

The process does not occur in the absence of a conscious observer. The universe—this plane of reality—exists only because we are here to observe it. If we did not exist there would be no defined present moments and only a continuum of nebulous possibilities would exist. The real world, the material world, would not exist if we were not here to observe and so define it. This is not a matter of mystical speculation. This is a case-hardened scientific fact, proven over and over again by laboratory experiments.[667]

We create reality, which is why we are here, and why we are alone. We exist as a sort of self-fulfilling prophecy. Consciousness is the Rip Van Winkle of probability. Notwithstanding its improbability, conscious matter is bound to wake up in that one actuality where it can exist. This universe is finely balanced on a razor's edge of coincidence. If certain natural forces varied in their values by the tiniest fractions, not even stars could exist.[668] In fact, in an infinite number of other possible universes, those values are different and stars don't exist. They exist here for the same reason we do, this is where they can exist.

We are alone for the same reason. We awaken in a universe where we can exist; our awakening in turn defines the universe.

Other intelligent life forms do not exist in this universe. This universe is ours by definition. It cannot accommodate another form of consciousness. A separate consciousness will define a separate universe, and we will be kept apart from it by the very nature of our reality.

This, of course, has some pretty hairy implications. We are certainly alone in the universe, but we've been getting used to that idea. The other implication is that our individual self-consciousness is really an illusion. Why, if consciousness defines reality, does your reality seem to coincide so closely with mine? Why isn't each of us off in his own private universe, like Zeyphod Beebelbrox?[669] The answer is that your consciousness and mine are not really separate. They are only different facets of the same universal consciousness. All Life in this universe is at present confined to this single planet. All Life together forms a single organism, possessed of a single unifying consciousness. We are all the same being.

**Atlantean Destiny**

Whether or not we are truly alone in the universe is really immaterial. We can debate the issue as an academic exercise, but it won't change one overwhelming fact: As far as we know, we are alone. And until we find out otherwise, it behooves us to proceed as if we were the only intelligent species in the whole space/time continuum.

This point of view should have a profound effect on our race. It is easy to gaze up at the night sky in wonder, and feel one's self to be an insignificant fly-speck. But it can be a little terrifying to look out on endless parsecs and realize that you are one of the most important things in it. It's enough to make your panic glasses go black.[670]

Being a solo act requires us to carry some very heavy baggage. Our responsibility to the Cosmos is absolute. In a very real sense, we must carry the weight of the universe on our shoulders. We are not just an insignificant species of semi-intelligent apes, charged only with the welfare of ourselves, or even of our little planet. Rather, we are the sole source of consciousness in an otherwise dead cosmos. It is all up to us. If we fail, Life, as a phenomenon in the universe, fails with us. Life never happened anywhere else before and it is unlikely to ever happen anywhere else again. If you believe in Life, if you believe in flowers and grass and trees, birds and whales and people; if you believe in children, then you must bear this Titanic burden. You must recognize yourself as one of the Olympians, one of but a tiny handful of the god-beings who inhabit this universe. For better or worse, we must accept the awesome implications of our place in the scheme of things—at the pinnacle of creation. It is our task to carry the torch of Promethean fire out into the frozen void, there to kindle the green flames of a billion billion living worlds. We few, we happy few, must decide the destiny of a universe.

## CHAPTER 8

# FOUNDATION

***I wish you'd stop what your doing,***
***and get on the case,***
***So we can blow this existence***
***Right out into space.***

**Wang Chung**

This is the last chapter of the *Millennial Project*, but it is about the first step in the Millennial Process. We have come far together: across the wide oceans, up the Bifrost Bridge, past the Moon and Mars, through the asteroids and comets, out to the stars themselves. We have seen the shining future, glimpsed man's cosmic destiny, and come to realize that the universe can be ours. The oceans, the planets, the stars; they are ripe fruits ready to drop into our hands—we need only reach for them.

Our species holds the keys to the Cosmos. Ours is the magic capacity to create any reality we can envision. The same force that built the Pyramids of Egypt and the cathedrals of Europe can create Aquarius and terraform Mars. These monuments to human resolution are carved into reality by the penultimate creative force in existence—a unified vision. Pyramids and cathedrals were built by individuals who coalesced themselves into purposeful groups. They worked and sacrificed, sweated and bled, to bring their visions to reality. We must do no less. The floating cities of Aquarius will be our pyramids, terraformed planets our cathedrals.

To enliven the universe is the destiny of mankind. To fulfill that destiny we must follow some path, either the one mapped here or some other. Our planetary culture is rapidly approaching

Kardashev Level One. For the first time in three billion years of evolutionary history, we have the stature to fulfill our role on the Cosmic stage. We have become the most potent force on the planet, dominating and changing even the climate. Our twin explosions of population and technology have brought us, with brutal suddenness, to this moment of decision. Hurtling over the precipice cannot now be avoided. As the brink approaches, we must choose: plummet to our doom, or fly!

**First Millennial Foundation**

One thing is certain: the existing social structures are not equipped to carry us to the stars any more than pigs have wings. We can't wait for NASA to build the Bifrost Bridge, or expect the United Nations to colonize the seas. If we wait for the existing institutions to propel us into the cosmos, we are doomed to failure. These organizations—agencies, nations, religions—were formed for different purposes; they were built on different foundations; they evolved in other ecologies of intent. To have any chance of expressing Life's Cosmic urge to break out into the larger universe, we must build a new human society ourselves. Together, we must begin immediately to construct a new civilization from scratch, one designed from the foundation up as a vehicle for the colonization of space. Laying this foundation is the first, and perhaps most important, step we will take on our long road to the stars.

Albert Einstein observed: "Everything has changed except the way we think."[671] At this critical juncture of evolutionary and human history, the time has come at last to change the way we think. It is essential to adopt a radically new paradigm of social organization. At present there exists no social system which offers the slightest hope of allowing us to fulfill our ultimate destiny in the Cosmos: capitalism, communism, socialism, monarchy, oligarchy, anarchy, theocracy, autocracy, democracy; all are bankrupt. They all evolved—not surprisingly—within the context of life on earth. Within that context, and until now, they have been more or less adequate. They will not, however, meet the needs of the future. Not only can we not take such clumsy and obsolete social systems with us into space, we can not even get there until we have freed ourselves from their constraints. Fulfillment of our destiny requires that we join together. We must coalesce, to become a cohesive organized society. Once unified, we will be capable of creating epic realities.

To convene, we must first transcend the superficial barriers of religion, race, politics, and geography that now separate us. In that transcendence, we can create a new nation; a nation not of territories and borders, but of concepts and ideals. This new nation will not be confined to any continent. It will exist as a network of free individuals. We need only reach out to each other—like neurons in the developing brain—extending our dendritic branches through the telecommunications web, forming a plexus of interconnected harmonic minds.

Together, we will form a new society founded on fresh principals and dedicated to the higher purpose of our mutual destiny. To that end we will together create the First Millennial Foundation.

Understanding what the Foundation is, requires first understanding what it is not: It is certainly not a church or a religion, although it has its roots in our numinous relationship with the Cosmos; it is not a political party, although it embodies the kernel of an entirely new approach to self-governance; it is not a country, although the people of the Foundation will be forming nothing less than a new nation; it is not a company, although the activities of the Foundation create net revenues which are used to further our purposes; it is not a charity, although people outside the structure of the organization contribute to it, and many in need will benefit from it; it is not a commune, although people in its core live and work in close, if not to say organic cooperation; it is not anything which social scientists have heretofore captured, chloroformed, pinned, and classified. It is an entirely new species of social butterfly.

Just as the vacuum will demand new approaches to the physical challenges of space, it will also demand new approaches to the social challenges. The Foundation will be our response. This will be the first social organization of its kind. If one is inclined to believe in mythic destinies and self-fulfilling prophecies, one might go so far as to call it: "First Foundation".[672]

**The Network is the Nation**

The Foundation is a network of free individuals, and so does not depend on geography for its structure. Our unity is forged within the realm of human consciousness—a union founded on philosophy, not plate tectonics. Electronic media will allow colonists all over a planet—this one or any other—to unite on common ground. A network is not limited, like a country, by the random conditions of the land it occupies. Nor is a network limited in its scope by any ethnic or geographic constraints.

Nations have heretofore always had to play in the zero-sum game of geography. Expanding required taking someone else's land. This fact has lead to oceanic spillings of blood throughout human history.

In the past, unity was based on physical proximity, there being no other way to get together. This is no longer true. The advent of electronic media has annihilated space. I can sit in my mountain retreat in Colorado and be 'with' someone in Rome. Today, on the telephone, I can be with them as certainly as if we were in the same room, separated only by a curtain. Tomorrow, on the videophone, the curtain will be removed. Short of arm-wrestling or making love, a videophone can effectively put us in the same room with anyone anywhere. So far, the dissolution of distance has had little or no effect on the basic structure of societies. Nations are still based on the proposition of geographic coincidence, and people are still unified by proximity, but this is going to change. New nations will not depend on geophysical or racial coincidence. Members of these new unions are just as likely to reside in Katmandu as Kokomo.

**Author's Disclaimer**

The intricate details of a fully developed society must be evolved. The specific blueprint for a new community can't and shouldn't spring fully developed from the mind of one man—like Athena leaping full-grown, her armor ringing round her, from the brow of Zeus—what a headache! It is desirable and feasible, however, to outline the rough shape that such a culture might take—like the medieval architect sketching out on the ground the foundation perimeter for a cathedral that will be built by thousands of people over many generations.

One of the most exciting things about space colonization is the chance to start over. The opportunity to carve new social sculptures from virgin stone is a compelling motivation to assail these new frontiers. Even if there were no other inducements, just the chance to build a new society from scratch is reason enough to go forth into space.

**We the People**

The Foundation will differ radically from other large-scale social systems in that it will be entirely apolitical. In all other systems, control is inevitably concentrated in the hands of a few individuals. The universal result is a travesty of governance. The body politic inevitably becomes a tool for the governors rather than a vehicle for the governed. The Foundation will eliminate this short-coming by adopting a system of pure, which is to say, true, democracy.

Present 'democracies' are democratic in name only. "Representative democracy" is an oxymoron, not unlike "congressional leadership". In theory, the people elect their representatives and so remain the ultimate source of power. In practice, the incumbents end up fixing the rules of the game, twisting them to their own ends. Eventually, they become untouchable by the electorate. The American Congressional elections of 1990 are a prime example. The populace was incensed by everything the legislators did, especially raising their own salaries. There was serious talk of a revolt, or at least a mass lynching. In spite of universal outrage, 98.5% of the incumbents running for reelection were returned to their seats. That is a higher ratio than that which prevailed in the Supreme Soviet before Perestroika!

The problem with any representative system is that it exerts its own sort of twisted Darwinian selection on the choice of candidates. Only those who thirst for power are likely to seek a place at the well head. Unfortunately, the more burning his thirst, the more likely is a man to be drawn to the well, and once there the more deeply he will drink. This unfortunate tendency has caused mankind, almost universally, to labor under the yoke of the power drunk. This is the fatal short-coming of any social system based on representative politics. [673]

The technology of electronic media has annihilated space and made possible the unity of segregate peoples. The same technology, allied with the power of the computer, has made it feasible for the people so unified to govern themselves. No longer is the intermediate agency of governors necessary for the regulation and direction of the body politic. The telephone, television, and computer, all combined into a single easy to use appliance, will make it possible for each person to function as his own legislator. In the bygone era of animal-powered transportation, it was essential for our forefathers to adopt a system of representative democracy. It was just not possible for every person to be in the legislature every time there was an important decision to be made. Now it is.

In the Foundation, there will be no elected representatives. The legislature will exist only in the air-waves and in the optical fibers of the telecommunications network. When a decision is to be made, it will be made by the people as a whole. Politics and its politicians are as cumbersome and obsolete as countries and their borders. With the technologies we already have in hand, they can be eliminated as easily.

**Bully Pulpit**

In your own brain, decisions are made through a holistic interaction of the entire dendritic network. There is no hierarchical structure, with a few "master" cells, dictating the decisions and conclusions of the rest. This interactivity is the key to our intelligence, and it will be the key to true democracy.

At first glance, one might think that a political system based on pure democracy is unworkable. Pre-conditioned by our past experience of the world, our prejudices throw up objections like flak over Baghdad: "No one will be in charge." "Who will make the decisions?" "Everyone and their dog will get involved in the process." "It will be utter chaos!" Exactly. And that is the point.

Pure democracy allows the body politic to tap into the ultimate driving engine of Cosmos—Chaos. It seems a contradiction in terms, but it is not.[674] The flip side of Chaos is Cosmos; underlying disorder is a weird symmetry, and the most highly complex and intricately ordered systems appear to be utterly chaotic.

This is easy to see in economic systems. Visit the bureau of central planning for pork production in Beijing: Everything is calm and perfectly ordered; neat rows of bureaucrats sit at their desks quietly filling out reams of paperwork; the party apparatus works with apparent machine-like efficiency as it allocates pork quotas to the lucky inhabitants of a worker's paradise. Now visit the floor of the Chicago commodities exchange: The pit is something out of Dante's Inferno. A crowd of sweaty traders are screaming in apparent apoplexy, veins starting out of their necks, arms waving, bodies jerking in frantic electric spasms—utter bedlam. It is a classic example of order versus chaos. Under the highly "ordered" system in communist China you can't get a piece of bacon for love nor money. In the U.S. I can tap into the "chaos" of the futures market by phone and buy six tons of pork bellies in two minutes.[675]

The most highly ordered systems depend for their order upon processes that appear chaotic. This will be the case with true democracy. To an outside observer it will seem to be nothing but an electronic riot; hundreds of people all babbling endlessly about the "issues" on dozens of TV channels, most of which aren't seen by more than a handful of watchers. Anyone with an idea, opinion, obsession, or pet peeve, will have access to the network and can put his two cents in. It will seem a cacophony of talking heads. But within the system, certain things will quickly happen.

The populace will rapidly begin to recognize certain speakers as being people with clear heads, persuasive tongues, and good ideas. They will tend to be listened to, especially by the people who agree with them. Of course, equally clear-headed and persuasive commentators with their own large followings will emerge at other points along the spectrum of opinion. These people, without being elected or appointed or chosen in any other way, will evolve into leaders. These 'leaders' will not be able to dictate anything to anyone any more than Rush Limbaugh can order an air strike on Libya. The leaders' only authority will be the power to persuade. Out of the wrangle, on each and every issue, will emerge an eventual consensus—a decision evolved by the body politic.

With a system of true democracy in place, representative politics, and most of the evils attending it can be eliminated. Each individual serves as his own legislator. The executive office is relegated to a largely administrative and leadership role, endowed only with the opportunity to persuade and not the power to coerce. The Foundation will be built on the bedrock of a constitution with deep roots in the philosophy of individual rights. Thus underpinned, the Foundation will have sufficient structural integrity to support a great society. True democracy will endow the Foundation with enough flexibility to maintain cohesion even while spread over many worlds.

**King Me**

As long as the old representative political systems hold sway, concentrating power in a few hands, humans will never achieve their cosmic potential. If we are to meet our destiny, then we must accept and develop the principle of individual sovereignty.

In the past, societies organized themselves by endowing one, or at most a few, individuals with sovereign power. The rule of these individuals was law, and the rights and freedoms of all other individuals were subordinate to their will. This primitive and obsolete model must be discarded. In the Foundation, each and every individual is endowed with absolute sovereign power. The limitations of this power are defined only by the corresponding and equal powers of every other individual sovereign. The Foundation is a society of kings.

This philosophy of government carries a trainload of ramifications. It means that every sane adult is endowed by the society with all the powers that in other societies are reserved to special individuals. In the Foundation, the individual is enormously powerful; he is also utterly accountable.

There are profound reasons for adopting a system of individual sovereignty. As we discover ever deeper secrets of the cosmos we find that empowering the individual is the surest way to tap into the high order harmonies inherent in chaotic systems. To evolve a new system of social order that taps into these symmetries we must leave power at its source—in the hands of individuals themselves. As we evolve technically, tools become available that make the previously unattainable goal of the personal imperium not only possible, but even practical

Industrial technology demanded that crops, products, and people, all be processed en masse.[676] The mass production techniques of the factory came to dominate all institutions in the industrial era. Humans passing through such institutions tended to be treated like so many identical soda bottles, to be filled up with a uniform mixture of sugar water and gas, to be labeled, capped, boxed and shipped to market.

Information technology eliminates the need for this mass treatment of goods and people. In the information age, the individual is once again preeminent in a way he hasn't been since before the advent of agriculture. The Foundation is a human movement of the information age. The tools of the agricultural era were the beast of burden and the plow; the tools of the industrial era were the factory and the assembly line; the tools of the information age are the telephone and the computer. Industrialism could not tolerate the individual; informationism can not function without him.

Humans are like specialized cells in a macro-organic superorganism. We have been highly evolved to enable the Cosmos to perceive, contemplate, and become self aware. We are Cosmic brain cells. This has important implications for the function and preeminence of individuals in the Foundation.

In most types of body tissue, muscle for example, the individual cells are less important than the aggregate mass. In muscles, all the cells must function in rigid synchrony. They are like soldiers, marching in cadence, each cell subordinating itself to the demands of the mass. This is most evident in the heart, where specialized muscle cells pulse reflexively in endless repetition. There is no decision making, no freedom of action, no creative latitude for the muscle cells. Theirs is a simple universe of stimulus and response.

In brain tissue, however, the function of the individual cell is altogether different. In the cerebral cortex, (the outer layer of the brain where consciousness resides) individual neurons possess thousands of dendrites which branch out in all directions. Each of these dendrites connects to other cells.[677] Each brain cell is in contact with thousands of others. Every neuron is connected, directly or indirectly, to every other cell in the network.

When a thought flashes through the brain, the activities of each individual cell are entirely problematic. There is no definite program, followed rigidly under every circumstance. When stimulated by other cells' dendrites, a neuron may or may not respond, depending on a number of other factors, especially the input it has received from other brain cells in its network.[678] Activity in the brain is fluid, spontaneous and unique. Each individual brain cell is free to respond to stimuli in its own way. If our brains lacked this flexibility, then our behavior would be like a frog's, limited to nothing more than conditioned reflex. We would essentially be unconscious. It is the freedom of the individual cells which allows the human brain to perform its magic.

Heart muscle is like a brigade of red-army soldiers, all in rigid lock-step with each other. The brain is more like a happening at Woodstock. Both groups respond to stimuli holistically. But the soldiers' response is uniform and invariable, while the audience's response is unique and unpredictable.

This spontaneity—based on freedom of the individual—is as critical to the higher functions of Cosmic consciousness as it is to our own. When a group of free individuals pulses in synchrony, responding by coincident choice in the same way to the same stimulus, then Cosmic consciousness attains a high order of harmony. This can be manifested as simply as in the beat at a rock and roll concert, or as subtly as in the common vision of a self-selected society.

Each person is entitled to live his life with perfect autonomy. The individual must be free of coercion and interference, free to choose his or her own ways and means, and free to express their own unique character. If the Foundation is built and maintained with this guiding principal in mind, we can tap the chaotic well-spring of Cosmic power. Once that power is focused and directed, we can carve out any future reality we desire.

The Foundation's common vision is of humankind carrying Life to the stars. The realization of that vision depends on Cosmic consciousness. The consciousness of the Cosmos depends on the freedom of the individual. Therefore, individual liberty must be the cornerstone of our foundation.

**The Human Laser**

In a light-bulb, the atoms jump around at random, sending off different waves of light in every possible direction. In existing societies, the people behave exactly the same way. They all buzz around chaotically, like electrified tungsten atoms. Everyone sends out his own ever-changing ripples into the fabric of time. Everyone's ripples are completely random with respect to everyone else's. The waves of chance generated by an uncoordinated society are incoherent, just like white light. An incoherent society bathes the future in a uniform glow of unfocused chance. Such a society is just as likely to actualize one potential future as another. The only thing that can be foretold with certainty about such a society is that its future is unpredictable.

By contrast, the Foundation will be a Human Laser. The people in the Foundation will act in synchronous harmony, like the molecules in a lasing crystal. By coordinating our actions and sharing a common focus we can create a coherent beam of intent. With this powerful ray of directed purpose, we can carve the future into a specific desired reality.

Waves, whether waves of intent in the probability field of the future, or waves of light in the fabric of space, share certain mysterious properties. One such property is the way that waves can combine to reinforce or negate their effective powers. For example: one hundred waves of white light, randomly interfering with each other, are only ten times as powerful as a single wave. A hundred waves of coherent light, however, are 100 times as powerful as a single wave. Coherent radiation, like that from a laser, is free from destructive interference. Therefore, lased energy is ten times as effective as the same amount of incoherent energy.[679] Coherent waves are additive in their power, while incoherent waves increase in power only by the square root of their total. Therefore, the larger the numbers, the more highly amplified is the effect of lased waves.

A person in a social laser can be orders of magnitude more powerful in his ability to actualize chosen futures than someone outside such a structure. As the number of people in the laser increases, so too does the individual's relative power. A thousand coherent people, operating inside a social laser, can have as much effect on the future as a million people buzzing randomly in the outside society.

As independent beings, we strive to make a difference in our world; we struggle to leave some lasting impact, some evidence of our passing. We are all concerned with the common fate of Life on planet Earth, and we are all trying to make the world a better place. But we live in incoherent societies where the positive individual is deprived of power. He is countered at every turn by the random, all too often negative, inputs of people around him. Too often, we are frustrated because we find that most of our efforts to do some good are neutralized. The background noise of negative static overwhelms our positive input.

In the coherent society, by contrast, we find ourselves suddenly empowered as individuals. Our energy is not being negated by conflicting vibrations. Our actions are coordinated and in harmony with the people around us. As our coherent society grows, we find that not only does our collective power grow, but so too does our individual power. Where we were frustrated and negated in the incoherent society, we find ourselves fulfilled and empowered by the social laser.

As the number of empowered individuals increases, their collective effect grows exponentially. There are five billion people in the world, but they radiate incoherently. Due to their mutual interference, the net effect of the world's population on the future probability field is equal only to the square root of five billion. By the turn of the Millennium, the global population is projected to be just over six billion. The square root of the human family in the year 2000 A.D. will be the magic number: 77,777. When the Foundation attains a coherent number at that mystical threshold we will have achieved critical mass.[680] At that point, the Foundation will have a greater effect on the future than the rest of the world combined. From then on, the shape of the future will be determined by the Foundation's focus. Once our human laser has reached this critical power level, the future of the living universe is assured.

**Focal Points**

It is generally assumed that unpredictability is one of the characteristic properties of the future. But this is not necessarily so. Complete predictability is, of course, out of the question, but we routinely predict the future for ourselves, and then actualize our prediction. It is such a routine part of life that we seldom appreciate the miracle.

I say to my brother: "I'll meet you under the Eiffel Tower at noon on August 10." At the appointed time, I find myself under the Eiffel Tower. Lo and behold, my brother appears. What a coincidence! If I had simply been wandering around the world, and happened to run into Dan, we would both have been speechless with amazement. But we <u>intended</u> to be there; so the only amazing thing was that neither of us was late.

By intending to rendezvous in Paris, my brother and I dramatically changed the shape of our future probability fields. Once we had formed that mutual intention, we created a sharp peak of probability under the Eiffel Tower. All the other potentialities for our mutual location on that day remained depressed. The chances of our meeting each other at the South Pole, for example, stayed remote.

As individuals, we do this all the time. It is the very essence of personal power. We say to ourselves at 16, "I am going to become a famous brain surgeon and live on the beach at Malibu." At 32, if you have great will power, you will be enjoying a banana daiquiri and a Pacific sunset after a hard day's surgery. If you lack motivation, you will end up waiting tables in New Jersey, wondering where you went wrong.

It is this ability to reach defined goals that distinguishes our species, and makes us the ideal instrument to fulfill Cosmic destiny.

Intention alone is not enough, however. I can intend to be in Paris till I'm blue in the face, but unless I get up and actually go there it will never happen. Just so, you can have the greatest will in the world to be a brain surgeon, but unless you continuously actualize that reality by working your ass off in medical school, you'll just end up waiting tables. A group of individuals can share a common vision. But unless they actually <u>do</u> something they will accomplish nothing.

For example: In the 60's there was a mass march on the Pentagon. Something like 50,000 people sat around attempting to levitate the monolith through their massed will. Needless to say, the Pentagon was unmoved. Now if those 50,000 people had attacked the problem with picks and shovels, they could have made a real impression.

Intention defines the point in the future probability field that you wish to actualize, but it is action that gets you there. A society without action is like a light-bulb without power. In order to shine, the atoms must vibrate.

**Motive Force**

The presence of human consciousness affords the universe a feed-back loop through which it can implement choices about the future it desires. Because of us, the fabric of future reality is flexible. The presence of consciousness in the universe changes the rules of the game. Instead of the Newtonian, clock-work universe, where the past pre-ordains the future, Life has created an organic, indeterminate universe with an unpredictable future.

There is an infinite probability field of potential futures. In many of those future universes, the Earth lies dormant—a wasted desert skeleton. There are many broad avenues of the future that lead to an Earth devoid of flowers, trees, animals and people. In fact these paths already occupy most of the available future space. For every thousand possible futures, there are 900 in which our planet lies dead, choked by smog, poisoned by pollutants, stripped to the bone by hungry billions, or burnt to cinders in atomic wars. As we progress into the determinant present, those potential futures grow, becoming ever more probable. But they do not occupy the entire probability field, only most of it. While a disastrous future is already quite likely, it is by no means certain.

There are other avenues into the future which lead to very different destinies for mankind and Mother Earth. There are alternate futures in which our planet reclaims her glory, to be clad again in verdant forests, regaining her crystal raiment of clear skies and clean waters. There are futures in which Life not only survives on Earth, but goes on to fertilize our sister worlds of Moon and Mars. And there is that glorious future in which Life goes on to fulfill its ultimate destiny and spread among the stars. As the embodiment of Cosmic Consciousness, we must insure the realization of this particular future.

**All Together Now**

First Foundation's human laser is the implement needed to carve this potential reality out of the blank surface of the future. Like a laser, the Foundation's organization has two main components: the cladding and the core. The cladding contains the great majority of people in the Foundation. The cladding is amorphous—like glass. There is no definite structure, people come and go at random. Those in the cladding go about their daily lives, and support the goals of the Foundation in any way they can. They attend lasings, read and contribute to Foundation publications, and donate goods, services, and funds. For those in the cladding, the Foundation is simply an element in the mix of their lives, one of many things in which they are involved. The function of the cladding is to pump power into the core.

The core is very different. The core contains only a small minority of the people in the Foundation. But the core is highly structured—like crystal. For people in the core, the Foundation is a way of life. Core members are sacred warriors in the Cosmic struggle for the universe. Their main focus is the salvation of Life, and its dissemination into space. These people live, eat, and breathe Cosmic destiny. They spend their time and invest their energies toward achieving the Foundation's ultimate goal: enlivening the universe.

Like the crystals in a laser, the core members of the Foundation focus the forces that can create new realities. The cladding pumps raw power into the core. The core transforms that power into a coherent beam of focused resolution. The people of the cladding make new realities possible; the people of the core make new realities happen. Often they are the same people, for individuals are apt to come and go between the glassy cladding and the crystalline core.

Undoubtedly, people of the core will be a unique breed. Core colonists are the people willing to dedicate themselves unreservedly to a higher calling. Like pioneers, they are capable of enduring hardships to attain great goals. They are as rare, precious, and tough as diamonds. With a critical mass of core members, we can drill our way out of this Malthusian gravity well that threatens to become Life's grave.

A lone spark may ignite a conflagration; leviathan has his genesis in a single cell; even mighty oaks begin as little nuts. So too, must the Foundation have its humble beginnings as a germinal idea. The Foundation will grow from a seed, precipitating atoms out of an amorphous solution, like a crystal. Charged ions will aggregate on the seed, incorporating themselves into the crystalline lattice of the Foundation's organization. The sea of humanity is the dilute solution, and those who are energized by a sense of cosmic destiny are the charged dilithium ions. These electrified 'persons' will precipitate themselves on the appropriate cathode. Eventually, these dilithium crystals, will grow large enough to power starships.

Galvanizing people into action requires the flow of ideas—the electric current of motive purpose. To grow the powerful crystal needed for the core of the Foundation's human laser we must pump high voltage ideas into the common consciousness. In its early phases, therefore, the Foundation must be an information dynamo.

***New Millennium***

After growing the first seed crystal, we will crank up the media engine. The Society's first media venture will be to launch its own magazine. The magazine—*New Millennium*—will be designed to compete directly on the open market with popular magazines like *Omni* and *Discover*. Using the new generation of desk-top publishing computers and software, a small cadre of hard-working people can put out a professional quality monthly periodical. *New Millennium* will emulate the flag-ship publication of another Society, dedicated to its own brand of exploration—*National Geographic.*

The purpose of *New Millennium* will be to present the Foundation's cosmic point of view to as wide an audience as possible. The magazine will cover relevant issues from the Foundation's unique perspective. Feature articles will detail particular aspects of the future: marine colonies, Lunar colonies, terraforming Mars, starship design. Regular departments will chronicle the growth and activities of Foundation crystals all over the world. *New Millennium* will be designed and edited with the intent of attracting as many people as possible to the idea of fulfilling man's destiny through the colonization of space.

The magazine will become the flagship of a growing media empire. In time, the Foundation will have enough power to activate that paragon of communications—television. TV has become the single most powerful means for expressing ideas that man has ever devised. The power of the tube is awesome in its potential, and we must harness that power to the chariot of our purpose.

The revolution in computer technology that makes New Millennium possible, is even now beginning to make its impact felt in the world of video. The age of "desktop broadcasting" is not far off. Armed with the powerful video tools already becoming available, we will be able to present the Foundation's ideas in ways that are clear, compelling, and viscerally exciting. Humans are visual creatures. TV, the visual medium, has impact like no other. Ideas like space colonies and star ships are more easily shown than explained. A series of polished programs about Aquarius, Asgard, and the other steps toward our destiny could do more to move people than any number of words.

Ultimately, we will establish a global television network to carry the Foundation's message to the world. CCN (the Cosmic Communications Network) will act primarily as a vehicle for Foundation broadcasts. As the Foundation grows, the network will become the extended nervous system through which colonists around the world communicate. Riding the rising wave of the cable and satellite TV industries, CCN will expand itself to incorporate a worldwide audience, ultimately numbering in the tens of millions.

We will also utilize other media formats: radio, seminars, concerts, records, computer networks, movies, even comic books—all will become vehicles for the Foundation's message. As we broaden the number and depth of our media outlets, they will begin to reinforce each other in positive feed-back loops, developing an exponentially growing synergy. Each medium will generate its own storm of associated activity, all forming reinforcing waves to pound the shores of common consciousness. As we continue to pump in power from the human laser, the Foundation's vision will begin to gel in the universal mind, like a hologram slowly being etched on a piece of film.

**Anlagen**

Once we have attracted to our banner a critical mass of hard-core cosmic warriors, we will begin to ramp up toward construction of Aquarius. The first chapter of this book describes the construction of Aquarius, but that construction cannot begin until the foundation has been laid. Building up to Aquarius requires first the formation of a world wide organization, bound together by the fiber optic threads of a telecommunications network. Once that organization has sufficiently matured, the first experimental space colonies can be established. When these test-bed colonies have ripened, then and only then, can we go on to the actual construction of Aquarius. Building a small demonstration space colony—Anlagen— is the first sub-step in that direction.[681]

| **Anlage**: "The foundation of a subsequent development; the first recognizable commencement of a developing part or organ in an embryo; an egg." |
|---|

This first colony will serve as a showcase, demonstrating how to live in symbiotic harmony with an environment. Whether that environment is the surface of Mars or the tropical sea, the lessons of Anlagen will be the same.

Two operative philosophies guide the development of all Millennial space colonies, including Anlagen. The first philosophy, is to live lightly—minimizing consumption. The second philosophy is to live symbiotically—existing in self-sustaining harmony with an ecosphere. Both of these principles are essential. When fully developed they will allow us to live harmoniously anywhere in space—including on our home planet.

Achieving a fully developed life-style of 'light symbiosis' will be of vital importance, not only to space colonists, but to humanity in general. Minimum consumption and maximum recycling are essential prerequisites on hostile worlds like the Moon and Mars. Light symbiosis is equally vital to the continued survival of humanity here on Earth. Ideally, the light lifestyle, and symbiotic techniques we develop in our space colonies will spread out into the global culture. These techniques for living can help to ameliorate our species' impact on all planetary ecospheres.

Living lightly means minimizing consumption while optimizing existence. Simply put—less is more. To live lightly is to live crisply, cleanly, with a streamlined life-style, disentangled from the debris of materialism. Space colonists need few possessions and must minimize their material attachments. Life in the colony is simple and uncluttered, but not Spartan. Great attention is paid to personal comfort and individual development. A minimalist approach will engender an elegance in our homes, our clothes, our architecture, and by extrapolation—our attitudes. Separating ourselves from the boxes of useless bric-a-brac that haunt our terrestrial existence is important. It frees us from material bondage; it is more economical of space and resources; and it makes it easy to pack—an important consideration for an upwardly mobile culture, on its way to the stars.

Living lightly is an essential life-style for space colonists. Like Bedouin, Eskimos, and other nomadic people, we will, by necessity, carry an absolute minimum of possessions. When moving into space, or living there, every single pound of inanimate matter, even air and water, carries a high price-tag. As a mobile, dynamic, outwardly expanding people, we will not be able to drag along U-Haul™ loads of garage-sale merchandise. Minimizing our encumbrances is an inevitable necessity. We may as well get used to it now.

Symbiosis means living in ecological balance with our environment. This is achieved by closing process-loops, creating systems in which wastes become feed-stocks. Symbiosis shares and complements many of light living's desirable attributes: Symbiosis is an objective moral good. It reduces the pollutants we emit to the outside environment to nearly zero. The gross entropy of our living system is thereby minimized. Symbiosis has great potential economies. And it is essential to a space-faring society.

The first Millennial colony will incorporate many of the features which will later be common to all space colonies. It will be designed to minimize costs of all kinds: energy, environmental, and economic. The colony will be as self-sufficient in food, energy, and services as possible. This first colony will serve as both test-bed and prototype to develop and demonstrate techniques to be used in space. This prototype colony will be established in an area where climatic, economic, and political conditions are optimal.

Eventually, we will build our own planets from scratch, but we must start somewhere. Therefore, we will choose the most benign space available for our first colony.

Accordingly, Anlagen, will be located somewhere in the Caribbean Sea. In a Caribbean location we will be able to develop the life style that we will carry with us to ecospheres throughout space. The advantage, of course, is that in the warm waters of the West Indies, there is no need to construct an ecosphere. Earth has already provided the environmental basics.

The political structure of the region is reasonably stable, and the Caribbean is close enough to the United States to be easily accessible. Accessibility is an important market consideration. Steady streams of visitors are essential to the economic self-sufficiency of all space colonies. Anlagen will be built from the ground up with guests in mind. It will be our intention to attract as many visitors as possible and to make their sojourn with us a memorable and enjoyable introduction to the Millennial life-style.

Visitors are extremely important for reasons other than cash flow. Every person who returns to the wider world from a visit to Anlagen becomes an ambassador carrying our message. If that individual is favorably impressed with what she sees, she will spread positive ripples through her own society. The negative is equally true, making visitors crucial to our long-term success.

Each guest is also a potential colonist. Anlagen, will be a place to demonstrate the technical and social workings of a space colony. If people are attracted by what they see, so much the better. The manipulative techniques of hucksters and cultists are of no use to us. We need clear-headed, hard-nosed realists who have made up their minds for themselves. If, after careful and skeptical reflection, a person decides to join us, then hot damn! Another noble warrior casts her lance with the cosmic legion.

**The Incubator**

Anlagen will provide a laboratory to test-fly social and cultural innovations. As an experimental colony, Anlagen's charter will specifically sanction social improvisation. From this living laboratory will eventually emerge a flexible template on which future space colonies can be modeled.

Solutions to the many problems faced by a new society cannot be imposed by bureaucratic fiat; they must evolve openly from organic interactions. Anlagen will provide a secure stage on which multiple scenarios can be played out. In the nurturing environment of a tropical colony, many alternatives can be tested. Free of the harsh forces that will dominate in the black ocean of space, nascent social systems can be fostered. Eventually, trial and error will winnow out the impractical, the uneconomical, and the inimical. The distillation process will ultimately leave us with workable and adaptable interactive systems, capable of coping with every facet of socio-cultural life.

We will be aided in this process by new technologies. The capacities of data processing and telecommunications hardware and software are expanding at breathtaking rates. By the year 2000, about the time we are ready to inaugurate Anlagen, individuals will have the power of supercomputers at their fingertips. This has innumerable ramifications, many of which are synergistic and so can't be accurately predicted. From the standpoint of an emerging new society, the implications are clear. These tools will allow us to institute social and cultural techniques dramatically different from any others that have ever been possible—the advent of 'real' as opposed to 'representative' democracy being only the most obvious. The new technologies can be used to radically transform fundamental organic systems like economics. With the massive computational powers at our disposal, the contribution of each individual to the economic well being of the colony can be accounted for with unprecedented exactness. This may allow evolution of a truly communal economic system, in which the individual is recognized and rewarded. Economic activity in an "intelligent environment" promises to be far more efficient. Just as mass-production techniques revolutionized the industrial era, ubiquitous computing will revolutionize the information era. With an interactive environment to track and evaluate productive activity, most layers of management can be eliminated entirely.

Significant new advances in software will enable a colony to enjoy the advantages of ubiquitous computing, without risking the dangers of "Big Brother". New systems of unforgeable encryption, allow individuals to enter electronic transactions while maintaining complete anonymity. The new techniques eliminate the possibility of compiling electronic dossiers, and put individuals firmly in control of all their personal data.[682]

The new tools at our disposal will allow us to function as nothing less than social genetic engineers. In essence, the cultural matrix is the DNA of a social organism. With powerful new tools, operating in the living laboratory of Anlagen, we can compress a thousand years of social evolution into a few decades.

Once Anlagen has fully ripened into a viable colony, we will be ready for the next strategic step—New Eden.

**New Eden**

New Eden is an intermediate step in the Millennial plan. It is a step designed to provide a bridge between the demonstration colony of Anlagen and the marine colony of Aquarius.

The idea is simply to find the best space in the universe to colonize. Since we are already on Earth, and for the time being are confined to this planet, we will limit our search to Gaia's surface. The tropical oceans immediately recommend themselves as habitable space. Aquarius needs to be near the coast of Africa to provide power for the Bifrost Bridge. We, therefore, turn our search to the equatorial regions of the Indian Ocean. There we find the Seychelles Islands.

New Eden takes its name from a curious historical footnote. General Charles "Chinese" Gordon, the famous defender of Khartoum, visited the islands around 1881. While there, he formulated the interesting hypothesis that the Seychelles were the flooded remnants of the original Garden of Eden. Chinese Gordon even purported to have located the Tree of Life in a remote valley on one of the main islands. That the Seychelles are the original home of primordial man is a doubtful hypothesis at best, but their idyllic splendor is undeniable. As space colonists it behooves us to survey the known universe for the most suitable habitats for our first colonies. We may have to search this galaxy for a very long time before we find a more ideal place than the Seychelles.

In New Eden, we will construct a true space colony. Built along the lines of Anlagen, New Eden will derive its power from the sea. A shore-based OTEC facility will provide all of the colony's power, water and food.[683] Surplus energy from the OTEC will be converted into hydrogen to fuel the colony's ships, planes, and dirigibles. The OTEC will also produce a steady flow of fresh water for irrigation of crops and fruit trees, for swimming pools, fountains, and waterfalls. Using the nutrient rich water from the depths, we will establish a variety of aquacultural enterprises to produce food and revenues for the colony. Hydroponic greenhouses will provide the bulk of the produce, with thousands of fruit and coconut trees being cultivated on the island.

The nutrient-rich cold water from the OTEC will be discharged into the shallow waters around the island. The nutrients thus released will stimulate the overall food-chain. The result will be a dramatic increase in fish production. Intensive aquaculture operations will be established in the shallow waters of the inter-island lagoons. In these areas, we will raise enough food for the colony, for the resort facility, and for export. In New Eden we will develop the beginnings of the sea-based industries that will later help support Aquarius.

The OTECs create a colony self-sufficient in energy, water, and food, allowing us to colonize any of the thousands of desert islands throughout the world. We will start with an uninhabited waterless island. We will then transplant a stable seed colony from Anlagen to this apparently desolate location. Then, using water and resources provided by the OTEC, seed colonists will transform the desert island into a little bit of paradise.

Unlike most tropical islands, the Seychelles are not of volcanic origin. The islands are granitic extrusions from the ancient super continent of Gondwanaland. There are some 90 islands in the group, about 30 of which are granitic. The rest are coral atolls. Each type of island has advantages and disadvantages. For our purposes, one of the desert coral atolls will probably be best. Such atolls are prefabricated Aquarian foundation platforms, already anchored in place, and with preformed mariculture containment lagoons. With our OTECs, we can transform barren slabs of dry coral into vibrant garden colonies. Once we have refined the process, it will be possible for other colonists to take the techniques and transform desert islands all over the world into viable colonies. Learning how to terraform desert atolls is our first baby step along the highway to the stars.

The Foundation will secure a long term lease to one of the islands in the group from the Seychelles government. This ought not to be insurmountably difficult, since most of the islands in the group are without fresh water and consequently uninhabited. Since our OTECs produce rivers of fresh water, we can have our pick of splendid islands no one else wants. There are 50 coral islands in the group and almost all of these are unoccupied. There is an international jet port in Victoria, the capital city on the largest island of Mahe'. This will provide good access to the islands, not only for ourselves, but for the many visitors we will attract.

Tourism will always provide one of the driving economic engines for the Foundation's expansion into space. Therefore, construction of a resort complex on the island will be one of our first major projects. The resort will be fashioned along the lines of super-resorts like the Hyatt Regency Waikoloa in Hawaii. These resorts are lavish fantasies of tropical luxury. There will be a multitude of activities for the guests. In addition to the usual leisure pursuits: scuba diving, submarine tours, sailing, surfing, water skiing, gambling, and golfing; it will also be possible for interested guests to study and participate in the activities of the colony.[684]

New Eden establishes a stable platform where we can perfect the means and build up the resources necessary for the construction of Aquarius.[685] Until we are ready to begin building Aquarius, we will make preparations in Eden. Research will be conducted on cybergenic construction techniques, like the accretion of galvanic sea cement. Detailed design and engineering of Aquarius will be completed. The marine seed ship—a mobile OTEC platform—will be constructed. Research will be done on the culture of algae from cold deep sea water. And economic reserves will be built up. Once we are ready, the actual construction of Aquarius can begin.

**Coming Full Circle**

After a long and winding journey, we have arrived at the beginning—again. This strange circular route is necessary because the Project is a process. It is an evolutionary process, and each succeeding stage is organically dependent on the stage preceding it. Aquarius can not spring spontaneously into being with the wave of a magic wand. Creating the Foundation is prerequisite to building Aquarius. But the catalyst needed to coalesce the foundation is a vision, and a plan for realizing that vision. Such has been the purpose of this book.

# Prelude

So there it is. The plan for the Millennial Project is planted in your mind—a spark, a germinal seed. You can see the whole thing, built up in layers like a Ziggurat. You can see it all: the floating islands of Aquarius, churning out enough food and clean power to save the planet; the Bifrost Bridge, lighting our rainbow road to space; the golden living bubbles of Asgard and Avallon, homes in sterile spaces to people and Life of all kinds; terraformed Mars, swathed in the gossamer lace clouds of a living world; the glistening myriad ecospheres of Solaria, a civilization of astronomic magnificence; and far out in the depths of interplanetary space, an exponentially growing wave of seed ships engulfing the star fields of the Milky Way in a green explosion of Life. You comprehend it all, holding the fate of an entire galaxy within the compass of your mind.

**Moment of Decision**

Saving the Earth and enlivening the universe are not questions of engineering, or physics, or even economics; they are matters of will power. We have the power. Now we must find the will.

What happens next is entirely up to you. The spark can whither and die, as if falling on cold iron. Or the spark can explode, as if igniting jet fuel. Extinction or explosion? You alone determine the result. Whatever is happening in your mind at this moment will be played out a million times over in the minds of others. Look now into the workings of your own consciousness, and see reflected there the same thoughts in a million other human minds. If you are enthralled, excited, stimulated, even mildly interested; then all is well. If you are irritated, alienated, disinterested, or bored; then welcome to the catastrophe. Whatever is happening inside your head will happen too in those million other brains. If this work has succeeded in making you want to create the shining future of a living galaxy, then it will succeed a million, a billion times more; and these things, and things more wondrous than we can imagine, will all come true. If it has failed, then the task must await some other more persuasive bard than I.

If your answer is *No*, that negative will echo a million times. Such an answer may negate the only viable future available, for our species, for our planet, perhaps for Life itself. If you decide against us, if you turn away, then throw this worthless pile of paper in the trash, and lament! For you will know that Life, the great experiment of Cosmos, is lost; and never, not in a billion eons, will such a chance come again. If your mind is cold, if these ideas have struck in you no flame of desire to grasp our Cosmic destiny, if you do not yearn to reach out and clutch by handfuls the myriad stars—then I have failed. Unless some other, more epic bard can sometime stir up your ashes to a growing fire, then all of us are lost: the whales, the butterflies, the flowers and the trees, you and me and all humanity yet unborn; all lost.

If, however, your answer is *Yes,* that affirmative will amplify a million times. A million minds can be set alight by that single spark. Shining futures can be made real; the planet can be saved; Life in the Cosmos can fulfill its shining destiny. If your mind is hot, if you feel the seed of this idea growing in your fertile brain, then there is cause for ultimate hope. If a fever of good purpose is rising in your head, then we are saved. If your soul takes flame with desire to carry these things forward then they shall indeed come true!

Now is the ultimate moment in the history of the universe. The fate of a million million worlds hangs here in the balance. An infinity of space awaits, hushed, afraid even to breathe, anticipating the next, the crucial moment—your decision.

### A Call to Arms

A million years, a thousand millennia from now, our progeny will look back in space and time to the glimmering dawn of galactic history. At the base of the awesome branching tree that is their living galaxy they will perceive a tiny blue-green seed. That morsel of Life, they know as the Mother of all planets. From that single cosmic micro-spore will have sprung a pan-galactic riot of Life. Throughout the coruscating star clouds will pulse a blizzard of animate matter; vibrant and alive, defiant of entropy. A hundred billion stars saturate the velvet sky with the virescent color of hope. Lovers, poets, and children beyond numbering, giggle, chortle, and laugh; filling the once silent voids with this most sacred music of consciousness. Macroscopic minds of a hundred billion hundred billions contemplate planes of mentality unknowable by us; yet for all their prodigious majesty, these demigods will reflect on our time with wonder.

A million years from now, the warriors of Troy in their horse-hair crested helms of bronze, and the heroes of Mars in their antennaed helmets of boro-silicate glass will all seem of a piece. Telescoped by unfathomable reaches of space and time, the few thousand years of Man's early history will seem but a moment. Our race will appear to have erupted from a long evolutionary gestation; transformed from gibbering apes to starmen in a single explosive event. Pre-stellar human history will look like the detonation of a runaway chain reaction. Our starships will be like shards of shrapnel, riding the shock wave out into the empty galaxy. Behind them will spread the expanding sphere of Life, as the real "Genesis Effect" transforms the universe.

Ours is the most mythic of all human ages. We stand poised at the march of a great transition. Behind us is the solid road of all history; from the creation of the universe to the present moment, all leading to this pregnant time of focus. Beyond, lie the realms of infinite choice, divided into two great camps: the universe as it has always been, utterly empty and devoid of life, and the universe as it can be, transfigured and alive! The time is now and the choice is ours. The destiny of the Cosmos is in our hands.

If we fail, then the living galaxy will never be born. To destroy the living universe we need only do nothing. The gateway to heaven has only just swung open. Very soon it will reach its widest margin, and then it will slam shut again, possibly forever.

We can sit around here on our collective ass and do nothing, watching, while the sole opportunity of this eternity slips by. Or, we can seize the day, grasp destiny by the hand, and leap forth to fill our place in the Cosmic order. It is up to us to enliven the universe. The time has come to arise and unite, ye warriors of Cosmos!

Let us form of ourselves an army of starship troopers, gird on our tungsten armor, and go forth to do battle with Chaos and the minions of darkness. Let us unfurl the green and golden banners of our cosmic crusade, storm the bastions of vacuum, sow our sacred seeds among the stars, and take this galaxy in the name of Life!

There are but a handful of us to wage this titanic battle, this *Gottermorgendammerung*, this Dawning of the Gods. Look now into the inner reaches of your soul, there you will find a creature of destiny, a glowing being, endowed by Cosmos with the fire of a sacred purpose. You were created in order to catalyze the transformation of this universe. You live so that the Cosmos may live. You are one of the god-beings of this universe. One of only a few existent things capable of translating your will into reality.

The time has come for you to use your awesome powers for the end they were intended: creation of a living universe.

Now your course is simple—join us. Throw your shoulder to the wheel. We have far to go, much to do, and little time. Together we can bring this dead universe to life. Come with us, add your power to the energy flux of the Foundation's laser. Beam up with us. We are going to the stars.

**The Beginning...**

# FIRST MILLENNIAL FOUNDATION

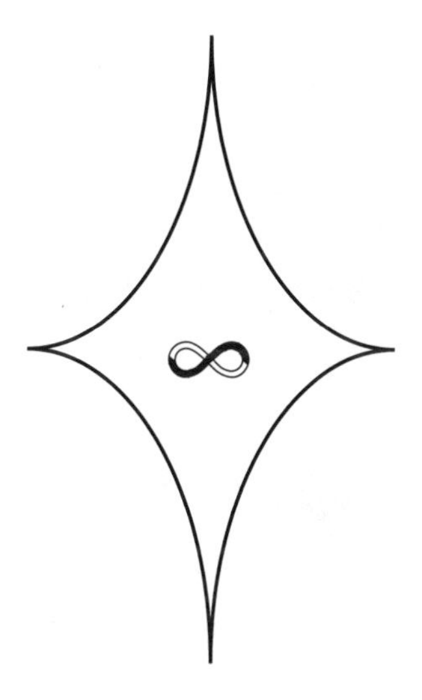

**P.O. Box 347**
**Rifle, Colorado 81650**
**Email: 73163,3612@CompuServe.com**
**Internet: mtsavage@pipeline.com**
**BBS: (303) 625-3273**
**(303) 625-2815**

# *APPENDICES*

## APPENDIX 1
## Aquarius

### 1.1 Oceanic Thermal Resources

There are more than a million cubic miles of warm surface water in the tropical oceans.[686] In one year, ten thousand sea colonies will cool 64,000 cubic miles of sea water.[687] It would take 150,000 colonies to fully deplete the available supply. Ten thousand colonies would absorb less than two percent of the energy available in the equatorial seas. To rewarm 64,000 cubic miles of water to 85° F. requires energy equal to a trillion barrels of oil. At the equator, solar energy input to the ocean is 1860 BTU per square foot per day—the equivalent of three million barrels of oil in each square mile every year.[688] The annual solar energy input to the equatorial warm-water zone is equivalent to 60 trillion barrels of oil. World oil reserves are two trillion barrels.[689]

### 1.2 - Reductions in Floor-space Requirements

In Aquarius all manufacturing operations will be carried out on floating platforms separate from the main structure of the colony. Therefore, factory space, 23.5 $ft^2$/capita, is not needed in the surface structure. Goods are distributed directly from storage so retail space, 44.6 $ft^2$/capita, is unnecessary. Commercial space used as residential space and vacant space, 24.8 $ft^2$/capita, do not count toward the area needed by the colonists either. These areas, amounting to 93 $ft^2$/capita, can provide 40% of the total commercial space requirements. Subtracted from the 225 square foot requirement, this still leaves the colonists with the American average for schools, offices, and other public buildings—102 square feet. Some activities which normally require additional separate space will be performed in the home. Many Aquarians will choose to work in their residences, telecommuting through their personal computers, reducing office space requirements. Part of the educational process takes place through interaction with a computer terminal; accordingly many students, especially at the university level, will choose to attend classes at home.

The subsurface platforms also provide an abundance of open floor space. While the bottom quarter of the support platforms is flooded to provide reserve buoyancy, the rest is dry. Altogether, there will be 13.5 million square feet of floor space available inside the buoyancy platforms able to house many large components: circulatory machinery, waste treatment facilities, bulk stores, warehouses, etc. Just by moving warehouses into the support platforms, 30 $ft^2$/capita can be subtracted from the area needs of colonists in the surface structure.

### 1.3 - Breakwater Barrier Specs.

The breakwater will be constructed in hexagonal sections of the same dimensions as the residence towers, using the same construction templates and techniques. Four concentric rings of platforms will provide a total surface area of 95 million square feet. The outermost ring with 350 platforms will form the primary breakwater.

The outer ring of platforms is exposed to the open sea and will take the brunt of its punishment. The forward sixth of each of these outer platforms will actually be under water. Every ten feet there is a three foot rise in the level of the platform, so the outer edge is 24 feet under water and each subsequent step is three feet closer to the surface. The distance from the outer edge, to the point where the platform surfaces is 82 feet. This submerged stair-step formation will cause waves to break as they approach the surface portion of the platform. This stair-step formation will continue, for another 82 feet laterally to the middle of the platform. At this point, the platform will be 24 feet above the water. Each of the steps above water is planted with a row of palm trees. There will be eight rows in all, each row staggered ten feet behind the row in front of it. From the crest to the back of the platform is an identical set of stair-step platforms, thickly planted with tropical fruit trees. This outer ring of platforms forms a sea barrier with an effective height of 48 feet, and a width of 328 feet. Not even the most powerful hurricane driven waves are likely to surmount such a barrier.

Inside this outer breakwater ring, are three more rings of platforms. These platforms, which serve primarily as agricultural and residential space, are only three or four feet high. These three rings provide altogether a surface area of around 70 million square feet. In this area, are the colony's hydroponic greenhouses, gardens, and fruit orchards. In this space, all of the colony's fresh fruits and vegetables will be grown.

### Appendix 1.4 - Anchoring

Mooring a floating structure the size of Aquarius will be a significant challenge. The design's modularity makes it feasible, however. Each OTEC cylinder and each supporting float will be individually anchored by a cable attached to the middle of the platform. The anchoring scheme is not dissimilar from those used to hold tension-leg oil platforms in place.

The anchoring cables will be of high performance Kevlar with a breaking strength many times that of steel. Each cable will have a diameter of nine inches with a breaking strength of 25 million pounds.[690] In addition to these vertical anchoring cables there will be three catenary cables attached to the outer edge of the surface structure. These catenary cables run in long curves to the sea bottom, where they are anchored. The catenary cables provide backup for the vertical cables and buffer twisting and drifting motions of the structure.

A modular approach to the anchor lines will give the composite effect of a single anchor cable 153 inches in diameter. Such a cable would have a breaking strength of around 30 billion pounds.[691] The fact that each cable is stronger than it needs to be and that it shares its load with many other cables builds tremendous redundancy into the system.

Kevlar is not an especially elastic material, but it will stretch 3-4% without any difficulty. The cables securing Aquarius will be a minimum of 3500 feet long and possibly as much as 15,000 feet long. With a 3% stretch in a 3500 foot cable, Aquarius can rise and fall as much as 105 feet. This is an ample allowance for the vertical action of tides.

The life of steel cable in sea water is limited, even when galvanized and protected with grease or tar. For this reason, the anchor cables must be of engineering plastics. In the future, there will be cheaper alternatives than Kevlar available. Recent advances in plastics technology have made it possible to produce cheap materials with more than six times the tensile strength of steel. By compressing and stretching an inexpensive grade of plastic like polyethylene, it is possible to force its long chain polymer molecules into long filamentous alignments, with many cross-linked carbon-carbon bonds. It is theoretically possible, using polymer cross-linking, to produce plastics with up to 40 times the strength of steel, on a pound for pound basis. It is quite likely that by the time we are ready to deploy Aquarius, plastics with 25 to 50% of this theoretical maximum will be available. The result will be a material that is light, strong, inexpensive and corrosion proof. Engineered plastic cables with a breaking strength 10 to 20 times that of steel will make it feasible to economically anchor marine colonies, even in very deep water.

The anchors themselves will be of sea cement accreted in place. A single dead-weight anchor will rest in the sediment, directly beneath each platform. Each anchor will weigh 11,500 tons.

## Appendix 1.5 - Nutrients in Deep Sea Water

Deep waters are also a rich source of nutrients other than nitrogen. Deep water also contains a relatively high concentration of phosphorus. Typically, there is ten times as much phosphorus in sea water at a depth of 3000 feet as is found at the surface. There are only .2 milligrams of phosphorus per liter in most surface waters, while at 3300 feet in the Indian Ocean there are as many as 9 milligrams per liter.[692] Other micro-nutrients like iron, copper, and manganese, are at twice the concentration in deep water as that found in surface waters.[693]

The other key nutrient for plant growth is $CO_2$. Carbon dioxide provides plants with a source of carbon. When plants undergo photosynthesis, the energy of sunlight is used to combine carbon dioxide with water to form sugar and oxygen, in the chemical reaction: $6CO_2 + 6H_2O \rightarrow C_6H_{12}O_6 + 6O_2$. On land, the minute concentration of $CO_2$ in the atmosphere often serves as a growth limiting factor. This is seldom the case at sea because the concentration of $CO_2$ in sea water is many times that in air. Sea water, due to its slight alkalinity, will hold as much as 45 cc/liter (4.5%) of carbon dioxide dissolved in solution.[694] Compare this to the concentration of $CO_2$ in the atmosphere—a mere .03%. There is 150 times as much carbon dioxide dissolved in sea water as there is available to plants in the air. For this reason, aquatic and marine plants always have enough carbon available to them. In fact, there is little difference between the amount of $CO_2$ in surface sea water and in the intermediate and deep waters.

## Appendix 1.6 - Spirulina Farming

To provide the initial seed population, a small volume of water, containing a high concentration of algal cells, will be piped back to the point where cold water enters the containment ponds. There, the algae will be injected into the cold water stream in a steady and controlled flow. Since each algal cell is a seed, this method provides a simple way to sow the original crop.

Phytoplankton are able to double their population three and a half times a day when there is an abundance of available nutrients and there are no other growth limiting conditions.[695] A single algal cell can multiply itself by 1000 times in less than three days. Some rapidly reproducing strains of Chlorella can double their populations in two hours and thirty six minutes—nine times a day.[696]

The discharge temperature of the deep water, 45° F., need not be a fatal obstacle to the rapid growth of algae. Annual plankton blooms occur in the polar seas at temperatures as low as 33° F.[697] Optimal growth temperature for Chlorella in sea water is

53.6°F.[698] Spirulina prefers higher temperatures and greater salinity, but the prospects for creating hybrid algae strains through genetic engineering are very good. Ideally, we will be able to produce a strain of Spirulina that, like Chlorella is able to reproduce rapidly in cold sea water. Discharge of the warm, surface water into the containment ponds raises the average temperature to nearly 62°F. Further heat gains from fresh water production, air-conditioning, cold-bed horticulture, and insolation, raise the pond temperature by a further few degrees.

After the cold deep water is discharged into the containment ponds it will flow slowly toward the perimeter. As the water travels, the algae will use the available nutrients to fuel their growth. After 24 hours, the algae saturated water will reach the outer perimeter wall. Three large floating processing factories sit along the perimeter wall of each segment. Continuous flow diatom nets will reel in the rich harvest of spirulina.

Algae, scraped continuously from the nets, will form a wet paste, comprising 70 - 90% water.[699] This paste is first pasteurized at 160° F. for about 20 minutes. The hot paste is then injected at high pressure into a steel chamber kept at 390° F. 93% of the moisture is removed as vapor, without overheating the algae or degrading fragile nutrients. The atomized Spirulina mist flash dries in the chamber to a fine powder which collects at the bottom of the vessel. This powder is removed for further processing.

After the water has been pumped through the algae strainers, it will be discharged into the outer containments. There, seaweed strips most of the remaining nitrogen, and purifies any metabolites produced by mariculture. The water is drained out the bottom of the outer containments at a depth of 100 feet. Once released, the water, now at about 65°F., will sink to its equilibrium depth.

### Appendix 1.7 - Capital Costs

Compared to standard construction techniques, growing Aquarius by cybergenic means is faster, simpler, and cheaper. In colonial economics, there is no cost of capital, so the costs of growing Aquarius are: 1- the incremental cost of power produced by the OTECs, 2- the life-support costs of the construction crews, and 3- the machinery.

Once the capital costs of an OTEC power plant are discounted, the cost of electrical power is incremental, since the fuel cost is zero. Without capital costs, electrical power can be produced from an OTEC for 1/2¢/kwh. However, half this cost is labor which is already accounted for during construction, so the actual cost for the power to grow Aquarius cybergenically is only a quarter of a cent per kilowatt hour. The total amount of power needed to accrete all of Aquarius's components is around eight billion kwh, at a cost of $20 million.

Sea cement is a lot like OTEC power in that the cost of the raw materials is zero. The finished product's costs are limited to capital, energy, and labor costs so total costs of finished sea cement are extremely low.

Millennial colonies are by design largely self-sufficient in terms of life-support, so the colonists costs of living are minimal. All Millennial colonies provide their own food, energy, and internal services like medical care. Cash expenditures for outside resources to support a crew of Millennial colonists will not exceed $10 per day per capita. The construction crew, making extensive use of robotics and cybergenic control systems, will number under 10,000, including on-site support personnel. Construction time needed to bring Aquarius to full completion is around 6 years. Overall costs for the construction crew will be roughly $220 million.

At $10/ft$^2$, finishing costs inside the surface structure, will be about $1 billion. Interior finish is extremely simple. The free-form sea-cement walls will be covered with a thin layer of plaster, derived from sea shells. The sea cement floor is left bare, and is finished by robotically operated hydraulic milling machines which even out irregularities and create a non-slip textured surface.

Internal service infrastructure is estimated to cost no more than $10,000 per inhabitant, or $1 billion. This figure would cover the cost of central food processing facilities, a pneumatic delivery system, sewage treatment, plumbing, electrical wiring, etc.

Goods and toys are estimated at another $10,000 per inhabitant. This figure would supply communal resources like speedboats, wind surfers, video cameras, etc.

Mariculture structures will cost around $120 million and the OTEC machinery will cost $1.1 billion. Miscellaneous items, like tools, ships, supplies, transport, etc. will all run about 25% of other costs, or $900 million. A tourist resort will cost about $350 million. A medical center will be built for about $700 million. This will be a state of the art facility designed not only to meet the medical needs of the colony, but also to attract a world-wide clientele for special therapies that are apt to be unavailable any where else. Research into questions of longevity and development of novel treatments to prevent disuse diseases associated with aging, will be an area of particular emphasis. The hospital will become a major source of employment and revenue for the colony. The computer utility will cost around $1.2 billion The total cost for Aquarius will be in the neighborhood of $8 billion dollars.

**Table 1.7.1**
**Capital Costs of Aquarius**

| **Item** | **Cost** $X10^6 |
|---|---|
| OTECs | 1,102 |
| OTEC shells | 83 |
| Aquarius shell | 275 |
| Interior finish | 1400 |
| Utilities | 1000 |
| Goods and Toys | 1000 |
| Mariculture Ops. | 120 |
| Resort | 350 |
| Hospital | 700 |
| Central computer | 600 |
| Data facility | 600 |
| Misc. costs | 770 |
| **TOTAL COSTS** | **$8,000** |

# APPENDIX 2
# Bifrost

### 2.1 - Rocket Science

The 'fuel' in a typical solid propellant rocket is 51% nitrocellulose (smokeless gunpowder) and 43% nitroglycerine. The solid-fuel boosters on the space shuttle are really just dynamite sticks the size of grain silos (45.5 meters long by 3.7 meters in diameter).

A rocket's ability to propel payloads to high velocities is directly proportional to the energy of the reaction driving it. In the case of chemical rockets, this is strictly limited by the propellants involved, hence the reliance on high explosives. Rocketry is Newton's third law in action: "*To every action force there is always opposed an equal reaction force.*" Propellant blasts out the back end, and the rocket moves skyward. That forward motion is in exact proportion to the action of the propellant. If the rocket and fuel weigh 1000 kg. and the rocket expels 1 kg. at 1000 mps (meters per second), then the rocket will move forward at one mps. The situation is always in balance, with the speed of the vehicle dependent only on the velocity of the propellant and the mass of the rocket.

In real rockets the situation is more complex. Propellant is continuously expelled, so the weight of the rocket is always changing. Because of this, the propellant expended at the end of a rocket's burn has vastly more influence on the rocket's final velocity than does the first propellant expended.

To express this changing relationship, there is a fundamental equation used in rocketry; called, after the great Russian visionary, 'Tsiolkovsky's equation': V = v *log* e M/m. The velocity of the rocket (V), is equal to the exhaust velocity of the propellant (v), times the natural logarithm of the mass ratio (M/m). The exhaust velocity is the speed at which the propellant is expelled from the reaction chamber of the rocket. The mass ratio is the ratio of the total weight of the rocket as it sits on the launch pad to the weight of the rocket after all the fuel has been burned. Since the natural logarithm of 2.718 is 1, the final velocity of a rocket with this mass ratio will equal the exhaust velocity. In short, the final velocity of a rocket, for a given mass ratio, is directly dependent on the exhaust velocity of the propellant.

Exhaust velocity is in turn directly dependent on the energy of the reaction which expels the propellant from the rocket. In chemical rockets, this is first limited by the energy available in the reactions involved, and is further limited by the heat produced by such reactions. Very few materials retain much structural integrity at temperatures over a couple of thousand degrees.

Today's rockets all use chemical energy. This is the same energy that is involved in all organic processes, it is the energy of life and chemistry, it is the energy of fire. In the hands of man, explosives are the ultimate expression of chemical power. As long as we are confined to chemical energies, rockets are hopelessly primitive. The principle of rocket propulsion is absolutely basic, it is grounded in the laws of physics and embedded in the very fabric of space, so we cannot discard the principle of rocketry. We must however, advance beyond chemical sources of propulsive energy.

In chemical reactions, the latent energy of the reaction is often expressed in the form of heat. Heat is just a measure of random molecular motion. The higher the degree of that motion, as expressed by the mean average velocity of the molecules, the higher the temperature. Heat energy has enormous disadvantages. The main problem is that the motion of the particles in the propellant is entirely random—explosions explode. Somehow the explosion must be contained, both physically and thermally. Unfortunately, chemical elements can only provide us materials that will remain solid at a few thousand degrees. Accordingly, most rocket motors have to be artificially cooled. This takes an extremely simple engineering principal, and turns it into a plumbing nightmare of epic proportions. Operating temperatures of some modern rockets have been raised as high as 2800° C.

(5000° F.), but only by means of the most heroic feats of plumbing ever conceived by the mind of man. Fireclay brick melts at 1750° C., and even zirconia brick melts at 2600° C. There is no available material which can withstand even these relatively moderate temperatures.

Engineers have been able to achieve high temperatures in the rocket's reaction chamber only by actively cooling the exhaust nozzle. This is achieved by an arrangement of tubes, pipes, and pumps of almost mind boggling complexity. In modern liquid fueled rockets, the walls of the nozzle and reaction chamber are built of bundles of bronze or copper tubing. When the rocket is ignited, the liquid fuel is pumped through these tubes at very high speed, thereby cooling the walls of the chamber and keeping the thin metal from melting. To make such a system work requires dozens of high speed turbo pumps, spinning at tens of thousands of rpms. These turbo-pumps are miracles of engineering and craftsmanship. The pump blades are carved by hand from blocks of surgical stainless steel by the best machinists on earth. Each blade is delicately curved as a rose petal and sharp as a scalpel. When spun up to full speed these pumps impel thousands of gallons of liquid fuel at enormous velocities and pressures through a spaghetti of tubing. The consequences of a single failure in this elaborate construction of high pressure tubing, high velocity liquid explosives, madly spinning pumps, and hellish temperatures are as instantaneous and disastrous as they are obvious. The only thing more amazing than the fact that we can build such elaborate mechanisms at all is the fact that having used them once we routinely throw them away.

The formula expressing the relationship between rocket performance and exhaust temperature is: $v \approx k\sqrt{t/m}$. The exhaust velocity (v), is approximately equal to .25 (k), times the square root of the temperature of the exhaust in degrees Kelvin (t), divided by the molecular weight of the propellant (m). As the temperature of a given propellant goes up so does its exhaust velocity. If the rocket fuel is liquid hydrogen with liquid oxygen as the oxidizer, the propellant will be water with a molecular weight of about 18, and if the combustion temperature is three thousand degrees Kelvin (2727° C.), then the exhaust velocity will be about 3 kps (kilometers per second). Chemical fuels are limited to maximum exhaust velocities of only 4-5 kps. If the temperature of the same propellant could be raised out of the realm of chemical fires and into the higher domain of thermonuclear reactions—around 100,000°—then the exhaust velocity could be increased to upwards of 40 kps.[700]

Orbital velocity around the earth is 7.77 kps (17,380 mph). It would seem an impossibility therefore to achieve orbit with a chemical rocket.[701] While most rockets cannot achieve exhaust velocities of more than about 2 kps—the space shuttle, for example, which is considered to be pushing the state of the art, can achieve an exhaust velocity of three kps—some rockets can achieve higher final velocities in two ways.[702]

The first and most obvious method is to increase the mass ratio. A mass ratio of 2.7 will enable a rocket to accelerate to a speed which matches its exhaust velocity. If the mass ratio is doubled, the final velocity is also doubled. However, this is probably the maximum practical limit that can be achieved by changing the mass ratio. The amount of fuel needed increases exponentially with the ratio of the final velocity V, to the exhaust velocity v. In our initial example, where V = v, this ratio is 1, and 17,183 kgs. of fuel were needed to accelerate 10,000 kg. of payload and rocket to the speed of the exhaust–2 kps. To accelerate the same payload and rocket to 4 kps., the amount of fuel must be increased to 64,000 kg.[703] A point of diminishing returns is quickly approached, where it takes more and more fuel to get smaller and smaller increases in final velocity.

Even carrying four times as much fuel will not, however, succeed in getting a rocket with an exhaust velocity of 2 kps into orbit. To achieve this, the rocket scientist must resort to the second gimmick in his bag of tricks—the step rocket. The basic idea behind a step rocket is to use a big rocket to launch a smaller rocket which in turn can launch a smaller rocket, etc. With a series of rockets, each with a mass ratio of 2.7 and an exhaust velocity of 2 kps, it requires four steps for the payload at the top of the rocket to be accelerated to orbital velocity at 8 kps.

While this provides a solution, it is not without its penalties, the main one being a kind of technological gigantism. The rocket at the top of the steps, the 'fourth stage' carries the payload, which in our example is 2000 pounds, but since the rocket itself weighs 8000 pounds and there are 17,183 kg. of fuel, the gross weight of the fourth stage is 27,183 kg. Since the mass ratio of each stage is the same—2.7—then the fourth stage is the 'payload' of the third stage. With structure and fuel the third stage must weigh 388,329 kg. This massive rocket is in turn the payload for the second stage, which must weigh 5.5 million kg. The first stage then must weigh 79 million kg. As you can see, the step rocket is an exercise in reducing an engineering question to an absurdity.

The most famous step rocket of all time was the mighty Saturn V, which in the 60's propelled man to the moon. The Saturn V was a more efficient rocket than our example, having a higher exhaust velocity and mass ratio, but it still weighed in at 6.4 million pounds. Of that great mass, its effective payload, that is the mass it could send to the moon, was only 45 tons, 1.4% of its total weight. The rest was fuel and the skyscraper sized tanks needed to hold it. The amount of fuel and its rate of consumption boggle the mind. The gigantic first stage which had five rocket motors, burned 15 tons of propellant every second. Accelerating at nearly 60 feet per second per second, the Saturn V burned 60 tons of liquid oxygen and kerosene just getting clear of the gantry. That is a mass of fuel one and a half times the weight of the total payload the rocket sent to the moon. To put such a machine in perspective, it is like having a car that burns fifty gallons of gasoline just getting out of the garage. If you owned such a monster you would need a tank truck just to drive down to the corner grocery.

The Saturn V, and the space shuttle are engineering dinosaurs. They are too massive, too cumbersome, too slow, too unreliable and too expensive as vehicles for getting around in our celestial neighborhood. As magnificent as these machines were—expressions of the ultimate engineering technology of their day—they have no place in the New Millennium.

There are chemical fuels that could theoretically be manufactured to yield much higher exhaust velocities than those available today. Monatomic hydrogen, with a specific impulse (Isp) of 15 kps; and triplet helium, with a Isp of 27 kps, are such "super-fuels".[704] Monatomic hydrogen, in the form of metallic hydrogen, can only be formed at pressures in excess of two million atmospheres. The Russians claim to have already produced samples of metallic hydrogen, by using a press capable of exerting pressures up to three megabars.[705] Triplet helium—at present a purely theoretical material—would be formed by exciting helium atoms to a high orbit and then unpairing them to form a 'triplet' state. In this state, the atoms would possess enormous potential energy. Velocities of 32 kps could be achieved with such rocket fuels.

**2.2 - Bifrost's Costs**

The costs of building the Bifrost Bridge can only be roughly estimated, since no such system has ever been assembled. Cost estimates are, however, available for most of the components

The motive force in the launch tube will be provided by massive super-conducting electromagnets called Homo-Polar Generators (HPGs). The HPGs for the launch tube will have a capital cost of $1000/MJ (mega-joule).[706] One MJ is equal to .28 kwh. The launch tube must be able to handle up to 70,000 kwh, or 250,000 MJ. Each HPG unit will be rated at 56 MJ, and will cost $56,000. Four thousand five hundred HPG's will be spaced along the track so they provide an even thrust of 1.4 million Newtons to the launch capsule. The capital cost of the HPGs will be around $250 million. With labor and other factors, their installed cost will be in the neighborhood of $750 million.

The super-conducting ring capacitors, needed to store electrical energy, will also be a significant cost item. Large systems cost $200/kwh. The cost of a single capacitor for the launch tube will be $14 million.[707] Multiple capacitors will allow rapid launches in succession. Installed capacitor cost: $44 million.

The super-conducting HPGs of the launch tube will be cooled with liquid nitrogen. Total costs for the nitrogen plant will be a billion dollars.[708] Other costs for the launch tube are detailed in Table 2.2.1 below.[709]

**Table 2.2.1**
**Launch Tube Costs**

| Item | Cost $X 10$^6$ |
|---|---:|
| Launch Tube Tunnel | $625 |
| Launch Tube Bed | 204 |
| Tunnel Liner | 1,500 |
| Homo-Polar Generators (HPGs) | 750 |
| Inductors and Switches | 140 |
| HPG Supports | 38 |
| Ring Capacitors | 44 |
| Liquid Nitrogen Plant | 1,000 |
| Other | 1,700 |
| **TOTAL** | **$6,000** |

Costs for the laser system are primarily in the lasers themselves. Free Electron Lasers in the gigawatt range will cost $10 per watt.[710] This cost includes the electron accelerator, the "wiggler", the initiating laser, the aiming mirror, and other attendant components of the system. Costs break down to $5/watt for the laser and another $5/watt for optics and related equipment.[711] Total cost of the lasers: $15 billion.

**Table 2.2.2**
**Bifrost Bridge Capital Costs**

| Item | Cost $X10$^9$ |
|---|---:|
| Lasers | 15.00 |
| Launch Tube | 6.00 |
| Laser Capacitors | 0.15 |
| Power Lines | 0.25 |
| Launch Capsules | 0.25 |
| Guidance & Control System | 0.25 |
| Support Facilities | 0.23 |
| Pre-boost system | 0.10 |
| Other | 2.77 |
| **TOTAL** | **$25.00** |

The Bridge itself will begin generating revenues even before it is completed. If we charge commercial customers, and governments half what it costs them to launch with their own rockets, we can charge $3300 a kilogram to put their payloads in orbit. Since our own launch costs amount to only a few dollars per kilogram, we can realize an enormous net profit on this business. If we launched just 1500 tons of commercial payload for outside customers each year, we could see net revenues of $5 billion a year from the launch facility.

# Appendix 3
# Asgard

### 3.1 - Bubble Membrane

The exact composition of the membrane cannot at present be known, but we can postulate a membrane that could be built with today's materials. An outer layer of very thin metallic foil would protect the membrane from ultraviolet light, atomic oxygen, and micrometeorites. The membrane itself will most probably be a sandwich of various polymers, predominately of the silicone family.

Virtually all polymers contain additives that impart special properties. Every layer of the membrane will be specially treated to enhance its particular characteristics, with special care being paid to resistance to ultraviolet light, oxygen, and fire. Polymer chemistry has been one of the most rapidly advancing fields for decades, and there is no evidence that progress will slow any time soon. The advice given to Dustin Hoffman's *Graduate*—"Plastics"—is just as valid today as it was then. By the time we build Asgard, we will undoubtedly have even better materials to choose from.

It may be desirable to bond the gold film to a sacrificial layer of very thin clear plastic. This outer membrane would be held out several inches from the surface of the membrane holding the water shield. The spacing between the two membranes could be maintained by mechanical spacers, or by very low gas pressure—water vapor could serve in this role. This outer film would serve both as a UV screen to protect the plastic membrane of the shield, and as a bumper, causing micrometeoroids to vaporize themselves. The inner membrane would then be subject only to strikes by larger, sand-grain sized, meteoroids.

The outer membrane would face degradation from bombardment by meteoric dust, intense UV and other radiation, atomic oxygen, and other space hazards. The membrane and its gold coating could be replaced and recycled every few years. In actual practice the sacrificial membrane would undergo a continuous process of replacement.

### 3.2 - Cable Nets

The stresses on bubble membranes increase with their radius, so very large pneumatic spheres must be reinforced with cable nets. The best cable nets form a hexagonal pattern, in which the membrane areas between the cables are uniform.

The strongest net will be required in the inner sphere of the outer shell. The cables in this net, made of titanium alloy, will need to be about a centimeter and a half thick, and spaced every 400 centimeters, giving the net a safety factor of three.

The outermost cable net can be less robust, since it must only provide enough counter pressure to maintain the water shield in its liquid form, a pressure of about one psi. The cables in this net therefore need be only three quarters of a centimeter thick. The inner spheres are normally under very little stress, since their internal and external pressures are nearly the same. Enough pressure is maintained inside the inner spheres to keep them inflated, but in zero g, this is very slight. Each of the inner spheres is also reinforced with a cable net so they can maintain a full 3 psi, against vacuum, should the outer spheres fail completely. Altogether the cable nets weigh 800 tons.

### 3.3- Algae System

To provide all of the food and oxygen needs for a colony of 100,000 will require a through-put of 600,000 liters per day. Clear plastic tubing, 6 centimeters in diameter is run on the inside of the water shield. The tubing is arranged in rows, with 720 tubes side by side, in two tiers, each twenty meters wide. These bands of algae-filled tubing run along the outer sphere's equator. Air, laden with carbon dioxide from the colonists respiration, is pumped into the tubing, where it is absorbed by the algae. The algae utilize the carbon in the $CO_2$, together with sunlight, water and nutrients, to produce carbohydrates, fats, and proteins. Oxygen is released as a byproduct. The outer surface of the clear tubing is perforated with many small holes. The inside surface is coated with a membrane, like Gore-Tex, which allows the passage of gases, but not liquids. The oxygen passes out of the algae tubing and is captured in a concentric tube; from there it is pumped through sunlit purification filters, and back into the ecosphere. The through-put of the system allows all the air in the ecosphere to be circulated twice a day. This keeps the concentration of $CO_2$ and oxygen in the atmosphere in balance. The whole system is closed and self-sustaining, so that the colonists need no outside inputs other than electrical energy to run the pumps and lights.

### 3.4 - Ecospheric Economics

One of the compelling forces that will attract humans into space in large numbers will be the economic advantage that space dwellers will have over ground pounders. To live on the Earth is enormously expensive. Everything we do here requires the expenditure of scarce resources: fossil fuels, metals, top-soil, and real estate, all of which are non-renewable.

Sitting in Denny's restaurant, having a cup of coffee and a piece of cake while contemplating the expenses inherent in our earth-bound way of life, it is all too obvious that everything has a real cost associated with it. The wheat that went into the flour, had to be planted and harvested by machines burning gasoline; the cocoa in the chocolate had to be gathered by workers in some distant jungle, and then bought and sold half a dozen times; it had to be moved half way across a planet to end up on my plate. The same is true for the sugar and for the coffee. The little dollop of cream that went into my coffee was in the past few days somewhere in the body of a cow, from whence it had to be extracted, then packaged and transported, probably hundreds of miles, all while being constantly refrigerated. The little plastic container holding that dollop of cream had to be injection molded out of petroleum feed stocks, and now, after having served its one and only purpose, it must be carted off to a land-fill where it will lie in state, serenely cluttering its bit of limited dump space for the next thousand years. The only thing more astonishing than the fact that all these disparate bits of industry can come together at all, is that they can do so without costing a king's ransom.

All of these resources are already limited, and in the future they will become even more so. As the human population burgeons, and the poor people of the world gain greater economic power, the demands on scarce resources will inexorably increase. These remorseless economic forces have one inevitable result: an increase in the price of cake. Not just an inflationary increase, but a real increase in the underlying costs.

Compare my situation at Denny's to the corresponding situation in Asgard. The circumstances are the same, it is late at 'night' and I am having coffee and cake with a friend. Now, however, the whole process has been simplified. The cake, the coffee, the chocolate, even the dollop of cream all come from the same source—the algae tubes. A food synthesizer orchestrates the combination of flavors and textures to produce the same sensations in my mouth, but the global network of industry and commerce has been replaced by a single integrated facility, located at most a few hundred meters away.

In the ecosphere, the costs of disposing of waste products is identical to the cost of producing feedstocks; in fact, it is the same cost. In Asgard, wastes are feedstocks. The cost of disposing of the cream container—which involves sending it to the central plant where it is melted down—is the same as the cost of producing the feedstock needed to produce its replacement. The same is true for organic wastes. They are disposed of in the SCWO where they become the feedstocks for algae production. Wastes become feedstocks in a simple closed loop. The only thing that enters the cycle is energy in the form of electricity; the only thing that leaves is energy in the form of waste heat. In such a system, the only costs involved are the energy costs, and replacement costs of the mechanisms. The machinery is built to last for generations, and the energy comes from the sun. The resources in the loop are never used up, and so have virtually no cost associated with them. The result is the production of consumables at a tiny fraction of their equivalent cost on Earth.

Ecospheric economics are clearer when considered within the hypothetical context of a one-man ecosphere. This one-man bubble would be just twenty meters across, and like Asgard would be in geosynchronous orbit around the Earth. (A one-person ecosphere is not a practical, or even desirable, space habitat, but it serves to demonstrate certain fundamentals.) Assume this tiny ecosphere is fully equipped with algae tubes, computer controls, photovoltaics, a SCWO, and a food synthesizer. Given such a micro-habitat, the question is: What would it cost to actually live in one's own personal ecosphere? All life support needs are provided. Wastes are broken down to provide feedstocks for food production. Air and water are continually repurified and renewed by the algae as they produce food. All energy needs are provided by photovoltaic cells. Absolutely no inputs are required from the outside, except sunlight to nourish the algae and provide power. As long as the sun shines, one can live with complete self-sufficiency. You could float through the ocean of space as self contained and independent as any diatom.[712]

What is true for a one-man microsphere is equally true for a large ecosphere like Asgard. Millennial ecospheres will be based on a deep symbiosis between man and algae. Each life form will produce wastes which form the feedstocks of the other's metabolism. The same tactic is used by certain fungi; living symbiotically with algae they form lichens. Not surprisingly, lichen is especially adaptable to harsh environments.

The economic advantages of the closed-loop cycle are almost incalculable. For an inhabitant of Asgard, the main costs of daily living—food, rent, utilities, and transportation—will be trivial. The price of each person's consumables: air, water, and food; amounts only to the cost of running the raw molecules through the process loop. This can be done with a few watts of solar power. Similarly, the costs of housing amount to the cost of bubble volume, amortized over the life of the inhabitant. Given the magic of cube root geometry, and the economies of tensile architecture, this is a minuscule per capita cost. The only big dollar items are for shield mass and transport costs to and from planetary surfaces. The Bifrost Bridge has already addressed transportation costs, and shield costs per capita can also be reduced to insignificance.

Life support costs in a Millennial ecosphere will be minimal, but productivity will be high. Asgardians, for example will engage in a variety of high-end manufacturing enterprises: growing ultra-pure crystals, producing genetically engineered substances, and fabricating high-density computer chips through molecular beam epitaxy; but this will not be the main business of the colony. Just as in Aquarius, most people in Asgard will work in some capacity as data managers. Tapping Asgard's telecommunications potential, coupled with the massive computer utility on Aquarius, Asgardians will be tightly interconnected with the entire data network of the Earth. They can work in data professions just as surely from orbit as they could in downtown Manhattan, with considerably greater ease, safety and convenience.

The difference between an Asgardian's gross production and his gross consumption will be his net productivity. Even counting transport charges and shield mass, the average space colonist will be able to live comfortably on a few dollars a day. The annual cost of supporting a person in space will be only a few thousand dollars, but her gross production will be equal to or greater than that of a data professional on Earth—tens of thousands of dollars. The net productivity of a colony like Asgard will therefore be enormous. Whereas, conventional earth-based civilizations can at best manage a net productivity of a few percent, the net production of a closed-loop civilization could easily be ten or twenty times as much.

Individuals who make the leap to space will be amply rewarded. On Earth, with its staggering life-support costs, people manage to save only a few percent of their income.[713] Average net worth in America, after a life time of work is around $75,000.[714] In Asgard, the average individual will realize an increase in net worth of around $10,000 per year. Compounded at 10% over a working life time of 40 years, the net worth of Asgardians could exceed $4,000,000 per capita. In Asgard, Millennial life-extension techniques and the gravity-free environment will extend the working life time to 80 years. If annual net worth increases of $10,000 are compounded annually for 60 years, the net worth of the average individual soars to an astounding $30 million. If for no other reason, people will migrate to space for the most basic of all motives—to get rich.

### 3.5 - Economy of Scale

A sphere has the remarkable property of possessing a volume which is related to its radius by the cube root. Small increases in a sphere's diameter lead to large increases in its volume. Economies of scale have profound implications for Asgard and other space colonies. For example, a one-man ecosphere would require a water shield with a mass of 3700 tons. By comparison, residents of Asgard each require only 60 tons of shield water. Thus, a one man ecosphere requires 60 times more mass per person than Asgard.

Doubling the radius of an ecosphere increases its internal volume by a factor of ten, and cuts the shield mass required per inhabitant in half. If we expanded Asgard to accommodate a million people, it would require a sphere with a volume of a billion cubic meters. The volume would increase by a factor of ten, but diameter would only be doubled. This relationship would decrease the shield mass required per inhabitant to 30 tons.

There is no reason to assume that we can't take advantage of this geometric economy up to very large sphere dimensions. Gerard O'Neill calculates the ultimate dimensions of one of his cylindrical space colonies as 25.75 km. long by 6.4 km. in diameter.[715] Professor O'Neill carefully and conservatively limits himself to the use of relatively primitive materials like steel cable with a working stress of 70,000 - 80,000 lbs./in$^2$. His calculations are based on a design which must support a meter and a half of dirt, buildings, and other loads, amounting to 7.5 tons per square meter. These masses must be supported against the stresses of artificial gravity. To bear these tremendous loads, O'Neill incorporates steel cables 150 centimeters (5 ft.) in diameter into his design. These cables are bundled in groups of four and spaced every 14 meters around the circumference of the cylinder to carry the loads imposed by the spin of the colony.

The materials available in the next Millennium will certainly outperform even the best materials available today. Kevlar 49 has a tensile strength of 270,000 psi, but super fibers of the next century are apt to push that figure well beyond 400,000 lbs/in$^2$.[716] Ecospheres like Asgard are subject only to the stress of the internal atmospheric pressure of 3 psi. Therefore, such ecospheres could be as much as 90 km. in diameter.[717] An ecosphere of this size would require less than 1000 lbs. of shield water for each inhabitant.[718]

An ecosphere 90 km. across could, in theory, provide a permanent home for over 380 billion people—75 times the population of our already crowded little planet. Seventy five times the Earth's population, housed in comfortable, spacious surroundings, with every individual's physical needs for energy, food, and shelter met, while occupying a volume of space, which on the scale of the solar system, can only be described as microscopic.

## 3.6 - Radiation Hazards

### Van Allen Radiation

Van Allen radiation is a product of the Earth's magnetic field and occurs in two broad bands at altitudes of 3000 and 22,000 kilometers. These donut shaped belts of radiation are comprised of electrons and protons—mostly of hydrogen—that have been ejected by the sun in the solar wind and subsequently trapped by the Earth's magnetic field. Earth's molten iron core gives our planet a magnetic field extending out thousands of miles. Charged particles entering the planet's magnetic field follow the field lines and are channeled to the poles. There they bombard the atmosphere causing the eerie glow of the northern lights.

The inner Van Allen belt, which does not extend beyond about 17,000 km., is made up mostly of protons with high energies of up to 100 million electron volts. The outer belt, which extends beyond the Moon, is made up of electrons with much lower energies of around 40 thousand electron volts.

The electron volt is a convenient yardstick for measuring the amount of energy carried by a moving atomic particle. Since atomic particles are small so are the units of energy they carry. The usual ways we use to measure power are units like the BTU (British Thermal Unit), which is the amount of energy it takes to raise a pound of water by one degree F. The calorie, the metric equivalent of the BTU, is the amount of energy needed to raise the temperature of one gram of water one degree C. (In the measurement of food energy people commonly use the kilogram-calorie which is the amount of energy needed to raise a kg. of water 1° C.) Looked at as an equivalent of heat energy, the electron volt doesn't amount to much, about $1.5 \times 10^{-22}$ BTU. So it would take 100 trillion 'high energy' protons to raise the temperature of a pound and a half of water by 1°. The peak flux rate of high energy protons in the inner Van Allen belt is around 10,000 protons per square centimeter per second. If a cubic centimeter of water were exposed to this dose of radiant energy it would raise its temperature by only a few hundred millionths of a degree.

So the amounts of energy involved in radiation hazards, at least in the Van Allen belts are extremely small. The flux rates are high though, peaking at distances of 6500 and 22,500 km. At a distance of 36,000 km.—geosynchronous orbit—the flux rate drops off to about 500 protons per square centimeter per second.[719] While such a flux rate could cause serious radiation exposure problems, the low energy of these particles make them easy to block with minimal shielding.

**Solar Flares**

While the Van Allen radiation belts are permanent fixtures of space around Earth, solar flares are periodic phenomena, erupting usually during times of peak sunspot activity. During a solar flare, huge quantities of ionized plasma are blown into space. A person who finds himself in the path of a flare in an unshielded environment can be fried to a figurative crisp in a matter of minutes. Solar flares create the most intense radiative fluxes we will encounter in space. Intense solar flare fluxes can send up to a hundred million high energy protons ripping through each square centimeter of space per second. On February 23, 1956, there erupted a Class 4 solar flare. It has since been estimated that this flare was so intense that an unshielded astronaut would have received 1000 times the fatal dose of radiation in less than thirty minutes.[720]

Most schemes for protecting astronauts and space colonists from solar flares involve some form of early warning system. By monitoring the sun visually, flares can be detected eight minutes after they erupt—the transit time for light from the sun. When a flare is detected a warning is given and people can move into especially hardened radiation shelters, until the danger has passed. While much of the radiation flux travels slowly enough to give space travelers an hour's warning to take shelter, some of the radiation travels in the form of high energy protons which move at a high percentage of the speed of light and so are apt to arrive within ten minutes of a solar flare's being sighted.

**Cosmic Rays**

Cosmic rays are a mysterious form of radiation about which little is known. They consist almost entirely—99%—of protons moving at close to the speed of light.[721] It is known that our sun generates some cosmic rays, particularly during violent episodes like solar flares. The balance seem to come from outside the solar system, but their origin is unknown. The energy flux of cosmic rays is extremely low—about the same as the energy flux of starlight. Despite this low energy density, the individual cosmic ray particles themselves can do enormous biologic damage.

Unlike Van Allen Belt radiation, cosmic rays are 'hard'. 'Soft' radiation, like the particles trapped in the Van Allen Belts have energies on the order of 100 million electron volts. Particles with these energies are fairly easy to stop with thin passive shields. Solar flare radiation can also be hard, with energies in excess of 10 billion electron volts being produced in extreme events, every three or four years. But these are periodic phenomena which announce themselves, allowing time to take shelter in especially hardened sanctuaries. Cosmic rays, by contrast, are both hard and continuous, with a flux rate of around .3 high energy particles per $cm^2$/sec. This radiation is both continuous and isotropic, meaning it comes from all directions. Sheltering from cosmic radiation requires a system to protect the entire space colony.

Most cosmic rays are primary—Hydrogen—protons traveling through space as individual particles. While these solo protons represent 90% of the cosmic ray particles, they contribute only about 20% of the total energy in the radiant flux. The balance of the energy, and consequently the major portion of the risk associated with cosmic rays comes from heavy nuclei. In these heavier "Z" particles, the whole nucleus of a heavier element, stripped of its electrons, is whizzing through space at a high percentage of the speed of light. These high energy particles pack an incredible wallop and are the most dangerous component of cosmic rays. Chief among these are the high energy nuclei of iron atoms.

$Fe^{56}$ is a natural isotope of iron containing 26 protons and 30 neutrons, and it is this element which comprises the deadliest part of the cosmic ray bombardment. If protons and electrons at a hundred million electron volts are like atomic bullets, then $Fe^{56}$ nuclei, at a billion electron volts, are like cannon balls. When these microscopic mortars crash into the cells of a human being they do terrible damage. A single iron nucleus will cause a microlesion 10 cells deep, and several cells wide. In the region around the microlesion—the penumbra—the collision will generate intense secondary radiation exposing nearby cells to radiation doses of up to 300 rad. In an organism the size of a human being, the death of a few hundred cells is not serious, but continuous exposure to such damage can be dangerous.

The flux rate of $Fe^{56}$ nucleons with energies between 100 million and a billion electron volts is around 55/$cm^2$/day. While this is a minute flux rate, compared with Van Allen electron flux rates of 10,000/second, the damage that can be done by these nuclei is not trivial. After just 90 days in geostationary orbit, an unshielded colonist would have accumulated 27,000 microlesions in every cubic centimeter of his body tissue. In sensitive areas, like the optic nerves, such damage could be mortal. The total radiation dosage from Cosmic rays amounts to about 100 rem per year, 20 times the maximum safe exposure.[722] If we are to live in space, then we must devise some way to cope with cosmic radiation.

Ironically, the slower moving, and hence less energetic cosmic ray particles tend to do the most damage. This is because they spend a relatively longer period of time passing through the tissue than faster particles, and so collide with more organic molecules. This seemingly upside-down property makes the shielding problem particularly difficult. If only enough shielding is provided to stop the most dangerous slow-moving particles, one may only succeed in slowing down the faster moving particles, thus making them more dangerous. The wrong amount of shielding can be worse than no shielding at all.

Improper shielding also produces secondary, or 'Bremsstrahlung' radiation. Such radiation results when a high energy nucleon collides with the nucleus of another atom. The result is the characteristic star-shaped explosion, familiar from bubble-chamber photographs taken in particle accelerators. Up to a certain threshold level, thicker passive shields actually make the radiation behind them worse, due to secondary particle production. This is one of the most intractable problems associated with passive shielding solutions to the galactic cosmic ray problem.

### 3.7 - Active Radiation Shields

Active force shields are a real possibility, but they pose enormous challenges. To construct an electrostatic force field able to deflect cosmic rays, the field must have the same energy as the particles to be repelled. In the case of cosmic rays, energies of several billion electron volts are required. Applying such a charge to the surface of a space habitat creates an intractable secondary problem. In order to repel the positively charged nuclei that comprise the most dangerous part of the cosmic rays, the charge on the habitat must also be positive. This positive charge then attracts any free electrons in space to the surface of the habitat. The charged habitat would accelerate the electrons towards itself with energies equivalent to its positive charge, that is, in the billion electron volt (BeV) range. This flux of incoming electrons would then effectively fry anyone inside the charged habitat.

Negating this effect requires a second sphere, surrounding the habitat sphere, with a negative charge of several BeV. The two spheres, oppositely charged, with a combined potential of many billions of electron volts must be separated from each other enough to prevent an electrical arc from discharging them. Since the dielectric constant of vacuum is 30,000 volts per centimeter, the two spheres would have to be separated from each other by a distance of at least two kilometers.

To protect ourselves from the dangers of ionizing radiation in space, we could take a cue from Mother Earth and provide ourselves with a magnetic shield. A magnetic force shield will require a field density of 377 Gauss.[723] To induce a field of this intensity would require an electrical current of 3 x $10^5$ amps. The entire field energy would be around 20 million joules. The conduits would need to be kept inside shielded vacuum tubes to remain cold. The conduits would be run longitudinally above the surface of a spherical habitat, converging at the axes, and passing in a bundle down a central tube running the length of the sphere's axis. At the center of the habitat would be a shielded target of lead. The charged cosmic rays and other dangerous particles would follow the magnetic force lines around the electrical conduits. The particles would spiral around the conduits and into the central axis. There they would smash into the lead target. Any secondary radiation would be absorbed in the surrounding shielding.

The main problems with active magnetic shields are the physical stresses created by the intense magnetic fields. These stresses must be contained by massive reinforcing cables and beams. It is not clear whether the mass of such components would be much less than that of passive shields.

Once such a magnetic field had been established, it would require little energy to maintain other than the losses occasioned by collisions with the charged particles themselves. Since this total energy flux is only about that of starlight, the energy requirements of the shield would be minimal. Eventually it will become practical to construct active magnetic force shields. In the near-term however, we will use passive water shields.

### 3.8 - Earth-Moon Network

Full development of Asgard will require the construction of an extensive support network. The network will have four major components, not counting Asgard itself: 1) a space station in low Earth orbit, 2) a space station positioned above the Moon at $L_2$, 3) a lunar station on the Moon, and 4) a station on one of the near-earth asteroids. Together this network of manned stations will provide the transportation, communication, and mining infrastructure needed to construct and support Asgard.

Valhalla will be the first of the network's components to be built. In Valhalla, we will develop the systems needed to live successfully in space for the rest of time. These include, zero-*g* algae farming, long term low-*g* metabolic adaptations, solar bubble power systems, and closed-loop ecologies. In this respect, Valhalla will be an experimental station—a toehold in space.

From Valhalla, we will be in position to construct the second link in the network—Camelot. Camelot will be similar in size and function to Valhalla. Instead of being in orbit around the moon, however, Camelot will be positioned at $L_2$ where it will remain fixed relative to the Moon and Earth.

Working from Camelot, we will be able to establish a base on the lunar surface in the crater Landsberg. At the lunar base, we will construct a mass launcher and mining and refining operations to produce materials needed for Asgard.

Once the lunar base is operating, we will send materials from the Moon into Earth orbit for the construction of the initial core of Asgard. At this stage, Asgard will be very small, and the necessary volatiles, like hydrogen, will be brought up from Earth.

As we prepare to expand Asgard to its ultimate size, the final link in the network will be forged. This will be a manned base on the AA asteroid Eros. A mass launcher will be constructed in space next to Eros, and tethered to one of its poles. Volatiles, mined and refined on Eros will then be fired into orbits designed to rendezvous with Asgard. From Eros will come all of the hydrogen and most of the carbon and nitrogen needed in Asgard.

With the support network in place, we will continue to expand Asgard, over a period of years, until the colony is completed. The modular construction of Asgard will make it feasible to expand from an initial small core to the final fully developed colony of 100,000 people. The first central bubble can grow from a very small start, until it reaches its final dimensions—200M in diameter. At this point, work will begin on a new colony, a few kilometers away in geosynchronous orbit.

Spaced every couple of kilometers around the circumference of the Clarke orbit, there is room for about a hundred thousand colonies like Asgard. If this many colonies were built, they could house twice the present population of the Earth.

# Appendix 6
# Solaria

### 6.1 - Population Dynamics

By the year 2030 there will be ten billion passengers riding Spaceship Earth—most of them traveling deck class. It is usually assumed that population growth will then halt, or at most continue to a maximum population of around 14 billion.

Population stability is generally thought to result from one of two scenarios. The first and most obvious is that death rates will rise due to famines and plagues caused by overcrowding and pollution. This is a scenario that admits hitting the Malthusian wall at full speed. The attendant holocaust in human and environmental terms is too horrible to contemplate. In the other scenario, the poor people of the world increase their standards of living enough to forgo having large families, as is now the case in the Western world. Either life is made so marginal and miserable that women are unwilling to bring children into such a disgusting world, as in the XUSSR, or the babies just die, as in Ethiopia; alternatively, people become so rich they can't afford to have children, as in the United States and Japan.

We can support a larger world population, at least up to a global population of 15 billion, by expanding out onto the oceans. The reserves of food and energy we can harvest from the seas, at least in the near term, are great enough to sustain even a very populated world. This presents no permanent solution however, since we cannot remove the Malthusian barrier, we can only push it back. Even with the bountiful oceans providing food and energy for 15 billion people, we will be back in crisis in just another 35 years. Not even the oceans can sustain a global population of 30 billion for very long.

The alternative scenario—increasing everyone's standard of living to that of North Americans—is attainable, but may not be desirable. The typical American uses 310 million BTU of energy in all forms every year. If a world population of 15 billion were using as much energy per capita as the typical American, world energy consumption would be 4500 quadrillion BTU.[724]

Not even the marine colonies could provide such a huge amount of energy indefinitely. The world's oceans represent a renewable power base of over 200 million megawatts.[725] The OTECS of Aquarius and the other marine colonies can eventually produce 50 trillion kilowatt hours of electrical power per year. That is enough electricity to meet the total energy demands of only 550 million American type consumers. The balance of the energy will have to come from land-based solar power, from fusion, and from other sources.

Fulfilling the development scenario would require raising the average standard of living in the world to that of the industrialized West in just 35 years. The average per capita income in the world in 1980 was around $2500.[726] For two billion of the world's poorest people the average annual wage is a mere $400. This is an astonishing figure to most of us in the First World. How, we wonder, can anyone keep body and soul together for a year on what most of us consider to be a modest weekly income? For the world's poorest people, $400 per year would actually be a big step up. In Bangladesh the average annual income is just over $100—28¢ a day. In Cambodia it is even less.

The average per capita income in the United States is around $15,000, 30 times the world average. Surprisingly, the world average can be brought up to the American average in 35 years at a growth rate of just 6%. In many places in the world the rate of industrial development is well in excess of 10%.

So this is an achievable scenario. There is enough energy available, and the rate of growth is or could be high enough to bring the world up to the American standard in just 35 years. What kind of world would it be though, with 10 billion people all consuming energy and producing trash at the same rate as Americans? It is all too apparent that the planet cannot sustain even the demands placed on it now. What will happen to the ecosphere in 35 years if it must carry a load which is not just twice, but 16 times that of today. Even if we could manage to sustain our own species, such development must inevitably be an ecological disaster for the rest of the planet. Just try to imagine building suburban tract housing for all 20 million of the people who will reside in Mexico City in 35 years. Try to imagine the freeway system that will be required for the 30 million commuters trying to drive each day between Bombay and Poona. Even if all the cars are hydrogen fueled, they will still take up space and suck up resources. Even if such a development scenario is viable from a raw resource standpoint, it nevertheless demands changes in life-style, almost as great as those brought about by the industrialization itself.

Neither scenario which would lead to population stability is really viable. We cannot face the Malthusian alternative with its human suffering, or the development alternative with its environmental carnage. What then is the answer? We are faced with a dilemma caused by continuing growth inside a closed system. No method of halting that growth appears viable. Therefore the answer must be to open the system. This is exactly what space colonization accomplishes. It is a release valve for our species and so for the planet as a whole. If we can get a grip on space, and learn to live there comfortably and economically, then we can escape the seemingly inevitable doom facing us and our world.

Moving a significant fraction of the human race into space is an epic challenge. In just 35 years we need to be in a position that will allow the rate of migration into space to accommodate not only all of the human race's new growth, but will also reduce the existing surface population at a rate of about 2% per year. At a 2% rate of continued growth, the world population of ten billion will be producing 200 million new people every year, that is about the population of England, France, Germany, and Italy combined. To actually reduce the world population will require moving out another 100 to 200 million people per year, at which rate we could get back to today's planetary population in another 35 years. So the total job requires accommodating the emigration of around 350 million people a year into space. There is some room for them on the Moon and more room on Mars, but the real home for the majority of these millions will be in the asteroid belt.

How you motivate people to migrate into space is a difficult political question. After the initial pioneering wave of sophisticated educated people of the First World, the later waves of immigrants will be mostly people from the Third World. These people have a wide variety of cultural and religious traditions, none of which include space colonization. Motivating these people to take the ultimate leap and leave their home planet for a strange life in outer space will be difficult in the extreme. The only possible answer to the question is that the means may be unknown and difficult to find, but they nevertheless must be found.

The correct formula will inevitably be some combination of carrot and stick. The carrot will have to be supplied by our own space-faring civilization. We Millennialists must succeed in perfecting means of living in space which supply all of the physical and psychological comforts necessary to a desirable life. This is an enormous challenge, but it must be undertaken and it can be met. Space can be transformed, more easily and cheaply than usually imagined, into an environment which is not only suited to human habitation, but which is ideal for it. The idea of living in space, afloat among the stars will always be alien to those born on Earth. It can nevertheless be done in a way which is attractive and comfortable. Achieving that end is crucial to the fate of both humanity and Earth.

The second part of the equation is the stick. This is the easy part, since the stick will provide itself. Conditions on Earth are deteriorating even now; the rate of deterioration will only accelerate. The only possibility for stemming and reversing the planet's slide into ruin is the emigration of humans into space. If the human race resists that solution, the environmental pressures will mount. At some point the growing horror of conditions here will leave people with no choice but to take the leap into the unknown and blast off. As Millennialists, it will be up to us to facilitate that decision as much as we possibly can. Obviously, education will be our primary means; making people understand on the one hand how attractive life in space can be, and on the other hand how necessary it is that they make the choice to emigrate.

We will be greatly aided in this effort by the fact that most of the people we need to move from the planet will be very young. Since they will be at the beginning of their lives they will typically be flexible, impressionable, and hopeful. They will be compelled by the lack of viable futures on Earth, and this combination of attitudes will make it easier to convince the young that their best future is in space.

The economics of space migration are more problematic. The means whereby hundreds of millions of people can each year be lifted from the planet and supplied with all of their needs in space is at first difficult to apprehend. However, it is nonetheless not only possible, but in fact highly feasible. For one thing, the emigration of new populations into space relieves their earth-bound political units of considerable stress. Accordingly, we can expect that these political bodies, countries for the most part, will be highly motivated to subsidize this effort. Breakthroughs in the methodology of accessing and surviving in space will make space migration enormously cheaper and easier than at present. Lifting people into space takes energy, but as we learned when exploring the Bifrost Bridge, that energy cost can be a matter of pennies. If we were engaged in a space migration effort of the magnitude envisioned here, the capital costs might be enormous, but the per capita costs could be trivial.

The scale of the effort needed to lift hundreds of millions of people off the planet is more than a little daunting. We would need to average around a million people per day to achieve stabilization and eventual reduction of the earth-bound population. Putting a million people a day into space by any means is a staggering proposition. With rockets it would probably be an economic, if not a physical impossibility. It would be just possible with the Bifrost Bridge. Sending a million people into orbit means launching 100 thousand tons of payload. We will need to build a big-bore launch tube, ten times the size of the heavy launcher we built in the Asgard phase. It will have to fire a hundred times a day, about once every 15 minutes. There will eventually be many such tubes: in Africa on Kilimanjaro, on Mt. Kenya, on Mt. Ruwenzori, in the Andes of South America, even in New Guinea. Other, more exotic means of conveyance—like Jacob's Ladders—will eventually come on line, but these are too esoteric to be included here.

By the year 2128, human numbers will exceed 80 billion—65 billion of whom will live in space. The overwhelming majority of these, 55 billion will be living in the asteroid belt. If these people maintained the average net productivity of core Millennialists—the equivalent of $10,000 per year—their net economic power would amount to $550 trillion. There are very few tasks, no matter how gargantuan, that we could not undertake with resources of such magnitude. These numbers, ten times the present population of the Earth, 100 times the GNP of the United States, seem incredible, but they represent only the merest seedling of the mighty tree of life our solar system will ultimately support.

Accommodating growth of human numbers in space is far easier than getting them there in the first place. While the area on the Earth, Moon, and Mars is strictly limited, this is not the case with space itself. Planetary bodies occupy only an infinitesimal fraction of the space available. Together, Earth, Moon, and Mars provide a total surface area of less than 700 million square kilometers. At saturation densities, these three worlds together will probably be able to hold no more than 25 billion people. The human population will approach this level in just two doubling periods, 70 years—a single human life-time.

The main thrust of human settlement in space must be towards the free-floating ecospheres of habitats like Asgard. There is a limitless bounty of space and energy waiting for us, free from the gravitational tyranny of any planet. To accommodate the entire present population of the earth, 5 billion people, would require an ecosphere only 22 kilometers across.[727] It is hard to believe, but the entire human race could live comfortably inside a volume it would take less than 15 minutes to drive across. When such a tiny bubble is put up against the globe of the earth, it is vanishingly small.

There is an instinctive revulsion to this idea. The mind immediately imagines all the huddled masses of humanity packed together like rats in a box. But this won't be the case. Even a tiny, 22 kilometer, bubble would provide every person with an abundance of room. If all five billion people were spread evenly throughout the interior of such a sphere, no person would be any closer to another than five meters. People don't live spaced out that way of course, they bunch up for domestic, professional, recreational, and social reasons. But suppose a typical couple and their two children were all together, floating in front of the TV screen in their living space. Together, their aggregate volume would be 4000 cubic meters, and the nearest next family would be twenty meters away, in their own living space. It is hard to conceive of everyone in the world able to live collectively in such a tiny space, and yet still have such an abundance of room. But such is the nature and wonder of cubic geometry..

Whether it would actually be desirable from political, social, economic, and engineering perspectives to create single habitats as large as twenty kilometers is an open question. With the materials on hand today, ecospheres with a diameter of 45 kilometers are feasible. Such a sphere could accommodate almost 50 billion people in spacious comfort. The only question is whether or not people will want to live in such monolithic habitats.

The people who move to the sea colonies and into space will be the young—of all ages. People already settled in their places and their cultures will live out their lives on the home planet. In just a few generations, though, the young—and all increase in population is ultimately in the form of young people—will make the transition to the new environments. As the new generations leave the old world, drawn by fantastic new opportunities and impelled by deterioration at home, the old generations left behind will fade away, leaving the earth mostly free of the encrustation of cities now afflicting her.

This is not to say that the Earth will be uninhabited, far from it. Ten or fifteen billion souls will make the Mother World their home. But they will not live on the Earth the way we do now—as blood-sucking parasites, like fleas infesting the hide of a mammoth. We will bring back from space the ultimate commodity: knowledge; specifically knowledge of how to live symbiotically with our environment. To inhabit space we will be forced to learn to live in self-contained and self-sufficient ecospheres. Happily, this is simple and easy to do. It is also extremely cheap. That technology, and that way of life will be the most important exports space colonies ever send back to Earth.

Martians will exemplify this symbiotic life style. Mars will have a new and tender ecology just establishing itself on a barren planet. Five billion people will inhabit that world in harmony with its infant biology. Martians live the same way as the Lunatics, from whom they are descended. Living inside their self-contained ecospheres, Martians have virtually no effect on the outside environment.

Bubble ecospheres and closed-cycle systems are just as viable on Earth as they are on the Moon or Mars. Over the next 125 years we will gradually shift the human population over to the Millennial way of life. At least 10 of Earth's 15 billion inhabitants will be living in closed-cycle ecospheres—neither depleting nor polluting the earth's environment. By the end of the first quarter of the Third Millennium, 2250, everyone on Earth will live in a way that puts them in symbiotic harmony with the planet. Our period of parasitism will be at an end. At the end of the Martian Phase of our development we will have attained a state of equilibrium on the Mother planet, in which environmental degradation has been halted, and the total population has been stabilized.

In this final transitional phase, the old cities will be slowly abandoned. They will be replaced by Millennial ecospheres. The ecospheres, which are able to support life comfortably on the dark side of the Moon, will be able to accommodate people in any earth environment with ease. The adaptability of these habitats will allow us to position cities in any locale which suits us. The object will be to choose locations which will not interfere with Gaia's own ecosphere. The oceans will already be highly developed, so other sites will come into play. Ecospheres can be built high in the mountains or deep in hostile deserts, even on the

Antarctic ice cap. People can live comfortably in these environments without usurping prime habitat that could be better used by less adaptable species. We can also move ourselves underground, leaving the surface free for natural biomes. By the end of this transitional period, 250+ years after the dawn of the New Millennium, the Earth will at long last be free from the burden of tool-using man. We will still inhabit the planet, and in large numbers, but the planet's ecology will no longer suffer as a consequence of our presence.

# Appendix 8
# Foundation

### 8.1. - LASER

LASER is an acronym for Light Amplification by Stimulated Emission of Radiation. Electromagnetic radiation originates when electrons, which have been boosted to an orbit further away from the nucleus of an atom, spontaneously fall back to a lower orbit. When they make this 'jump' to a lower orbit, the extra energy the electrons held in the higher orbit is lost. The energy leaves the electron in the form of a photon—a particle of light.

Electron jumping is what illuminates a typical light bulb. The tungsten atoms in the bulb filament are excited by the passage of an electric current. Tungsten is a miserable conductor of electricity, so a lot of the energy in the current is transformed into heat by resistance. Heat is just the energetic bouncing around of atoms at high speed. This bouncing around causes electrons, energized by collisions with other electrons, to jump up to higher orbits. When the electrons 'fall' back to their lower orbits, as they inevitably do, they give off their energy in the form of light. In a typical light bulb filament, the atoms are all performing this crazy dance, jiggling around, jumping up and down at random all the time. The result is a wash of all kinds of photons going in every direction. We see this random energy as white light.

As we all learned in sixth grade science class, white light is really composed of all the colors of the rainbow. This is because each color is a specific frequency of light, and white light is a mix of various frequencies. A laser, by contrast, is one particular color because the light is all of one frequency. In a laser, atoms are first 'pumped' by an outside energy source which boosts the electrons into higher orbits, where they remain, until a catalyst causes them all to jump down together. Since all the atoms are the same, and they all have electrons in the same high orbits, all the electrons jump down the same distance and so release photons of the same frequency. The result is a burst of coherent radiation.

The laser photons are all contained in a reflective tube that has mirrors at both ends. The mirrors cause the waves of light to bounce back and forth, reinforcing each other—kind of like making tidal waves in the bathtub. When the waves reach a critical energy threshold, they break through the partially reflective front mirror as a collimated ray of coherent light—a laser beam.

# *Bibliography*


Abell, George. *Exploration of the Universe*. New York: Holt, Rinehart and Winston, 1973.

Abercrombie, Stanley. *Ferrocement: Building with Cement, Sand, and Wire Mesh*, New York: Schocken Books, 1977.

Adams, Douglas. *The Hithhiker's Guide to the Galaxy*, New York: Random House, 1979.

Adan, B. and E.W. Lee. "High Rate Algal Growth Pond Study Under Tropical Conditions." *Workshop on Wastewater Treatment and Resource Recovery, Singapore, 1980*. IDRC-154e. Ottawa: International Development Research Center, 1980.

*Air Supported and Frame Supported Structures Pre-Engineered or Custom Built for Multi-Industrial Uses*. Tappan: ASATI (Air Structures Air Tech. International) brochure.

Aldrin, Buzz. "The Mars Transit System," *Smithsonian Air & Space*, (Oct./Nov., 1990), pp. 40-47.

"Algae Fuel Petrol Research," *New Scientist* (24 Sept. 1987), p. 35.

Allen, F.G. "Magnetic Electrostatic Shielding", NASA CR-153755, (March, 1967).

Allman, William F. "Rediscovering Planet Earth," *U.S. News and World Report* (Oct. 31, 1988), pp. 56-68.

Ames, B.N., R. Magaw, and L.S. Gold. "Ranking Possible Carcinogenic Hazards," *Science*, No. 236, (April 17, 1987), pp. 271-280.

"An Active Radiation Shield for Cylindrically Shaped Vehicles," *Journal of Spacecraft and Rockets*, (July, 1971), pp. 773-776.

Anderla, Georges and Anthony Dunning. *Computer Strategies: 1990-99*. New York: John Wiley & Sons, 1987.

Anderson, Ian. "Australia Prepares to Measure the Rise and Rise of the Pacific," *New Scientist* (22 July 1989), p. 26.

Anderson, Poul. "Our Many Roads to the Stars," *Galaxy: The Best of My Years*, James Baen, ed., New York: Ace Books, 1980.

Anderson, Poul. Personal communication, 17, Feb. 1994.

Andrus, G. Merrill and George T. Gillies. "MAGLEV: Transportation for the 21st Century," *Civil Engineering*, (April, 1987), pp. 65-67.

Ante'bi, Elizabeth and David Fishlock. *Biotechnology: Strategies for Life*. Cambridge: MIT Press, 1985.

Argyle, Edward. "Chance and the Origin of Life," *Extraterrestrials: Where Are They?* New York: Pergamon Press, 1982.

Armstrong, David, et al. "Natural Flavors Produced by Biotechnological Processing," *Flavor Chemistry*: Trends and Developments, Roy

Teranishi, et al., eds., Washington: American Chemical Society, 1989.
Asimov, Isaac. *A Choice of Catastrophes.* New York: Simon and Schuster, 1979.
Asimov, Isaac. *Mars, The Red Planet.* New York: Lathrup, Lee & Shepard Co., 1977.
Averner, M.M. and R.D. Macelroy. *On The Habitability of Mars: An Approach to Planetary Ecosynthesis.* NASA SP-414, 1976.
"Baltic Freight Index." *The Journal of Commerce and Commercial Shipping* (March 21, 1989), p. 3B.
Baroika, Allen A. "The Chip," *National Geographic*, Vol. 162, No. 4, (October, 1982), pp. 421-457.
Barr, Roderick A., et al. "Theoretical Evaluation of the Sea-Keeping Behavior of Large OTEC Plant Platforms and Cold Water Pipe Configurations." *U.S. ERDA Cont. No. ET-76-C-02-2681*, COO-2681-3, (August, 1978).
*Basic Petroleum Data Book - Sept. '86.* "Petroleum Industry Statistics," Vol. VI, No. 3. American Petroleum Institute, 1986.
Bassett, L.S. et al. "Prevention of Disuse Osteoporosis in the Rat by Means of Pulsing Electromagnetic Fields," *Electrical Properties of Bone and Cartilage: Experimental Effects and Clinical Applications.* Carl T. Brighton, Jonathan Black and Soloman Pollack, eds. New York: Grove & Stratton, 1979.
Bates, D.R., ed. *Space Research and Exploration.* New York: William Sloan Assoc., 1958.
Beard, Jonathan "Balloon in Space Takes the Heat off Spacecraft," *New Scientist,* 14 Oct. 1989, p. 35.
Beatty, J. Kelly, Andrew Chaikin and Brian O'Leary, eds. *The New Solar System.* New York: Cambridge University Press, 1982.
Becker, Robert O. "Effects of Electrical Currents on Bone In Vivo," *Nature*, Vol. 204, (Nov. 14, 1964), pp. 652-654.
Becker, Robert O. "The Significance of Electrically Stimulated Osteogenesis," *Clinical Orthopaedics and Related Research*, Vol. 141, (June, 1979), pp. 266-274.
Bedding, James. "Money Down the Drain," *New Scientist*, (15 April 1989), pp. 34-38.
Bender, David F. "The Lunar Capture Phase of the Transfer of Asteroidal Material to Earth Orbit by Means of Gravity Assist Trajectories," *Journal of the British Interplanetary Society*, Vol. 40, (March, 1987), pp. 129-132.
Berkovsky, B. "Ocean Thermal Energy - Prospective for a Renewable Source of Power," U.S. DOE CONF-780236, *The Fifth Ocean Thermal Energy Conversion Conference*, (February, 1978).
Berry, Adrian. *The Iron Sun: Crossing the Universe Through Black Holes.* New York: E.P.P. Dutton, 1977.
Berry, Adrian. *The Next Ten Thousand Years.* New York: E.P. Dutton & Co., Inc., 1974.

Bickel, John O., ed. *Tunnel Engineering Handbook.* New York: Von Nostrand Reinhold Co., 1982.
Biddle, Wayne. "Two Faces of Catastrophe," *Smithsonian Air & Space,* (Aug./Sept., 1990), pp. 46-49.
Birch, Paul. "Radiation Shields for Ships and Settlements," *Journal of the British Interplanetary Society*, Vol. 35, (November, 1982), pp. 515-519.
Black, Jonathan. *Electrical Stimulation.* New York: Praeger, 1987.
Bloembergen, Nicolaas, and C. Kumar Npatel. "Strategic Defense and Directed Energy Weapons," *Scientific American*, Vol. 257. No. 3, (September, 1987), pp. 39-45.
Bluestone, Mimi. "Solar Power: Alive, Well - and Almost Making Money," *Business Week,* (July 18, 1988), pp. 132-133.
Bodin, Michael A. "Brief Human Vacuum Exposure In Relation to Space Rescue Operations," *Journal of the British Interplanetary Society*, Vol. 30, (February, 1977), pp. 55-62.
"Boeing Aerospace Facility to Aid SDI Decision on FEL Technology," *Aviation Week and Space Technology*, (Aug. 18, 1986), pp. 66-69.
Bond, Alan, and Anthony R. Martin. "A Conservative Estimate of the Number of Habitable Planets in the Galaxy," *Journal of the British Interplanetary Society*, Vol. 31, (November, 1978), pp. 411-415.
Bowman, Geoffrey. "Phobos and Deimos," *Spaceflight*, (October, 1980), pp. 303-311.
Bowman, Norman J. "The Food and Atmosphere Control Problem on Space Vessels," *Realities of Space Travel,* L.J. Carter. ed. New York: McGraw Hill Book Co., Inc., 1957.
Brand, Stewart. *The Media Lab: Inventing the Future at M.I.T.* New York: Penguin Books, 1988.
Brand, Stewart, ed. *Space Colonies.* New York: Penguin Books, 1977.
Brandt, Richard and Otis Port. "Intel: The Next Revolution," *Business Week,* (Sept. 26, 1988), pp. 74-80.
Brewer, J.H., et al. "Construction Feasibility of OTEC Platforms." *U.S. ERDA Contract No. ET-78-C-02-4931*, COO-4931-1.
Brighton, Carl T. "Bioelectrical Effects on Bone and Cartilage," *Clinical Orthopaedics and Related Research*, Vol. 124, (1977), pp. 2-4.
"Britain Pushes to the Fore in Growing Algae for Drugs," *New Scientist* (1 Oct. 1987), p. 35.
Broecker, Wallace S. "Unpleasant Surprises in the Greenhouse," *Nature*, Vol. 328 (1987), p.123.
Bruckner, A.P., and A. Hertzberg. *Ram Accelerator Direct Launch System for Space Cargo.* IAF-87-211, Seattle: University of Washington, 1987.
Bubenik, G. A., et al. "The Effect of Neurogenic Stimulation on the Development and Growth of Bony Tissues," *The Journal of Experimental Zoology*, Vol. 219, (1982), pp. 205-216.
"Building a Space Station," *U.S. News and World Report*, (Sept. 26, 1988), p. 60.

Bustin, Robert. "Determination of Hydrogen Abundance in Selected Lunar Soils," *NASA-N87-30239*, Oct. 31, 1987.

Campbell, Paul A. "Aeromedical and Biological Considerations of Flight Above the Atmosphere," *Realities of Space Travel*, L.J. Carter. ed. New York: McGraw Hill Book Co., Inc., 1957.

Casimati, Nina. *Guide to East Africa: Kenya, Tanzania, and the Seychelles*. London: Hippocrene Books, Inc, 1985.

Caulkins, David. "Raw Materials for Space Manufacturing - A Comparison of Terrestrial Practice and Lunar Availability," *Journal of the British Interplanetary Society*, Vol. 30, (August, 1977), pp. 314-316.

Chandler, William V., et al., "Energy for the Soviet Union, Eastern Europe, and China," *Scientific American*, (Sept. 1990).

Chapman, Clark R., and David Morrison. "Chicken Little Was Right," *Discover*, (May, 1991), pp. 40-43.

Chapman, V. J. *Seaweeds and Their Uses*. London: Methuen & Co. Ltd., 1970.

"Charged Up Muscles," *Prevention*, Vol. 41, (April, 1989), p. 17.

Charles, Dan, et al. "Space May be Too Dangerous for Human Beings," *New Scientist*, (3 March 1990), p. 24.

Chaum, David. "Achieving Electronic Privacy," *Scientific American*, (August 1992), pp. 96-101.

Churchill, R.R. and A.V. Lowe. *The Law of the Sea*. Manchester: Manchester University Press, 1983.

Clark, William C. "The $CO_2$ Question," *Science*, Vol. 223, (9 March 1984), p. 104.

Clark, W.C. and R.E. Munn, eds., *Sustainable Development of the Biosphere*. Cambridge: Cambridge University Press, 1986.

Clarke, Arthur C. *Ascent to Orbit: A Scientific Autobiography*. New York: John Wiley & Sons, 1984.

Clarke, Arthur C. "Electromagnetic Launching as a Major Contribution to Space-Flight," *Journal of the British Interplanetary Society*, Vol. 9, No. 6, (November, 1950), pp. 261-267.

Clarke, Arthur C. "Extra-Terrestrial Relays," *Wireless World*, (Oct. 1945), pp. 305-308.

Clarke, Arthur C. *Interplanetary Flight: An Introduction to Astronautics*. New York: Harper & Brothers, Pubs., 1960.

Cleator, P.E. *An Introduction to Space Travel*. New York: Pitman Publishing Corp., 1961.

Cleary, Daniel. "Warm Messages on an Atomic Scale," *New Scientist* (26 Jan. 1991), p. 31.

Coates, Andrew. "Surviving Radiation in Space," *New Scientist*, (21 July 1990), pp. 42-45.

Cole, H.S.D., et al., editors. *Models of Doom: A Critique of the Limits of Growth*. New York: Universe Books, 1973.

Cole, Stephen. "Space Station in the Balance," *Astronomy*, (May, 1989), pp. 25-31.

Collins, Michael. "Mission to Mars," *National Geographic*, Vol. 174, No. 5, (November, 1988), pp. 733-764.
"The Connection Machine." *Scientific American*, Vol. 256, No. 6, (June 1987).
Cook, James. "Solar Energy Getting Hotter," *Forbes,* (Dec. 12, 1988), pp. 95-96.
Cook, William J. "The New Frontiers," *U.S. News & World Report* (Sept. 26, 1988), pp. 50-62.
Couper, Alastair, ed. *The Times Atlas of the Oceans.* New York: Van Nostrand Reinhold Co., 1983.
Crawford, William P. *Mariner's Weather.* New York: W.W. Norton & Co., 1978.
Criswell, David R. "Lunar Materials for Construction of Space Manufacturing Facilities." *Princeton Conference on Space Manufacturing Facilities,* Princeton: (7-9 June, 1975).
Crone, Andrew and Peter Liss. "Carbon Dioxide, Climate and the Sea," *New Scientist,* (21 Nov. 1985), pp. 50-54.
Cross, Michael. "Battle to Win the World's Greatest Memory Contest," *New Scientist* (22 July 1989), p. 35.
Cross, Michael. "Do Computers Dream of Intelligent Humans?" *New Scientist,* (26 Nov. 1988), pp. 42-46.
Crowe, Devon G. "Laser Induced Pair Production as a Matter-Anti-Mater Source," *Journal of the British Interplanetary Society*, Vol. 36, (November, 1983), pp. 507-508.
Crutzen, Paul J. and T.E. Groedel. "The Role of Atmospheric Chemistry in the Environment - Development Interactions." *Sustainable Development of the Biosphere.* Cambridge: Cambridge University Press, 1986.
Cunningham, Clifford J. *Introduction to Asteroids.* Richmond: William-Bell, Inc., 1988.
Currier, D. P. "Muscular Strength Development by Electrical Stimulation in Healthy Individuals," *Physical Therapy*, Vol. 63, No. 6, (June 1983), pp. 915-921.
Damon, Thomas. *Introduction to Space: The Science of Spaceflight.* Malobar: Orbit Book Co., 1989.
Dantzig, George B., and Thomas L. Saaty. *Compact City.* San Francisco: W.H. Freeman, 1973.
"Data Storage Technologies for Advanced Computing," *Scientific American*, Vol. 257, No. 4, (October, 1987).
Davis, Ged R. "Energy for Planet Earth," *Scientific American*, Vol 263, No. 3, (September, 1990), pp. 55-62.
Davis, Kingsley. "The History of Birth and Death," *Bulletin of the Atomic Scientist*, Vol. 42, No. 3, (April 1986).
Davis, L. "Biological Diversity: Going... Going...?" *Science News*, (September 27, 1986), p. 202.
"Deciding to Colonize the Moon," *Spaceflight*, Vol. 31, (July, 1989), pp. 237-240.

DeKarne, James B. "Hydroponic Greenhouse Gardening," *Mother Earth News*, No. 29, (September, 1974), pp. 68-71.
Delitto, Anthony and Lynn Snyder-Mackler. "Two Theories of Muscle Strength Augmentation Using Percutaneous Electrical Stimulation," *Physical Therapy*, Vol. 70, No. 3, (March, 1990), pp. 158-164.
Delitto, Anthony and Andrew J. Robinson. "Electrical Stimulation of Muscle: Techniques and Applications," *Clinical Electrophysiology*. Baltimore: Williams & Wilkins, 1989.
Delsemre, A.H. "Chemical Composition of Nuclei," *Comets*, Laurel L. Wilkening, ed. Tuscon: University of Arizona Press, 1982.
DeSan, M.G. "The Ultimate Destiny of an Interstellar Species - Everlasting Nomadic Life in the Galaxy," *Journal of the British Interplanetary Society*, Vol. 34, (June, 1981), pp. 219-237.
Dickinson, R.E. "Impact of Human Activities on Climate - A Framework." *Sustainable Development of the Biosphere*. Cambridge: Cambridge University Press, 1986.
*Diet and Health: Implications for Reducing Chronic Disease Risks*. National Research Council, Washington: National Academy Press, 1989.
Dodge, Barnett F. "Fresh Water from Saline Waters: An Engineering Research Problem," *American Scientist*, Vol. 48, (1960).
Donnelly, Ignatius. *Atlantis: The Antediluvian World*. New York: Gramercy Pub. Co., 1949.
Dooling, David, Jr. "Closed Loop Life Support System," *Spaceflight*, Vol. 14, (April, 1972), pp. 134-139.
Dooling, David, Jr. "Controlled Thermonuclear Fusion for Space Propulsion," *Spaceflight*, Vol. 14, (January, 1972), pp. 26-27.
Drew, Phillip. *Frei Otto: Form and Structure*. Boulder: Westview Press.
Drexler, K. Eric. *Engines of Creation: The Coming Era of Nanotechnology*. New York: Doubleday, 1987.
Drexler, Eric. "Deep Space Material Sources," *Princeton Universtiy Conference on Space Manufacturing Facilities*, (Princeton, 1975).
Dunbar, Lyle E. "Market Potential for OTEC in Developing Nations," *8th Ocean Energy Conference*, Washington, June, 1981.
Dyson, Freeman. *Disturbing the Universe*. New York: Basic Books, Inc., 1979.
Dyson, Freeman. *Infinite in All Directions*. New York: Harper & Row, Publishers, 1988.
Dyson, Freeman J. "Search for Artificial Stellar Sources of Infra-red Radiation," *Science*, Vol. 131, (3 June 1960), p. 1667.
Eberhart, J. "An Inflatable U.S. Space Station," *Science News*, (July 18, 1987), p. 37.
Ehricke, Krafft A. "A Long-Range Perspective and Some Fundamental Aspects of Interstellar Evolution," *Journal of the British Interplanetary Society*, Vol. 28, (November, 1975), pp. 713-734.
Ehrlich, Paul R. and Anne H. Ehrlich. "World Population Crisis," *Bulletin of the Atomic Scientists*, Vol. 42, No. 3, (March, 1986), p. 13.

Eicher, Carl K. "Facing Up to Africa's Food Crisis," *Foreign Affairs*, Vol. 61 (1), (Fall, 1982), p. 151.
Elliott, Edward C. "Reinforced Silicone Resins," *Silicone Technology*, Paul F. Bruins, ed., Interscience Publishers, 1969, pp. 121-139.
Engel, K.H., et al. "Biosynthesis of Chiral Flavor and Aroma Compounds," *Flavor Chemistry Trends and Developments*, Roy Teranishi, et al. eds. Washington: American Chemical Society, 1989.
"Environmental Structures," Environmental Structures, Inc.
"Eros," *Journal of the British Interplanetary Society*, Vol. 30, (June, 1977), p. 234.
Evett, Arthur. *Understanding the Space-Time Concepts of Special Relativity*. New York: Halsted Press, 1982.
Farquhar, Robert and David Dunham. "Libration-Point Staging Concepts for Earth-Mars Transportation," *Manned Mars Missions Working Group Papers, Vol. I,* NASA, June 10, 1985.
Felten, James E. "Feasibility of Electrostatic Systems for Space Vehicle Radiation Shielding," *The Journal of Astronautical Sciences* (Spring, 1964), pp. 16-22.
Fennessy, Edward. "Telecommunications AD 2000," *Journal of the British Interplanetary Society*, Vol. 32, No. 10, (October, 1979), pp. 371-378.
Fielder, Judith and Nickolaus Leggett. "A Second Generation Lunar Agricultural System," *Journal of the British Interplanetary Society*, Vol. 41, (June, 1988), pp. 263-268.
Finney, Ben R. and Eric M. Jones, eds. *Interstellar Migration and the Human Experience*. Berkeley: University of California Press, 1985.
Fisher, Arthur. "Global Warming, Part I: Playing Dice With Earth's Climate," *Popular Science*, (August, 1989), pp. 51-58.
Fisher, Arthur. "Global Warming, Part II: Inside the Greenhouse," *Popular Science*, (September, 1989), pp. 63-70.
Fogg, G.E. *Algal Cultures and Phytoplankton Ecology*. Madison: University of Wisconsin Press, 1965.
Fogg, Martyn J. "Extra-Solar Planetary Systems II: Habitable Planets in the Galaxy," *Journal of the British Interplanetary Society*, Vol. 39, (March, 1989), pp. 99-109.
Fogg, Martyn J. "The Feasibility of Intergalactic Colonization and its Relevance to SETI," *Journal of the British Interplanetary Society*, Vol. 41, (November, 1988), pp. 491-496.
Fooland, Just, ed. *Population and the World Economy in the 21st Century*. New York: St. Martin's Press, 1982.
Forward, Robert L. "Antimatter Revealed," *Omni,* (November, 1979), pp. 45-48.
Forward, Robert L. "Feasibility of Interstellar Travel: A Review," *Journal of the British Interplanetary Society*, Vol. 39, (September, 1986), pp. 379-384.
Forward, Robert L. "A Programme for Interstellar Exploration," *Journal of the British Interplanetary Society*, Vol. 29, (October, 1976), pp. 611-632.

Forward, Robert L. and Joel Davis. "Ride a Laser to the Stars," *New Scientist,* (2 Oct. 1986), pp. 31-35.
Fox, Barry. "Portable Phones Ring the Globe by Satellite," *New Scientist,* (7 July 1990), p. 30.
Fox, Barry. "A Record for the Compact Disc," *New Scientist*, (22 July 1989).
Freeman, Michael. *Space Traveller's Handbook.* New York: Sovereign Books, 1979.
Freiherr, Greg. "The Orbiting Junkyard," *Final Frontier,* (Nov./Dec., 1990), pp. 41-47.
Freitas, Robert A., Jr. "Interstellar Probes: A New Approach to SETI," *Journal of the British Interplanetary Society*, Vol. 33, (March, 1980), pp. 95-100.
Freitas, Robert A., Jr. "Terraforming Mars and Venus Using Machine Self-Replicating Systems (SRS)," *Journal of the British Interplanetary Society*, Vol. 36, (March, 1983), pp. 139-142.
Friedman, Louis. *Starsailing*. New York: John Wiley & Sons, Inc., 1988.
Freund, Henry P. and Robert K. Parker. "Free Electron Lasers," *Scientific American,* (April, 1989), pp. 84-89.
Freundlich, Naomi J. "Rings of Power," *Popular Science,* (January, 1989), p. 66.
Fry, Ian V., et al. "Application of Photosynthetic N2-Fixing Cyanobacteria to the CELSS Program," *Controlled Ecological Life Support Systems*, N88-12257.
Gair, Thomas J. "The Spectrum of Silicone Applications," *Silicone Technology*, Paul F. Bruins, ed., Interscience Publishing, 1969, pp. 1-7.
Gall, Norman. "We Are Living Off Our Capital," *Forbes*, (Sept. 22, 1986), p. 62.
Gallant, Roy A. *National Geographic Picture Atlas of Our Universe.* Washington: National Geographic Society, 1986.
Gaubatz, W.A., P.L. Klevatt, J.A. Copper. "Single Stage Rocket Technology," 43rd Congress of the IAF, IAF-92-0854, Washington: 1992.
Gilfillan, Edward S., Jr. *Migration to the Stars: Never Again Enough People*, Washington: Robert B. Luce, Co., 1975.
Glaser, Peter E. "The Satellite Solar Power Station," *Princeton Conference on Space Manufacturing Facilities*, Princeton, 1975.
Gleick, James. *Chaos: Making a New Science.* New York: Viking Penguin, Inc., 1987.
Glenn, Jerome Clayton and George S. Robinson. *Space Trek: The Endless Migration.* New York: Warner Books, 1978.
Golden, Frederic. *Colonies in Space: The Next Giant Step*. New York: Harcourt Brace Jovanovich, 1977.
Goldsmid, H.J. *Thermoelectric Refrigeration.* New York: Plenum Press, 1964.
Gordon, J.E. *The Science of Structures and Materials.* New York: Scientific American Library, 1988.

Gore, Rick. "Extinctions," *National Geographic*, (June 1989), p. 664.
Greenberg, J.M. "What Are Comets Made of? A Model Based on Interstellar Dust," *Comets*, Laurel L. Wilkening, ed. Tuscon: University of Arizona Press, 1982.
Greenwald, John. "Tune In, Turn On, Sort Out," *Time*, (May 29, 1989), p. 68.
Gribbin, John, *Hothouse Earth: The Greenhouse Effect and Gaia*, New York: Grove Weideneld, 1990.
Gribbin, John. *In Search of Schroedinger's Cat*. New York: Bantam Books, 1984.
Gribbin, John. "Warmer Seas Increase Greenhouse Effect," *New Scientist,* (6 Jan. 1990), p. 31.
Grieve, Richard A.F. "Impact Cratering on the Earth," *Scientific American,* (April, 1990), pp. 66-73.
Gritton, E.C. et al. "Projected Engineering Cost Estimate for an Ocean Thermal Energy Conversion (OTEC) Central Station," *8th Ocean Energy Conference*, Washington, June, 1981.
Grossmann, John. "The Blue Collar Space Suit," *Smithsonian Air & Space* (Oct./Nov., 1989), pp. 58-67.
Groves, Donald G. and Lee M. Hunt. *Ocean World Encyclopaedia.* New York: McGraw-Hill Book Co., 1980.
Gupta, Vijay K. "An Overview of Ocean Thermal Energy Systems." *Alternative Energy Systems*. New York: Pergamon Press, 1984.
Gustan, E. and T. Vinopal. "Controlled Ecological Life Support System: A Transportation Analysis," *NASA-CR-166420*, NASA-Ames Research Center, Moffett Field, CA, 1982.
Haffner, James W. *Radiation and Shielding in Space*. New York: Academic Press, 1967.
Hagen, Arthur W. *Thermal Energy From the Sea*. Park Ridge: Noyes Data Corp., 1975.
Hall, A.J. *The Standard Handbook of Textiles*. London: Heywood Books, 1969.
Hallmark, Clayton L. and Delton T. Horn. *Lasers: The Light Fantastic, 2nd Ed.* Blue Ridge Summit: Tab Books, Inc., 1987.
Hamakawa, Yoshihiro. "Photovoltaic Power," *Scientific American*, Vol. 256, No. 4, (April 1987), pp. 86-92.
Hamilton, S.B., Jr. "Silicone Elastomers," *Silicone Technology*, Paul F. Bruins, ed., Interscience Publishing, 1969.
Hammonds, Keith H. "The Graphics Revolution," *Business Week,* (Nov. 28, 1988), pp. 142-156.
Hannah, Eric C. "Radiation Protection for Space Colonies," *Journal of the British Interplanetary Society*, Vol. 30, (August, 1977), pp. 310-313.
Hanrahan, James S. and David Bushnell. *Space Biology: The Human Factors in Space Flight*. New York: Science Editions, Inc., 1961.
Hanson, Joe A. *Open Sea Mariculture*. Stroudsburg: Dowden, Hutchinson & Ross, Inc., 1974.

Harding, Richard. *Survival in Space: Medical Problems of Manned Spaceflight*. New York: Routledge, 1989.
Harper, David E. "The Future of Telecommunications: Part I," *Telecommunications,* (January 1989), pp. 27-32.
Hart, Michael H. "Atmospheric Evolution, the Drake Equation, and DNA: Sparse Life in an Infinite Universe," *Extraterrestrials - Where Are They?* New York: Pergamon Press, 1982.
Harte, John and Robert H. Socolow. *Patient Earth.* New York: Holt, Rinehart and Winston, Inc., 1971.
Head, William and Jon Splane. *Fish Farming in Your Solar Greenhouse.* Eugene: Amity Foundation, 1979.
Hecht, Jeff. "Will We Catch a Falling Star?," *New Scientist*, (7 Sept. 1991), p. 48.
Hempsell, Mark. "Space Industrialization - A New Perspective," *Spaceflight*, Vol. 31, (July, 1989).
Henbest, Nigel. *The Exploding Universe.* New York: Macmillan Publishing, 1979.
Henbest, Nigel and Jeff Hecht. "Ice Dwarfs at the Edge of the Solar System," *New Scientist,* (13 July 1991), p. 24.
Heppenheimer, T.A. *Colonies In Space*. Harrisburg: Stackpole Books, 1977.
Heppenheimer, T.A. "Free Electron Lasers," *Popular Science* (December, 1987), p. 63.
Herzog, Thomas. *Pneumatic Structures: A Handbook of Inflatable Architecture*. New York: Oxford University Press, 1976.
Hilbertz, Wolf H. "Accretion Coating and Mineralization of Materials for Protection Against Biodegradation." *U.S. Patent No. 4,461,684*, (July 24, 1984).
Hilbertz, Wolf H. "Autochthonous Sea Architecture: Preliminary Experiments." *Man-Environment Systems*, Vol. 6, No. 1, (January 1976).
Hilbertz, Wolf H. "Electrodeposition of Minerals in Sea Water: Experiments and Applications." *IEEE Journal of Oceanic Engineering*, Vol. OE-4, No. 3, (July 1979).
Hilbertz, Wolf H. "Marine Architecture: An Alternative." *Architectural Science Review*, Vol. 19, No. 1, (December, 1976).
Hilbertz, Wolf H. "Mineral Accretion of Large Surface Structures, Building Components and Elements," *U.S. Patent #4,246,075*, (Jan. 20, 1981).
Hilbertz, Wolf H. "Repair of Reinforced Concrete Structures by Mineral Accretion." *U.S. Patent No. 4,440,605*, (April 3, 1984).
Hilbertz, Wolf H. "Toward Cybertecture." *Progressive Architecture*, (May 1970).
Hofele, W., et al. "Novel Integrated Energy Systems: the Case of Zero Emissions." *Sustainable Development of the Biosphere*, Cambridge: Cambridge University Press, 1986.

Holden, Constance. "New Annual Report on Global Deterioration," *Science*, Vol. 232, (1986), p. 822.
Holdren, John P. "Energy in Transition," *Scientific American*, (September, 1990), p. 160.
Hosternoth, Stefan and Peter J. Lamb. *Climatic Atlas of the Indian Ocean: Vol. I - Surface Climate and Atmospheric Circulation.* Madison: University of Wisconsin Press, 1979.
Houghton, Richard A. and George M. Woodwell. "Global Climatic Change," *Scientific American*, Vol. 260, No. 4, (April, 1989), pp. 36-44.
Huddleston, H.T., "Metal Extraction from Lunar Ore," *Journal of the British Interplanetary Society*, Vol. 32, No. 1, (January, 1979), pp. 27-31.
Huston, S.L. "Radiation Shielding for Lunar Bases Using Lunar Concrete," 43rd Congress of the IAF, IAF-92-0339, Washington: 1992.
"Induction-Type Laser Faces Critical Tests," *Aviation Week and Space Technology,* (August 18, 1986), pp. 54-59.
*Information Please Almanac: Atlas and Yearbook - 1988.* Boston: Houghton Mifflin Co., 1988.
*International Energy Annual—1991.* Energy Information Administration, U.S. Department of Energy, Washington: 1992.
Jirka, G.H. et al. "Intermediate Field Plumes From OTEC Plants: Predictions for Typical Site Conditions," *8th Ocean Energy Conference*, Washington, June, 1981.
Jones, Eric M. "Estimates of Expansion Time Scales," *Extraterrestrials - Where Are They?* New York: Pergamon Press, 1982.
Jones, Kenneth L., Louis W. Shainberg and Curtis O. Byer. *Foods, Diet, and Nutrition, 2nd Ed.* San Francisco: Canfield Press, 1975.
Kantrowitz, Arthur. "Laser Propulsion to Earth Orbit: Has Its Time Come?," *Proceedings: SDIO/DARPA Workshop on Laser Propulsion*, Vol. 2, CONF-860778.
Kantrowitz, Arthur. "Propulsion to Orbit by Ground-Based Lasers," *Astronautics and Aeronautics*, Vol. 10, No. 5, May 1972.
Kare, Jordin T. "Trajectory Simulation for Laser Launching," *SDIO/DARPA Workshop on Laser Propulsion*, Vol. 2, CONF-860778.
Kasting, James F., Owen B. Toon and James B. Pollack. "How Climate Evolved on the Terrestrial Planets," *Scientific American*, Vol. 258, No. 2, (February, 1988), p. 90.
Kasting, James F., Christopher P. McKay and Owen B. Toon, "Making Mars Habitable," *Nature*, Aug. 8, 1991, p.489.
Kavaler, Lucy. *Green Magic: Algae Rediscovered.* New York: Thomas Y. Crowell, 1983.
Kavaler, Lucy. *The Wonders of Algae.* New York: The John Day Co., 1961.
*Ken's Fish Hatchery, Vol. 6.* Alapaha: Ken's Fish Farms, Inc., 1983.
Keplinger, H.F. "The CRISIS in Energy is Here Now!," *World Oil*, (July 1986), p. 52.

Kerr, Richard A. "Carbon Dioxide and the Control of Ice Ages," *Science*, Vol. 223, (9 March 1984), p. 1053.

Kerr, Richard A. "Ocean's Deserts are Blooming," *Science*, (13 June 1986), p. 1345.

"Kevlar Aramid: The Fiber that Lets You Re-Think Strength and Weight." Dupont Co. brochure.

Keyfitz, Nathan. "The Growing Human Population," *Scientific American*, (September, 1989), p. 119.

Kimball, John W. *Biology*. 5th Ed., Reading: Addison Wesley Publishing Co., 1983.

Kincaide, William C. and John N. Murray. "Advanced Alkaline Electrolysis Systems for OTEC," *8th Ocean Energy Conference*, Washington, June, 1981.

Kirk, David, et al. *Biology: The Unity and Diversity of Life*. Belmont: Wodsworth Publishing Co. Inc., 1978.

Klass, Philip J. "Scientists Focus on FEL Technology to Counter ICBMs," *Aviation Week and Space Technology*, (August 18, 1986), pp. 40-45.

Kootz, Ronald B. *Cable: An Advertiser's Guide to the New Electronic Media*. Chicago: Crain Books, 1982.

La Marche, Valmore C., Jr. et al. "Increasing Atmospheric Carbon Dioxide: Tree Ring Evidence for Growth Enhancement in Natural Vegetation," *Science*, (7 Sept. 1984), p. 1019.

Lampe, David. "Grow Your Own Buildings," *Mother Earth News*, (March/April 1980).

Landsberg, Hans H., chairman. *Energy: The Next Twenty Years*. Cambridge: Ballinger Pub. Co., 1979.

Langone, John. *Superconductivity: The New Alchemy*. Chicago: Contemporary Books, 1989.

Langton, N.H., ed. *The Space Environment*. London: University of London Press, 1969.

Lawton, A.T. "Photometric Observation of Planets at Interstellar Distances," *Spaceflight*, Vol. 12, No. 9, (September, 1970), pp. 365-373.

Levinson, Alfred A. and S. Ross Taylor. *Moon Rocks and Minerals: Scientific Results of the Study of the Apollo 11 Lunar Samples with Preliminary Data on Apollo 12 Samples*. New York: Pergamon Press, 1971.

Lewin, Roger. "Damage to Tropical Forests or Why Were There so Many Kinds of Animals?," *Science*, Vol 234, (10 Oct. 1986), pp. 149-150.

"Life Against the Odds," *Discover*, (June, 1990), p. 15.

"Limited Occupancy," *Scientific American*, (June, 1987), p. 25.

*Longman Illustrated Dictionary of Astronomy & Astronautics: The Terminology of Space*. Harlow: York Press, 1987.

Lother, Michael et al. "Cost Targets for OTEC: The Impact of Innovative Financing," *8th Ocean Energy Conference*, Washington, June, 1981.

Lovelock, James. *The Ages of Gaia: A Biography of Our Living Earth*. New York: W.W. Norton, 1988.

Lovelock, J.E. *Gaia: A New Look at Life on Earth.* Oxford: Oxford University Press, 1979.

Lovelock, James and Michael Allaby. *The Greening of Mars.* New York: St. Martin's Press, 1984.

Lunan, Duncan. *Man and the Planets: The Resources of the Solar System.* Bath: Ashgrove Press, 1983.

"Lunar Water Process," *Spaceflight*, Vol. 13, No. 4, (April, 1971), p. 131.

Lynch, Wilfred. *Handbook of Silicone Rubber Fabrication.* New York: Van Nostrand Reinhold Co., 1978.

Mallove, Eugene and Gregory Matloff. *The Starflight Handbook: A Pioneer's Guide to Interstellar Travel.* New York: John Wiley & Sons, Inc., 1989.

Martin, Anthony R. "Space Resources and the Limits to Growth," *Journal of the British Interplanetary Society*, Vol. 38, (June, 1985), pp. 243-252.

"A Mass Extinction Without Asteroids," *Science*, Vol. 234 (1986), p. 14.

Matloff, Gregory L. "Cosmic-Ray Shielding for Manned Interstellar Arks and Mobile Habitats," *Journal of the British Interplanetary Society*, Vol. 30, (March, 1977), pp. 96-98.

Matloff, Gregory L. "World Ships and White Dwarfs," *Journal of the British Interplanetary Society*, Vol. 39, (March, 1986), pp. 114-115.

Matloff, Gregory L. and Alphonsus J. Fennelly. "Optical Techniques for the Detection of Extra-Solar Planets: A Critical Review," *Journal of the British Interplanetary Society*, Vol. 29, (July-August, 1976), pp. 471-481.

Matloff, Gregory L. and Eugene Mallove. "The Interstellar Solar Sail—Optimization and Further Analyses," *Journal of the British Interplanetary Society*, Vol. 36, (May, 1983), pp. 201-209.

Matloff, Gregory L. and Kelly Parks. "Interstellar Gravity Assist Propulsion: A Correction and New Application," *Journal of the British Interplanetary Society*, Vol. 41, (November, 1988), pp. 519-526.

Mazria, Edward. *The Passive Solar Energy Book*, Emmaus: Rodale Press, 1979.

McLarney, Bill. "Fish Farm with Cages," *Mother Earth News*, No. 81, (May/June, 1983), pp. 38-41.

McAleer, Neil. *The Cosmic Mind-Boggling Book.* New York: Warner Books, 1982.

McElroy, M.B. "Change in the Natural Environment of the Earth: The Historical Record." *Sustainable Development of the Biosphere*, Cambridge: Cambridge University Press, 1986.

McElroy, R.D. "CELSS (Closed Ecological Life Support Systems) and Regenerative Life Support for Manned Missions to Mars," *Manned Mars Mission Working Group Papers*, N87-17749.

McEvedy, Colin and Richard Janes. *Atlas of World Population History.* New York: Penguin Books, 1978.

Meals, Robert N. and Frederick M. Lewis. *Silicones.* New York: Reinhold Publishing Corp., 1959.

Meindle, James P. "Chips for Advanced Computing," *Scientific American*, Vol. 257, No. 4, (October, 1987), pp. 78-89.

Mendola, Dominick. "Aquaculture," *Energy Primer: Solar, Water, Wind, and Biofuels.* Richard Merrill and Thomas Gage, eds. New York: Dell Publishing Co., 1978.

*Military Standardization Handbook: Magnesium and Magnesium Alloys.* MIL-HDBK-693 [MR], Dept. of Defense, 30 Sept. 1964.

Miller, Julie Ann. "Diet for a Blue Planet," *Science News*, Vol. 127, (April 6, 1985), pp. 220-222.

Miller, William H. "Whatever Happened to Our Energy Alternatives?," *Industry Week*, (May 18, 1987), p. 32.

Millman, Joe. "Southern Discomfort," *Technology Review*, (18 July 1986), pp. 16-17.

Mogel, Leonard. *The Magazine*, 2nd ed. Chester: Globe Pequot Press, 1988.

Molton, P.M. "The Protection of Astronauts Against Solar Flares," *Spaceflight*, Vol. 13, (June, 1971), pp. 220-224.

"More Food for Thought," *Science News*, (Feb. 18, 1989), p. 111.

Morey, Emily R. and David J. Baylink. "Inhibition of Bone Formation During Space Flight," *Science*, Vol. 210, (22 Sept. 1978), pp. 1138-1141.

"Mother of Short Comets," *Discover,* (February, 1991), p. 8.

Muller, Robert A. and Theodore M. Oberlander. *Physical Geography Today.* New York: Random House, 1978.

Murphy, Jamie and Andrea Dorfman. "The Quiet Apocalypse," *Time*, (Oct. 13, 1986), p. 80.

Mutch, Thomas A., et al. *The Geology of Mars.* Princeton: Princeton University Press, 1976.

Myers, Norman, ed. *GAIA: An Atlas of Planet Management.* New York: Doubleday & Co., Inc., 1984.

Nadis, Steve. "Mars the Final Frontier," *New Scientist*, 5 Feb. 1994, pp. 28-31.

"Near Earth Asteroids Observed," *Spaceflight*, Vol. 31, (July, 1989), p. 227.

Newport, John Paul, Jr. "Get Ready for the Coming Oil Crisis," *Fortune*, (March 16, 1987), p. 47.

Nicholls, Peter, ed. *Ecology 2000.* New York: Beaufort Books, Inc., 1984.

Nicolson, Iain. *The Road to the Stars.* New York: William Morrow & Co., Inc., 1978.

Nieuwalt, S. *Tropical Climatology.* New York: John Wiley & Sons, 1977.

Nordhaus, W.D. and G.W. Yahe. "Future Paths of Energy and Carbon Dioxide Emissions," *Changing Climate: Report of the Carbon Dioxide Assessment Committee.* Washington: National Academy Press, 1983.

Nuttall, L.J. "Advanced Water Electrolysis Technology for Efficient Utilization of Ocean Thermal Energy," *Proceedings of the 8th Ocean Energy Conference*, USDOE, DOE/CONF-810622, (June, 1981), pp. 679-683.

Oberg, James E. *New Earths: Restructuring Earth and Other Planets*. Harrisburg: Stackpole Books, 1981.

Oberg, James E. and Alcestis R. Oberg. *Pioneering Space: Living on the Next Frontier*. New York: McGraw-Hill Book Co., 1986.

*Odds on Virtually Everything*. Editors of Heron House. New York: G.P. Putnam's Sons, 1980.

Oguchi, Mitsuo, et al. "Food Production and Gas Exchange System Using Blue-Green Alga (Spirulina) for CELSS," *Controlled Ecological Life Support System*, NASA Conf. Pub. 2480.

O'Keefe, M.T. "Seashells in the Seychelles," *Saturday Evening Post*, (September, 1988), p. 78.

O'Leary, Brian. *Mars 1999: Exclusive Preview of the U.S.—Soviet Manned Mission*. Harrisburg: Stackpole Books, 1987.

O'Leary, Brian. "Mining the Apollo and Amor Asteroids," *Science*, Vol. 197, (22 July 1977), pp. 363-366.

O'Leary, Brian. *Project Space Station*. Harrisburg: Stackpole Books, 1983.

Oleson, M. and R.L. Olson. "Controlled Ecological Life Support Systems Conceptual Design Option Study," NASA Contract NAS2-11806, (June, 1986).

"One Small Drop for Man," *Discover*, (March, 1988), pp. 20-22.

O'Neill, Gerard K. "The Colonization of Space," *Physics Today*, (September, 1974), pp.32-44.

O'Neill, Gerard K. *The High Frontier: Human Colonies in Space*. New York: William Morrow & Co., Inc., 1977.

Ovenden, W. "Meteor Hazards to Space Stations," *Realities of Space Travel*, L.J. Carter, ed. New York: McGraw Hill Book Co., Inc., 1957.

Palmeira, R.A. and G.F. Pieper. "Cosmic Rays," *Introduction to Space Science*, Wilmont N. Hess, ed. New York: Gordon and Breach, 1965.

Parbit, Michael. "Antarctic Meltdown," *Discover*, (September, 1989), pp. 38-47.

Park, C. and S.W. Bowen. "Ablation and Deceleration of Mass-Driver Launched Projectiles for Space Disposal of Nuclear Wastes," AIAA-81-0355.

Parkinson, Bob. "Superfuels," *Spaceflight*, Vol. 18, (October, 1976), pp. 348-350.

Parkinson, Gerald. "New Techniques May Squeeze More Chemicals from Algae," *Chemical Engineering*, (May 11, 1987), pp. 19-23.

Parkinson, R.C. "Cities on the Moon—A Lost Vision," *Spaceflight*, Vol. 31, (July, 1989), pp. 220-223.

Parkinson, R.C. "The Resources of the Solar System," *Spaceflight*, Vol. 17, (April, 1975), pp. 124-128.

Parkinson, R.C. "Take-Off Point for a Lunar Colony," *Spaceflight*, Vol. 16, (September, 1974), pp. 322-326.

Parliment, Thomas H. and Rodney Croteau, eds. *Biogeneration of Aromas*. Washington: American Chemical Society, 1986.

Parsons, Brian K. and Harold F. Link. "System Studies of Open-Cycle OTEC Components," *Ocean Engineering and the Environment, Oceans 85: Conference Record, IEEE Ocean Engineering Society, Vol. 2*, San Diego, November, 1985.

Parry, M.L. and T.R. Carter. "The Effect of Climatic Variations on Agricultural Risk." *Climatic Change*, Vol. 7 (1985), p. 95.

Parry, M.L. "Some Implications of Climatic Change for Human Development," *Sustainable Development of the Biosphere*, Cambridge: Cambridge University Press, 1986.

Pearce, Fred. "Blowing Hot and Cold in the Greenhouse," *New Scientist*, (11 Feb. 1989), pp. 32-33.

Pearce, Fred. "How to Stop the Greenhouse Effect," *New Scientist*, (18 Sept. 1986), p. 29.

Peled, Abraham. "The Next Computer Revolution," *Scientific American*, Vol. 257, No. 4, (October, 1987), pp. 57-64.

Pelton, Joseph. "The Satellite Communications Industry: New Doors Opening," *Telecommunications*, (February, 1989), pp. 59-60.

"Permanent Storage of Optical Disks Finds New Uses in Aero-Space, Defense," *Aviation Week & Space Technology*, (July 10, 1989), p. 51.

Pfeiffer, John. *The Cell*. New York: Time Inc., 1964.

Pimentel, D. et al. "Energy and Land Constraints in Food Protein Production," *Science*, (21 Nov. 1975), pp. 754-761.

*Pioneering the Space Frontier: The Report of the National Comission on Space*. New York: Bantam Books, 1986.

Pisias, Nicholas G. and Nicholas J. Shackleton. "Modelling the Global Climate: Response to Orbital Forcing and Atmospheric $CO_2$ Changes," *Nature*, Vol. 310, p. 757.

Pollack, Soloman et al. "Microelectrode Studies of Stress Potential in Bone," *Electrical Properties of Bone and Cartilage: Experimental Effects and Clinical Applications*. New York: Grove & Stratton, 1979.

Pournelle, Jerry E. *A Step Farther Out*. New York: Ace Books, 1979.

Pournelle, Jerry. "That Buck Rogers Stuff," *Galaxy: The Best of My Years*, James Baen, ed. New York: Ace Books, 1980.

Prescott, J.R.V. *The Political Geography of the Oceans*. New York: John Wiley & Sons, 1975.

"Producing Metallic Hydrogen," *Spaceflight*, Vol. 19, (May, 1977), p. 175.

*Properties of Mazlo Magnesium Products*. Cleveland: American Magnesium Corp., 1941.

"Radio Frequency FELs May Win Role in SDI," *Aviation Week and Space Technology*, (Aug. 18, 1986), pp. 59-63.

Ramirez, Anthony. "A Warming World," *Fortune*, (July 4, 1988).

Ransome, T. "Lagrange Points and Applications," *Spaceflight*, Vol. 12, No. 12, (December, 1970), pp. 488-490.

Reisch, Marc. "Biotech Firm Readies Microalgae Products," *Chemical and Engineering News*, (July 27, 1987), p. 14.

Richmond, A., et al., "Quantative Assessment of the Major Limitations on Productivity of *Spirulina platensis* in Open Raceways," *Journal of Applied Phycology*, Vol. 2, No. 3, pp.195-206.

Ridpath, Ian, ed. *The Illustrated Encyclopedia of Astronomy and Space*, Revised Ed. New York: Thomas Y. Crowell, Publishers, 1979.

"Rocket Fuel Keeps Acid off Florida Oranges," *New Scientist*, (21 July 1990), p. 32.

Roels, O. A. "The Economic Contribution of Artificial Upwelling Mariculture to Sea-Thermal Power Generation," City University of New York.

Roels, O. A. "From the Deep Sea: Food, Energy, and Fresh Water," *Mechanical Engineering*, (June, 1980), pp. 37-43.

Roels, O. A. and S. Laurence. "Marine Pastures: A By-Product of Large (100 Megawatt or Larger) Floating Ocean Thermal Power Plants," *USERDA Contract No. E(11-1) 2581*, COO-2581-3.

Roels, O. A. and S. Laurence. "Potential Mariculture Yield of Sea Thermal Power Plants: Part 1 - General Statement," CONF-751235, American Geophysical Union, San Francisco, 1975.

Roels, O. A. and S. Laurence. "Potential Mariculture Yield of Sea Thermal Power Plants: Part 2 - Food Chain Efficiency," Port Aransas Marine Laboratory, University of Texas Marine Sciences Institute.

Roels, O. A. and Donald F. Othmer. "Power, Fresh Water and Food from Cold, Deep Sea Water," *Science*, Vol. 182, No. 4108, (12 Oct. 1973), pp. 121-125.

Roels, O. A. and Donald F. Othmer. "Power, Fresh Water and Food from the Sea," *Mechanical Engineering*, (September, 1976).

Rogers, Peter. "Water: Not as Cheap as You Think," *Technology Review*, (Nov./Dec., 1986).

Rothschild, Michael. *Bionomics: The Inevitability of Capitalism*. New York: Henry Holt and Co., 1990.

Rowan-Robinson, Michael. *Our Universe: An Armchair Guide*. New York: W.H. Freeman and Co., 1990.

Reid, Marvin and J. Richard Shoner. "Is Another 'Energy Crisis' Coming Or is Industry Simply Crying Wolf?" *National Petroleum News*, (March, 1987, p. 41).

Rice, E.E., L.A. Miller and R.W. Earhart. "Preliminary Feasibility Assessment for Earth-to-Space Electromagnetic (Railgun) Launchers," *NASA Report No. CR-167886*.

Richards, I.R. "A Closed Ecosystem for Space Colonies," *Journal of the British Interplanetary Society*, Vol. 34, (September, 1981), pp. 392-399.

Richards, I.R. and P.J. Parker. "Estimates of Crop Areas for Large Space Colonies," *Journal of the British Interplanetary Society*, Vol. 29, (December, 1976), pp. 769-774.

Roland, Conrad. *Frei Otto: Tension Structures*. New York: Praeger Publishers, 1970.

Ross, Jonathan M. and William A. Wood. "OTEC Mooring System Development: Recent Accomplishments," *NOAA Technical Report OTES-4*.

Roth, Reece, Warren Rayle and John Reinmann. "Fusion Power for Space Propulsion," *New Scientist,* (20 April 1972), pp. 125-127.

Roymont, John E.G. *Plankton and Productivity in the Oceans*. New York: Pergamon Press, 1963.

Rubas, Thomas J., J. Michael Wittig and Klemens Finsterwalder. "OTEC 100 MWe Alternate Power Systems Study," *6th OTEC Conference: Ocean Thermal Energy for the 80's*, USDOE CONF-790631, June, 1979.

Rucker, Rudy. *The 4th Dimension*. Boston: Houghton Mifflin Co., 1984.

Russell, Peter. *The Global Brain: Speculations on the Evolutionary Leap to Planetary Consciousness*. Los Angeles: J.P. Tarcher, 1983.

Russell-Hunter, W.D. *Aquatic Productivity*. New York: The Macmillan Co., 1970.

Ruzic, Neil P. *Where the Wind Sleeps: Man's Future on the Moon, a Projected History*. New York: Doubleday, 1970.

Sagan, Carl and Ann Druyan. *Comet*. New York: Random House, 1986.

Sagan, Carl. *COSMOS*. New York: Random House, 1980.

Sagan, Carl and Frank Drake. "The Search for Extraterrestrial Intelligence," *Scientific American Exploring Space*, Special Issue, Jonathan Piel, ed., Vol 2, No. 1, (1990).

Salbeld, Robert and Donald Patterson. *Space Transportation Systems: 1980-2000*. New York: American Institute of Astronautics, 1978.

Sampson, R. Neil. "ReLeaf for Global Warming," *American Forests*, (Nov./Dec. 1988), pp. 9-14.

Sandeman, T.F. "Cosmic Radiation and its Possible Biological Effects," *Spaceflight*, Vol. 1, No. 8, (July, 1958), pp. 291-296.

Sanger, Eugen. *Space Flight: Countdown for the Future*. New York: McGraw Hill Book Co., 1965.

Savage, Harry K. *The Rock That Burns: Oil from Shale*, Boulder: Pruett Press, 1967.

Savitz, Eric J. "Wave of Prosperity: Rising Rates, Ebbing Capacity Buoy Tankers," *Barron's*, (June 19, 1989).

Scuibba, C., Jr. "New Concepts Enhance Position of Open and Hybrid OTEC Power Cycles," *5th OTEC Conference*, USDOE CONF-780236, Miami Beach, September, 1978.

"SDI Experiments Will Explore Viability of Ground-Based Laser," *Aviation Week and Space Technology,* (Aug. 18, 1986), pp. 45-54.

Seargent, David A. *Comets: Vagabonds of Space*. Garden City: Doubleday & Co. Inc., 1982.

*The Seychelles Group*: 61AC061036, and *Mahe Island*: 61BHA61541. Clearfield: Defense Mapping Agency Depot.

*Shaded Relief Map of the Tharsis Quadrangle of Mars*, M5M 15/112 R, I-926(MC-9), Dept. of the Interior, USGS.

Shapley, Harlow. *The View From a Distant Star*. New York: Dell Publishing Co., Inc., 1963.
Sheaffer, Robert. "Are We Alone After All?" *Spaceflight,* (Nov./Dec., 1980), pp. 334-337.
Shelef, G. et al. "Waste Treatment and Nutrient Removal by High-Rate Algae Ponds," *Workshop on Wastewater Treatment and Resource Recovery, Singapore, 1980.* IDRC-154e. Ottawa: International Development Research Center, 1980.
Shepherd, L.R. "The Possibility of Cosmic Ray Hazards in High Altitude and Space Flight," *Realities of Space Travel*, L.J. Carter, ed. New York: McGraw Hill Book Co., Inc., 1957.
Sheppard, D.J. "Concrete on the Moon," *Spaceflight*, Vol. 17, No. 3, (March, 1975), p. 91.
Sherman, Irwin W. *Biology: A Human Approach.* 2nd Edition. Oxford: Oxford University Press, 1979.
Shimony, Abner. "The Reality of the Quantum World," *Scientific American,* (January, 1988), p. 46.
Singer, C.E. "Interstellar Propulsion Using a Pellet Stream for Momentum Transfer," *Journal of the British Interplanetary Society*, Vol. 33, (March, 1980), pp. 107-115.
Simon, Julian L. and Herman Kahn. *The Resourceful Earth.* New York: Basil Blockwell, Inc., 1984.
Sitwell, Nigel. "The Queen of Gems Comes Back," *Smithsonian,* Vol. 15, No. 10, (January, 1985), pp. 40-51.
"Sixty-Six Days on Algae," *Spaceflight*, Vol. 3, (May, 1961), p. 86.
Slavin, T. et al. "CELSS Physiochemical Waste Management Systems Evaluation," NASA, NAS2-11806, June, 1986.
Smoluchowski, Roman. *The Solar System: The Sun, Planets, and Life.* New York: Scientific American Library, 1983.
Soderblom, Laurence A. and Torrence V. Johnson. "The Moons of Saturn," *Scientific American, Exploring Space,* Special Issue, Jonathan Piel, ed., Vol 2, No. 1, (1990).
Soi, Fred T. "The Population Factor in Africa's Development Dilemma," *Science*, Vol. 226, p. 801.
Soviero, Marcelle M. "A Cure for Soggy Sandwiches," *Popular Science*, March 1992, p. 23.
Spaulding, Arthur O. "The Looming Energy Crisis," *Oil and Gas Journal*, (May 19, 1986), p. 89.
"Spirulina: Protein for the Future," *Mother Earth News*, No. 68, (March/April, 1981), pp. 180-181.
"Stanford University Prepares to Test New Storage-Ring FEL," *Aviation Week and Space Technology,* (Aug. 18, 1986), pp. 77-79.
Starke, Linda, ed. *State of the World: 1986*. New York: W.W. Norton and Co., 1986.
*Statistical Abstract of the United States: 1988.* Washington: U.S. Dept. of Commerce, Bureau of the Census, 1988.

Stephenson, David G. "Comets and Interstellar Travel," *Journal of the British Interplanetary Society*, Vol. 36, (May, 1983), pp. 210-214.

Stillwell, James D. (chairman). *Architectural Fabric Structures.* Washington: National Academy Press, 1985.

"Stimulating Athletes," *Discover,* (March, 1989), p. 14.

Stine, G. Harry. *Handbook for Space Colonists.* New York: Holt, Rinehart and Winston, 1985.

Stine, G. Harry. "Hardening Humans," *Analog,* (December 12, 1987), pp. 126-129.

Stine, G. Harry. "The Third Industrial Revolution: The Exploitation of the Space Environment," *Spaceflight*, Vol. 16, (September, 1974), pp. 327-334.

Stobaugh, Robert and Daniel Yergin, eds. *Energy Future.* New York: Random House, 1979.

Straack, L. Holmes. *Magnesium.* New York: Harper & Brothers, 1943.

Sullivan, Walter. *Black Holes.* Garden City: Anchor Press, 1979.

Sumner, Eric. "Telecommunications Technology in the 1990s," *Telecommunications,* (January, 1989), pp. 37-38.

"Superconductor Capability Gets Big Boost," *Boulder Daily Camera,* (Nov. 2, 1989), p. 3A.

"Supersmall Lasers May be the Key to Supersmart Computers." *Business Week*, (August 17, 1989), p. 72.

Sutcliffe, Harry and John Wilson, eds. *1983 Rapid Excavation and Tunneling Conference, Proceedings: Vol. 2.* New York: American Institute of Mining, Metallurgical and Petroleum Engineers, 1983.

Switzer, Larry. *Spirulina: The Whole Food Revolution.* New York: Bantam Books, 1982.

Talib, A. et al. "Hydrogen and Alternative Means of Energy Delivery From Ocean Thermal Energy Conversion (OTEC) Plants," *Symposium Papers: Hydrogen for Energy Distribution*, Institute of Gas Technology, Chicago, July, 1978.

Taylor, Stuart R. *Lunar Science: A Post Apollo View.* New York: Pergamon Press, Inc., 1975.

Taubes, Gary. "The Case of the Cosmic Rays," *Discover,* (September, 1989), pp. 52-60.

*Tension Span Structures.* Seaman Corp. brochure.

Thompson, D'Arcy. *On Growth and Form.* Cambridge: Cambridge University Press, 1971.

Thompson, Russell, "What's Going Wrong with the Weather?," *New Scientist*, (24 March 1988), p. 65.

Toffler, Alvin. *The Third Wave.* New York: Morrow, 1980.

Tooper, R.F. "Electromagnetic Shielding," ASD-TDR-63-194, May, 1963.

*Topographic Map of Mars*, 1:25,000,000 Topographic Series, M25M 3 RMS, U.S. Geological Survey, 1976.

Tortolano, F.W. "Strides in Technology Brighten the Future for Plastics," *Design News,* (Nov. 17, 1986), pp. 58-62.

Townsend, L.W. "HZE Particle Shielding Using Confined Magnetic Fields," *Journal of Spacecraft and Rockets*, Vol. 20, No. 6, (Nov./Dec., 1983), pp. 629-630.

Townsend, L.W. "Preliminary Estimates of Galactic Cosmic Ray Shielding Requirements for Manned Interplanetary Missions," NASA, NAS1.15:101516, October, 1988.

Trukhanou, K.A., et al. "Active Shielding of Spacecraft," U.S. Air Force, AD-742410.

"TRW Using Superconducting Materials to Boost Radio Frequency FEL Power," *Aviation Week and Space Technology,* (Aug. 18, 1986), pp. 73-75.

Turner, Frederick. "Life on Mars: Cultivating a Planet—and Ourselves," *Harper's*, Vol. 279, No. 1671, (August, 1989).

Tweedy, R.G., et al., eds. *Pesticide Residues and Food Safety: A Harvest of Viewpoints.* Washington: American Chemical Society, 1991.

Uehling, Mark D. "Tackling the Menace of Space Junk," *Popular Science,* (July, 1990), pp. 82-85.

Vaughan, Chris. "Hiroshima Study Shows Higher Risks of Low-Level Radiation," *New Scientist,* (6 Jan. 1990), p. 28.

Venkataraman, L.V. et al. "Freshwater Cultivation of Algae with Possibilities of Utilizing Rural Wastes in India," *Workshop on Wastewater Treatment and Resource Recovery, Singapore, 1980,* IDRC-154e, Ottawa: International Development Research Center, 1980.

Vertregt, M. *Principles of Astronautics.* New York: Elsevier Publishing Co., 1965.

"Vitamin A Saves Lives," *New Scientist,* (13 Oct. 1990), p. 13.

Vogler, Frank H. "Analysis of an Electrostatic Shield for Space Vehicles," *American Institute of Aeronautics and Astronautics (AIAA) Journal,* (May, 1964), p. 872.

Von Hoerner, Sebastian. "Population Explosion and Interstellar Expansion," *Journal of the British Interplanetary Society*, Vol. 28, (November, 1975), pp. 691-712.

Von Puttkamer, Jesco. "The Next 25 Years: Industrialization of Space: Rationale for Planning," *Journal of the British Interplanetary Society*, Vol. 30, No. 7, (July, 1977), pp. 257-264.

"Voyager Collects Three More Moons," *New Scientist,* (12 Aug. 1989), p.19.

Ward, Fred. "The Pearl," *National Geographic*, August 1985, pp. 193-223.

Watts, Susan. "Parallel Tracks to Standard Processing," *New Scientist,* (12 Aug. 1989), pp. 44-47.

Weaver, Kenneth F. "Meteorites: Invaders From Space," *National Geographic,* (September, 1986), pp. 390-418.

Webb, Paul. "The Space Activity Suit: An Elastic Leotard for Extra-Vehicular Activity," *Aerospace Medicine*, Vol. 39, (April, 1968), pp. 376-382.

Weisburd, Stefi. "Waiting for the Warming: The Catch-22 of $CO_2$," *Science News*, Vol. 128, (Sept. 14, 1985), p. 170.

Whipple, Fred L. *The Mystery of Comets*. Washington: Smithsonian Institution Press, 1985.

Whitmire, D.P. and A.A. Jackson. "Interstellar Laser Powered Ram Jet," *Journal of the British Interplanetary Society*, Vol. 30, (1977), pp. 223-226.

Whitmore, William F. "OTEC: Electricity From the Ocean," *Technology Review*, (October, 1978), pp. 58-63.

Wigley, T.M.L., P.D. James and P.M. Kelly. "Scenario for a Warm, High $CO_2$ World," *Nature*, Vol. 283, (1980), p. 17.

Wiley, John P., Jr. "Worst Mass Extinction Since the Dinosaurs," *Smithsonian*, (November, 1986), p. 42.

Winterberg, F. "Launching of Large Payloads Into Earth Orbit by Intense Relativistic Electron Beams," *Journal of the British Interplanetary Society*, Vol. 31, (September, 1978), pp. 339-343.

Wolff, William A. "OTEC World Thermal Resource," *Proceedings of the 6th OTEC Conference - Ocean Thermal Energy for the 80's*, USDOE CONF - 790631, June, 1979.

Wood-Kaczmar, Barbara. "The Junkyard in the Sky," *New Scientist*, (13 Oct. 1990), pp. 37-40.

Woodcock, G.R. and D.L. Gregory. "Derivation of a Total Satellite Energy System," *Princeton Conference on Space Manufacturing Facilities*, Princeton, May 9, 1975.

Woodwell, G.M. et al. "The Biota and the World Carbon Budget," *Science*, Vol. 199, 1978, p. 141.

Work, Clemens P. "Wiring the Global Village," *U.S. News and World Report*, (Feb. 26, 1990), pp. 44-46.

Yam, Phillip. "Atomic Turn On," *Scientific American*, (November, 1991), p. 20.

Zahl, Paul A. "Algae: The Life-Givers," *National Geographic*, (March, 1974), pp. 361-377.

"Zap!: Coil Guns Offer to Orbit Small Cargoes on a Regular Schedule," *Scientific American*, (April, 1990), pp. 20-22.

Zellner, B. "Asteroid Taxonomy and the Distribution of the Compositional Types," *Asteroids*, Tom Gehrels, ed. Tucson: University of Arizona Press, 1979.

# *End Notes*

# Notes

[1]Vijay K. Gupta, "An Overview of Ocean Thermal Energy Systems," *Alternative Energy Systems*, (New York, 1984), p. 116.
[2]Hans H. Landsberg (Chairman), *Energy: The Next Twenty Years*, (Cambridge, 1979), p. 476. 40,000 BTU (British Thermal Units) of latent heat. Assumes lowering sea temperature by 10° C., from 20° to 10° C.
[3]Gasoline- 20,000 BTU/lb. Sea water is about 1.5 times as dense as gasoline, so a gallon of gasoline has about 660 times as much energy as a gallon of sea water. Average depth of the world ocean is 2.3 miles, 12,500 feet.
[4]Assumes an average depth of the warm water layer of 1000 feet over 140 million square miles of ocean surface.
[5]Mass of oceans = $1.4 \times 10^{24}$ grams.
[6]O.A Roels and S. Laurence, "Marine Pastures: A By-Product of Large (100 Megawatt or Larger) Floating Ocean Thermal Power Plants," *U.S. ERDA Contract No. E(11-1) 2581,* COO-2581-2, p. 36. Total planetary biomass is 911 billion tons. Nitrogen is 1.7% of biomass on average. Total world biomass nitrogen = 15.5 billion tons.
[7]Linda Starke, ed., *State of the World - 1986*, (New York, 1986), p. 87.
[8]311,000 kcal/kg. for feedlot beef. X 136 million kg. = $4.23 \times 10^{13}$ kcal. = $1.68 \times 10^{14}$ BTU @ $6 \times 10^{6}$ BTU/bbl. = 28 million barrels of oil.
[9]920,000 kcal/kg. X 136 million kg. = $1.25 \times 10^{14}$ kcal = $4.94 \times 10^{14}$BTU.
[10]David Pimentel, et al., "Energy and Land Constraints in Food Protein Production," *Science*, Vol. 190, (November, 1975), p. 756. Zebu cattle produce .76 kg. of animal protein per hectare of grazing land with zero input of outside energy.
[11]*Ibid.*, p. 758. Exceptional rangeland, like that in Texas, receiving 27.5 inches of rain per year, can produce 2.2 kg. of animal protein per hectare per year.
[12]Samuel W. Matthews, "Is the World Warming?," *National Geographic*, (October, 1990), p. 75.
[13] *International Energy Annual—1991*, Energy Information Administration, U.S. Dept. of Energy, Washington. pp. 21, 115 - 118.
[14]Crude oil typically contains 19,000 BTU/lb. Oil weighs 7.5 lbs./ gal. A barrel of oil contains 42 gallons and weighs 315 lbs. X 19,000 BTU = $6 \times 10^{6}$ BTU. One kwh contains 3413 BTU absolute. @ 40% conversion efficiency 1 kwh = 8533 BTU. $6 \times 10^{6}$ BTU/8533 = 703 kwh/bbl.
[15] Gupta, p. 117.
[16]William C. Kincaide and John N Murray, "Advanced Alkaline Electrolysis Systems for OTEC," *8th Ocean Energy Conference*,

---

(Washington, June, 1981), p. 690. Each sea colony will produce 7.2 million kilowatt hours of surplus power per day for export. The conversion of electricity to hydrogen is about 80% efficient, so each colony will convert 5.76 million kwh to hydrogen each day. Hydrogen liquefaction requires 17.4% of the energy in the hydrogen being liquified, or 1.1 million kwh. This leaves a balance of 4.75 million kwh in the form of liquified hydrogen. Total energy being converted to liquid hydrogen each day equals 16.2 billion BTU. The heating value of hydrogen is 242 BTU/ft$^3$.

[17]Feedback is an engineering term referring to what happens when a process creates one of its own inputs. In the case of 'positive' feedback, that input tends to amplify the process. The classic example of positive feedback is the screeching sound that occurs when a microphone feeds back on itself.

[18]John Gribbin, "Warmer Seas Increase Greenhouse Effect," *New Scientist*, (6 Jan. 1990), p. 31.

[19]John Gribbin, *Hothouse Earth: The Greenhouse Effect and Gaia.* (New York, 1990), p. 132.

[20]*Ibid.* Water vapor constitutes up to 4% of air, as opposed to .03% for $CO_2$. $CO_2$ absorbs IR with wavelengths only from 13-17 microns. Water vapor absorbs across the IR band, from 1 to 100 microns.

[21]Gribbin, "Warmer Seas."

[22]Including seaweed.

[23]Total additions to $CO_2$ are 6 billion tons, 5 from fossil fuels and one from deforestation. Dennis Normile, "Science Gets the $CO_2$ Out," *Popular Science*, (February, 1994), p. 67.

[24] William F. Whitmore, "OTEC: Electricity from the Ocean," *Technology Review,* (October, 1978), p. 60.

[25] *Ibid.*

[26]*Ibid.*

[27]Gupta, p. 118.

[28]The efficiency of a thermal engine can be calculated by use of the Carnot Formula. The total temperature difference is 80° - 41° = 39° , but the effective temperature difference is only 77° - 59° = 18°. The effective temperature difference is that which is available after heat exchange has taken place. Net power of 2.5% takes into account the overall efficiency of the power production facility—turbine and generator—of 75 - 80%.

[29]B. Berkovsky, "Ocean Thermal Energy - Prospective for a Renewable Source of Power," U.S. DOE CONF-780236, *The Fifth Ocean Thermal Energy Conversion Conference*, (February, 1978). p. II-193. This figure takes into account the 1000 year latency inherent in deep ocean turn-over.

[30]Ibid.

[31]If restricted to tropical seas, where the temperature difference is at least 40° F., the ultimate resource is 10,000 GW. Gupta, p. 118.

[32]Brian K. Parsons and Harold F. Link, "System Studies of Open-Cycle OTEC Components", *Ocean Engineering and the Environment, Oceans 85: Conference Record, Vol. 2*, (San Diego, November 1985), pp. 1230-1231. Warm water in @ 25°C., 77°F., warm water out at 20°C., 68° F. Cold water in @ 5°C., 41°F., cold water out at 7°C., 44.6° F.
[33]Donald F. Othmer and Oswald A. Roels, "Power, Fresh Water, and Food from Cold, Deep Sea Water," *Science*, Vol. 182, No. 4108, (12 October 1973), p.122.
[34] Whitmore, p. 63.
[35]Stefan Hosternoth and Peter J. Lamb, *Climatic Atlas of the Indian Ocean: Vol. I - Surface Climate and Atmospheric Circulation*, (Madison, 1979), Charts 50-61.
[36]William A. Wolff, et al., "OTEC World Thermal Resource," *Proceedings of the 6th OTEC Conference - Ocean Thermal Energy for the 80's*, (June 1979), U.S. Dept. of Energy, CONF-790631, p. 13.5.2.
[37]Lyle E. Dunbar, "Market Potential for OTEC in Developing Nations," *8th Ocean Energy Conference*, (Washington, June 1981), p. 948.
[38]Wolff, p. 13.5.2.
[39] Arthur W. Hagen, *Thermal Energy From the Sea.* (Park Ridge, 1975), p. 37.
[40]Thomas J. Rubas, et al., "OTEC 100 MWe Alternate Power System Study," *6th OTEC Conference: Ocean Thermal Energy for the 80's*, USDOE CONF-790631, (June, 1979), p. 9.1.11. Note that the figures given in the paper by Rubas, et al. are for a 100 MWe net OTEC. The design involved here is for a 100 MWe gross OTEC. Accordingly, the costs/kwh given by Rubas were factored by .6. The result is in 1977$, to approximate actual anticipated expenditures in the 2000 - 2010 time frame, these 1977$ amounts were then doubled. This assumes an average annual inflation rate of around 3%.
[41]Hagen, p. 7.
[42]This is an overly generous allocation of power to the colonists. Life in Aquarius will be very energy efficient. Habitats on the equatorial ocean require very little energy for heating or cooling. The real power requirements of the colonists are likely to be only a fraction of this amount.
[43]Dunbar, p. 951. Cost to the colony of the electricity (not counting ammortized capital costs) will be limited to the operating and maintenance costs of the system since the fuel is free and there are no external interest costs. Operating and Maintenance costs of OTECs are rated at 1.5% of total capital costs per year. Total capital costs of the OTECs are $1.1 billion, so annual operating costs will be $16.5 million. The OTEC facilities' net electrical output is 3.6 billion kilowatt hours per year; this brings the cost of power to the colony to under 1/2¢ per kwh. Ammortized capital costs amount to 1¢/kwh, over a 30 year plant life.

[44]Hagen, p. 37. $5.95 \times 10^5$ lbs. cold water per second to produce 100 MW gross power.
[45]Average annual discharge of the Nile at Aswan, 1870 to 1952 avg. = $92 \times 10^9$ $M^3$
[46]S. Laurence & O.A. Roels, "Potential Mariculture Yield of Sea Thermal Power Plants: Part 2 - Food Chain Efficiency," CONF-751235, (San Francisco, 1975). p. III-21.
[47]M.B. McElroy, "Change in the Natural Environment of the Earth: The Historical Record", *Sustainable Development of the Biosphere*, W.C. Clark and R.E. Munn, eds., (Cambridge, 1986), p. 205. The temperature difference between the surface waters at the poles and the equator fuels the same type of adiabatic circulation in the oceans as occurs in the atmosphere, though at a much slower rate. This slow circulation of oceanic water is called the 'thermohaline circulation'. The warm waters from the equator slowly migrate toward the poles where they cool and sink. This sinking zone comprises only about 10% of the ocean's surface, so most oceanic waters are in the process of slowly upwelling. In higher latitudes, where water temperatures average around 10° C., a thousand trillion cubic meters of water sink each year, continually recharging the nutrient rich deep waters. Sinking, the cold polar waters drift toward the equator, where they slowly warm and rise, making one round trip about every thousand years.
[48]Joe A. Hanson, ed., *Open Sea Mariculture*, (Stroudsburg, 1974), p. 87. The rate of circulation of deep water is very slow, about half a centimeter per second. In the Atlantic, the deep water turnover period is about 300 years, while in the broader Pacific it is more like 2000 years.
[49]Laurence & Roels, "Food Chain Efficiency",. p. III-21.
[50]Othmer & Roels, p. 76.
[51]Here the term 'nitrate' will be meant to include nitrogen in all of its fixed forms.
[52]John E. G. Roymont, *Plankton and Productivity in the Oceans*, (New York, 1963), p. 161.
[53]32.5 grams of total salts/liter = 8.7 tbs/gal.
[54]Robert A. Muller and Theodore M. Oberlander, *Physical Geography Today*, (New York, 1978), p. 155.
[55]"Ocean and Oceanography", *Encyclopaedia Britannica*, 1958, V. 16, p. 691. The polar downwelling zones provide the cold water for the thermohaline circulation, but the nitrogen is supplied by algal biomass produced globally. The concentration of nitrogen in deep waters may also depend on nitrate reserves in ocean sediments—a nearly inexhaustible warehouse.
[56]Donald G. Groves and Lee M. Hunt, *Ocean World Encyclopaedia*, (New York, 1980), p. 235.
[57]O.A Roels and S. Laurence, "Marine Pastures," p. 4.

[58]Laurence & Roels. "Food Chain Efficiency", p. III-23. Protein is 65% of the algae's dry weight. 15% of algae will be consumed by filter feeders.
[59]Larry Switzer, *Spirulina: The Whole Food Revolution*, (New York, 1982), p. 21.
[60]Donella H. Meadows, et al., *The Limits to Growth*, (New York, 1972), p. 47.
[61]Pimentel, p. 754.
[62]Lucy Kavoler, *Green Magic: Algae Rediscovered*, (New York, 1983), p. 45.
[63]*Ibid.*
[64]Paul A. Zahl, "Algae: The Life Givers", *National Geographic* , (March 1974), pp. 361 - 377.
[65]Kavoler, p. 11.
[66]Gerald Parkinson, "New Techniques May Squeeze More Chemicals from Algae," *Chemical Engineering*, (May 11, 1987), p. 19.
[67]Zahl, p. 361.
[68]Many algae are also capable of sexual reproduction.
[69]Spirulina flourishes in highly alkaline waters simply because no other plants can stand the harsh conditions, leaving it free from competition. While spirulina will grow in other media, its growth rate varies. Water discharged from the OTECs will be around 60° F. Desalination, fresh water production through atmospheric condensation, air conditioning, cold-bed horticulture and residence time in the containment ponds will all serve to increase this base temperature. Genetic selection, and eventually genetic engineering, will increase productivity. Strains of algae other than Spirulina may also be cultivated.
[70]The USRDA for $B_{12}$ is six micro-grams per day. Without some source of animal protein in the diet it is almost impossible for a human to get enough of this vital nutrient. Those forced to subsist on a purely vegetable diet generally suffer from pernicious anemia as a consequence of $B_{12}$ deficiency. A single tablespoon of Spirulina powder in the diet will provide 15 micrograms, or 250% of the USRDA for Vitamin $B_{12}$. A person taking enough Spirulina powder in his diet to provide half his protein needs—two tablespoons, 20 grams—would be supplied with 70% of the USRDA for Vitamin $B_1$ (Thiamine), 50% of $B_2$ (Riboflavin) and 12% of $B_3$.
[71]Algae are equivocal forms of life. Phytoplankton contain chlorophyll and are therefore generally considered to be plants. They nonetheless exhibit many animal characteristics: they are often motile, swimming actively from place to place; many propagate through cellular division; some even have 'eyes'. While, for practical purposes, they can be thought of and even rigorously defined as plants, algae are animal enough to produce complete proteins.
[72]Switzer, p. 23.

[73]Switzer, p. 22.
[74]"Vitamin A Saves Lives," *New Scientist*, (13 Oct. 1990), p. 13.
[75]Switzer, p. 28.
[76]Hanson, p. 13. Taking massive doses of pre-formed Vitamin A, which is fat soluble, can lead to a toxic build-up, but this doesn't happen with Vitamin A derived from spirulina. Like the Vitamin A found in carrots, spinach, and other vegetable sources, the Vitamin A in Spirulina is in the form of Beta Carotene. Beta Carotene is broken down to release Vitamin A only on an as needed basis by the body, so there is no accumulation and no toxicity.
[77]"Britain Pushes to the Fore in Growing Algae for Drugs," *New Scientist*, (1 Oct. 1987), p. 35.
[78]At such enormous rates of production, the market value of these micronutrients is bound to fall substantially, but they are apt to remain quite valuable.
[79]Marc Reisch, "Biotech Firm Readies Microalgae Products," *Chemical and Engineering News*, July 27, 1987, p. 14.
[80]Switzer, p. 244.
[81]Pimentel, p. 754. U.N. Food and Agriculture Organization recommends 41 grams/day of reference protein (egg) for a 155 pound man.
[82]Based on a design proposed by Airship Industries, Ltd.
[83]Laurence and Roels, "Potential Mariculture Yield - Pt. 2," p. III-23. Assuming 10% of the nitrogen in the deep water is fixed by the benthic algae, and that crabs exhibit the same 35% conversion efficiency as shellfish in transmuting algal protein to animal meat, then each 1000 gallons of sea water could produce .123 oz. of crab meat (wet weight). The 42 billion gallons of deep sea water discharged by Aquarius's OTECs each day, could produce 322,000 lbs., of crab, and other meats per day—58,900 tons/yr.
[84]M.M. Soviero, "A Cure for Soggy Sandwiches," *Popular Science*, March 23, 1992, p. 240. Per capita demand for paper and paper products in the U.S., including newsprint and packaging, is 700 pounds. Chitin from lobster and crab production can satisfy about 160 pounds of this demand. The balance will be supplied by paper produced from sea weed cellulose.
[85]Fred Ward, letter of April 20, 1992. Small pearls—2 to 4 mm.—retail at $15 to $25 apiece. Black pearls—14 to 18 mm—retail at $3000 and more, apiece. Pearls sold at the annual auction in Rangoon in 1992 sold at wholesale for $3000 apiece. A matched necklace can be worth $800,000.
[86] V.J. Chapman, *Seaweeds and Their Uses*, (London, 1970), p. 111. Nitrogen comprises 15% of the weight of protein and protein comprises 9% of the dry mass of seaweed (*Laminaria Digitata*).
[87]Chapman, p. 211. Algin ($C_6H_{10}O_7$) is a chain-like polymer molecule with a high molecular weight and many reactive side chains.

[88]*Ibid.* and "Silk and Sericulture", *Encyclopaedia Britannica*, Vol. 20, p. 670. Algin fibres weigh 1.23 grams/denier (9000 meters) compared to silk fibres at 1.8 to 3.8 g/denier.
[89]A.J. Hall, *The Standard Handbook of Textiles*, (London, 1969), p. 49.
[90]Chapman, p. 196.
[91]*Statistical Abstract of the United States - 1988*, U.S. Dept. of Commerce, Table No. 1268. Per capita U.S. demand for textiles for apparel and home furnishings in 1985 was 36 lbs.
[92]Pimentel, et al., p. 2. Assuming these fish average the same conversion efficiency as catfish, they will convert 10.5% of their feed—20 tons—into protein. Since the wet weight of fish is 22% protein, there will be a gross daily production of 90 tons of fish, shrimp, and lobster. Thirty tons of this will end up as scrap, and so be recycled in the feed pellets.
[93]Assumes wholesale price is reduced by one half.
[94]60,000 tons/yr. of grass algae grazers, 60% as crab, 30% as abalone, 5% as conch and 5% as turtle. 30% scrap.
[95]Crab meat wholesales @$7.50/lb., abalone, conch, and turtle the same.
[96]56,575 tons/yr of filter feeders, one culture per containment segment. 30% scrap.
[97]All shellfish at $2.50/lb.
[98]9636 tons per year of carnivores, one culture per containment segment. 20% scrap.
[99]Lobster and shrimp @ $7.50/lb.
[100]Mackerel, herring, and cod @ $2.00/lb
[101]Salmon @ $5/lb.
[102]According to KELCO, a division of Merck Pharmaceutical Co., raw alginate wholesales for $5 to $15 per pound, depending on purity and other factors. Here, the value of sea-silk is assumed to be $5/lb., the cost of the raw feedstock. This does not account for the value added in transforming the feedstock into finished textiles.
[103]Paper @$.25/lb.
[104]James Gleick, *Chaos: Making a New Science*, (New York, 1987), pp. 119-153. Strange attractors are points or areas of equilibrium in chaotic systems.
[105]The central OTEC platform, will support a column 550 feet high; each of the six OTECs in the first ring will support a column 340 feet high. The towers in the second ring, each 210 feet high, are supported by specially constructed floats that underpin them. All towers in subsequent rings are similarly supported. The weight of the central tower, 412,500 tons, exceeds the surplus bouyancy of the central OTEC platform. The central tower does not, however stand alone. Construction on the tower structure does not begin until after all seven OTEC platforms have been completed. Since all OTEC platforms are tied firmly together, they form a coherent bouyant platform with a total bouyant force of 2.5 million tons.

---

[106]In the Norse creation myth, Ymir the frost giant was licked out of the ice by a cow at the dawning of the universe. Ymir's sons, including the God Odin, later slew the giant and made the Earth out of his body.
[107]J.H. Brewer, et al, "Construction Feasibility of OTEC Platforms," *Contract No. ET-78-C-02-4931*, p. 35. The OTEC modules in Aquarius will each be 284 feet in diameter and 200 feet deep.
[108]Roderick A. Barr, et al., "Theoretical Evaluation of the Sea-keeping Behavior of Large OTEC Plant Platforms and Cold Water Pipe Configurations," *US ERDA Cont. No. ET-76-C-02-2681*, (August 1978), p. 84.
[109]69,878 sq. ft.
[110]There are 93 million square feet in the surface towers, 13.5 million in the subsurface flotation platforms, 15.2 million on the tower roofs, and 95 million on the breakwater.
[111]William U. Chandler, et al.., "Energy for the Soviet Union, Eastern Europe and China," *Scientific American*, (September 1990), p. 121.
[112]Jonathan Silvertown, "A Silent Spring in China", *New Scientist*, 1 July 1989, p. 55.
[113]*Statistical Abstract of the United States - 1988,* Tables 1237-1239.
[114]Includes space for schools, restaurants, stores, warehouses, hospitals, hotels, churches, and other commercial buildings. For all 'commercial' purposes, America has some four million buildings with 52 billion square feet of floor space available. For a population of 233.7 million, this amounts to a per capita allocation of commercial floor space of 225 square feet.
[115]Linda Leigh, and Kevin Fitzsimmons, "An Introduction to the Intensive Agriculture Biome of Biosphere II," *Space Manufacturing 6: Nonterrestrial Resources, Biosciences, and Space Engineering*, Proceedings of the Eighth Princeton/AIAA/SSI Conference, May 6-9, 1987, p.78.
[116]The flotation chambers underpinning the structure must be extended by about 15 to 20 feet each time five new stories are added to the structure.
[117]At ring eight, there are 217 modules, each covering 69,878 square feet. Total area per floor at the bottom of the structure, eight rings wide, is 15,162,875 square ft. Four new floors, eight rings wide, will provide 60,651,500 square feet of new space. If a ninth ring is added there will be 271 modules with 18,936125 square ft per floor. Five more floors, nine rings wide will add 94,680,625 square feet.
[118]Wolf H. Hilbertz, inventor, "Mineral Accretion of Large Surface Structures, Building Components and Elements," *U.S. Patent No. 4,246,075*, (Jan. 20, 1981). The other minerals are magnesium hydroxide (brucite), sodium chloride, and silicates. Of these, only brucite contributes significantly to the mass of the accreted material.

[119]Wolf Hilbertz, "Electrodeposition of Minerals in Sea Water: Experiments and Applications," *IEEE Journal of Oceanic Engineering*, Vol. OE-4, No. 3, (July 1979), p. 98.
[120]David Lampe, "Grow Your Own Buildings," *Mother Earth News*, (March/April, 1980), p. 118.
[121]Hilbertz, "Electrodeposition of Minerals", p. 109.
[122]Wolf Hilbertz, inventor, "Repair of Reinforced Concrete Structures by Mineral Accretion," *U.S. Patent No. 4,440,605*, Apr. 3, 1984.
[123]Wolf Hilbertz, *U.S. Pat nt No. 4,246,075*, p. 5.
[124]*Ibid.*
[125]Hilbertz, *U.S. Patent No. 4,440,605*, p. 5.
[126]Hilbertz, *U.S. Patent No. 4,246,075*, p. 5.
[127]*Ibid.*
[128]Watts = volts X amps. 1000 watts = 12 volts X 83.33 amps. 83.33 amps/ .189 amps/ft$^2$ = 440.9 ft$^2$.
[129]Seven 100 MW OTECS at the heart of Aquarius plus the 100 MW seed-ship = 472 MW net – 9 MW for surface uses = 463MW.
[130]Hilbertz, *U.S. Patent No. 4,246,075*, p. 1.
[131]The complete process is as follows: Sea water is first mixed with calcium oxide, CaO; magnesium exchanges places with the calcium to form magnesium hydroxide which precipitates out of solution and is then filtered out; this precipitate is then mixed with hydrochloric acid, being converted to magnesium chloride in the process, $Mg(OH)_2+2HCl = MgCl_2+2H_2O$; the magnesium chloride is then melted, and the molten compound is electrolyzed; the resultant products are 99.9% pure magnesium and chlorine gas. The chlorine is then combined with hydrogen to form hydrogen chloride which is then recombined with the process water to reconstitute hydrochloric acid; in this way neither hydrochloric acid nor chlorine gas, both deadly poisons, ever leave the closed process loop.
[132]Waste heat from the molten magnesium chloride tank can be recovered to provide some of the needed process heat.
[133]When extruded in bars, rods, and wires: 8.5% aluminum, .5% zinc, .3% silicon, .15% manganese, .05% copper.
[134]Anodizing involves electrolytically coating a metal surface with a protective film—usually of another metal. In Aquarius we will use mostly copper, cobalt, and nickel to anodize magnesium surfaces. Anodizing has the additional advantage of imbuing surfaces with a wide variety of metallic hues.
[135]Harold V. Thurman, *Introductory Oceanography*, (Columbus, 1975), p. 77. Manganese nodules are abundant in most of the deep ocean basins. There are major deposits of them in the equatorial regions of the Indian ocean, very near where Aquarius will be located. The nodules are typically 30% manganese dioxide, $MnO_2$, and 20% iron dioxide, $Fe_2O_3$.

Other metals, principally copper, cobalt, and nickel occur in concentrations of 1 to 2%.

[136]John H. Perry, *Chemical Engineer's Handbook*, 3rd Ed., (New York, 1950), p. 1811.

[137]Stanley Abercrombie, *Ferrocement: Building with Cement, Sand, and Wire Mesh*, (New York, 1977), p. 54. Ferrocement typically requires 840 pounds of steel per cubic yard.

[138]Each ton of magnesium requires roughly a ton of calcium carbonate. Production will accordingly be scaled to match the availability of CaCo from mariculture operations.

[139]*Statistical Abstract of the United States: 1988*, 108th ed., Table No. 1191. World primary production of magnesium in 1985 was 361,000 tons.

[140] On average, sea water contains 3.5% solids by weight. There are about 428 mg/kg. of calcium carbonate, and 1270 mg/kg. of magnesium in sea water. Each kg. of sea water therefore contains around 1700 mg. or 1.7 grams of potential building material. It requires some 6.5 trillion kg. of sea water to provide the amount of calcium carbonate and magnesium needed for the accretion of Aquarius and its attendant parts. At an average flow of 1 knot, 16 million cubic meters, 16 billion kg., of sea water will pass through Aquarius's growth zone every hour.

[141]Lampe, p. 119.

[142]Hilbertz, *U.S. Patent No. 4,461,684,* p. 3. Total dissolved solids entering oceans each year = 2.73 X $10^9$ metric tons = 2.73 trillion kg. X 5% = 136.5 billion kg./yr of building components.

[143]"Ocean and Oceanography," *Encyclopaedia Britannica*, Vol. 16, p. 685. Total dissolved salts = 5 X $10^{16}$ tonnes. Calcium carbonate and magnesium @ 5% = 2.5 X $10^{15}$ tonnes

[144]Aquarius is self-sufficient in both plastics and glass. Petrochemical feedstocks are derived from genetically engineered algae. Silicates—the basic raw material for glass—are available from the silaceous skeletons of diatoms which can be cultured along with other varieties of algae in huge quantities. In essence, the large scale cultivation of algae is a form of nanotechnology that will enable us to extract virtually any raw material we need from sea water.

[145]Motorola Inc., *1989 Annual Report*, p. 11. Circuit density has been doubling every two to three years since the advent of the integrated circuit. That trend is anticipated to continue, at least through the 1990s.

[146]James P. Meindl, "Chips for Advanced Computing," *Scientific American*, (October, 1987), Vol. 25, No. 4, p. 81.

[147]William F. Allman, *Apprentices of Wonder: Inside the Neural Network Revolution*, (New York, 1989).

[148]Intel Corporation, *1989 Annual Report*, p. 4.

[149]Michael Cross, "Do Computers Dream of Intelligent Humans?," *New Scientist*, (Nov. 26, 1988), p. 45.

[150]"Data Storage Technology for Advanced Computing", *Scientific American*, (October, 1987), p. 117.
[151]*Ibid.*
[152]State of the art optical storage systems, like Kodak's *System 6800* can store 1 terabyte, 1000 gigabytes, in 13 square feet, 77 gigabytes/sq.ft.
[153]According to the U.S. Library of Congress Planning Office, as of January 1988. Note that graphic media, photographs, films, etc. are enormously more demanding of data storage than print media.
[154]Charles Seiter, "Hard Disk Alternatives", *Macworld*, (July 1989), p. 122. Present optical data storage media costs as little as 3.3¢/megabyte. We can expect that data storage costs will fall by one or two more orders of magnitude in the near future.
[155]Daniel Clery, "Warm Message on an Atomic Scale," *New Scientist*, (26 Jan. 1991), p. 31. The Japanese electronics giant, Hitachi, has recently taken to engraving messages in atoms on the head of a pin. Hitachi scientist predict that their atomic scrimshaw will ultimately allow data to be recorded at a density of one billion bits per square micron. Laboratory lasers now have a focal resolution as small as a single atom—2.5 Ångstroms. (An Ångstrom is all of one ten-millionth of a millimeter.) The pits on today's optical disks are half a micron across, 5000 Ångstroms. A millimeter of optical track can accommodate only 2000 half-micron bits, while it will hold 4 million atom sized bits. In the next 25 years, we may build lasers and recording media which operate reliably on the atomic scale, allowing us to pack thousands of times as much data on a disk.
[156]Philip Yam, "Atomic Turn-on", *Scientific American*, (November, 1991), p. 20. Single atomic bits can probably be stored amid clusters of as little as a thousand other atoms, allowing the Library of Congress to be compressed onto a single 12 inch disk.
[157]Danny Hillis, "The Connection Machine," *Scientific American*, Vol. 256, No. 6, (June 1987).
[158]Ross and Wood, p. 14.
[159] Lyle E. Dunbar, "Market Potential for OTEC in Developing Nations," *8th Ocean Energy Conference,* Washington, D.C., (June, 1981), p. 950.
[160]William P. Crawford, *Mariner's Weather*, (New York, 1978), p. 185.
[161]S. Nieuwolt, *Tropical Climatology,* (New York, 1977), pp. 80-82.
[162]J.R.V. Prescott, *The Political Geography of the Oceans*, (New York, 1975), p. 209.
[163]Alastair Couper, ed., *The Times Atlas of the Oceans*, (New York, 1983), p. 241.
[164]Prescott, p. 105.
[165]*Statistical Abstract of the United States-1988*, Table No. 688.
[166]Shelter-$4316, household operations-$343, housekeeping supplies-$398, furnishings-$1021.

---

[167]Assumes 70% of population is 18 year and older, and 85%+ of adult population is working.
[168]One million visitors/year @ $1000 avg. exp.
[169]15,000 telecommuters @ $30,000 avg. annual income.
[170]Even if expenditures for each worker amounted to the same $28,000 as in our example, total expenditures for the colony would only be $1.7 billion.
[171]The Saturn V's first stage burned fuel at the rate of 15 tons per second. In its 2.5 minute burn, the massive first stage, generating 7.5 million pounds of thrust, would accelerate the rocket to 6000 miles per hour at an acceleration of less than 2 *g*. It took the Saturn V 4.5 seconds to clear the gantry during which time it burned 67.5 tons of fuel.
[172]Donald E. Hunton, "Shuttle Glow," *Exploring Space, Scientific American, Special Issue*, Jonathan Piel, ed., (1990), p. 34.
[173]*Proceedings: SDIO/DARPA Workshop on Laser Propulsion,* CONF-860778, Vol. 1. p. 5.
[174]The prospects for affordable access to space have improved dramatically of late, with the advent of the DC-X, an experimental Single Stage To Orbit (SSTO) rocket. SSTO technology could conceivably provide access to space for as little as $1,100 - $110.00 per kilogram. (Launch costs for full-scale SSTO rockets have been estimated at $10 million per launch, with costs conceivably as low as $1 million per launch, for a 9000 kg. payload. W.A. Gaubatz, et al., "Single Stage Rocket Technology", 43rd Congress of the IAF, IAF-92-0854, p. 4.) Colonizing space on a grand scale, however, will require a further reduction in cost by one or two orders of magnitude. To achieve this, something other than rockets will be required.
[175]The distance to the horizon depends on the height of one's observation point. To actually see the horizon at eight kilometers requires a raised vantage, eight kilometers is used here for purposes of illustration.
[176]Curvature of Earth is 4.877 meters in every eight kilometers.
[177]Gravitational acceleration at the Earth's surface is 9.8 meters per second per second. In the first second of free-fall a body will accelerate from 0 to 9.8 mps, at an average velocity of 4.9 mps.
[178]Acceleration of 10.197 $g$ = 100 M/sec$^2$. Acceleration to 5000 M/sec @ 100 M/sec = 50 seconds. Average velocity in the tube = 2.5 kps (5600 mph). 50 X 2.5kps = 125km.
[179]*g* is the symbol for the acceleration of gravity. One *g* is the same acceleration experienced by a falling body near the Earth's surface.
[180]Paul A. Campbell, "Aeromedical and Biological Considerations of Flight above the Atmosphere," *Realities of Space Flight*, L.J. Carter, ed., (New York,1957). p. 257.
[181]G. Harry Stine, *Handbook for Space Colonists*, (New York, 1985), p. 92.

[182]The radius of curvature of the track is ten kilometers. 255 *g*s seems extreme, but is not as bad as it sounds. You would experience the same degree of force if you fell on your back from a height of .3M (1 ft.).
[183]E.E. Rice, et al., "Preliminary Feasibility Assessment for Earth-to-Space Electromagnetic (Railgun) Launchers", *NASA Report No. CR-167886*, p. 6. Power transfer efficiency is 85%: Inductor Electrical to Rail Electrical. Note that Rice's study uses an electromechanical system for the hydraulic conversion of power. He indicates an overall efficiency of 66%. It is believed, that the application of superconducting technology throughout will minimize such conversion inefficiencies.
[184]Total including line losses.
[185]It may actually prove cheaper to convert power to hydrogen in Aquarius, and then use the hydrogen to fuel a power plant closer to the Bridge, but conceptually, the power comes directly from Aquarius.
[186]Naomi Freundlich, "Rings of Power", *Popular Science*, (January, 1989), p. 67.
[187]Timothy O. Bakke, "Satellite Catapult," *Popular Science*, (August, 1990), p. 73.
[188]By comparison, a Concorde SST weighs around 91,000 kg. or 1500 kg. per meter. In the case of the Concorde, however almost 25% of the total weight is taken up by its engines and landing gear, neither of which will encumber the launch capsules.
[189]Herman Krier, et al., "CW Laser Propulsion", *Proceedings: SDIO/DARPA Workshop on Laser Propulsion*, CONF-860778, Vol. 2, p. 27.
[190]The laser energy will actually be delivered in two distinct pulses. The first pulse, of relatively low power, serves to vaporize a quantity of the ice. The second pulse, of much higher amplitude, then causes the steam layer to detonate, forming a shock wave which drives the capsule forward. The pulses are only microseconds apart, so to the passenger or cargo it cannot be distinguished from continuous thrust. This pulsed laser technique is oten referred to as LSD (Laser Supported Detonation). The timing of the pulses, about 50 nanoseconds apart, is about the same as that produced by RF (Radio Frequency) driven FELs. *Proceedings: SDIO/DARPA Workshop on Laser Propulsion*, CONF-860778, Vol. 2, p. 2.
[191]Arthur C. Clarke, *Interplanetary Flight: An Introduction to Astronautics,* (New York, 1960), p. 17.
[192]In space the specific impulse of LHOX is 441 seconds.
[193]Jordin T. Kare, "Trajectory Simulation for Laser Launching," *Proceedings: SDIO/DARPA Workshop on Laser Propulsion*, CONF-860778, Vol. 2, p. 64.
[194]*Proceedings: SDIO/DARPA Workshop on Laser Propulsion*, CONF-860778, Vol. 1-Executive Summary, p. 1.
[195]*Ibid.*

---

[196]The lower the molecular weight of a propellant the greater its relative thrust.
[197] Neil McAleer, "The Light Stuff: Laser Propulsion, *Space World*, (July 1987), p. 11.
[198]A better propellant might be "liquid water with some small proportion of solid material to bind it"—the world's first "watermelon booster". Poul Anderson, personal communication, 17 Feb. 1994, p. 2.
[199] Arthur Kantrowitz, "Laser Propulsion to Earth Orbit: Has Its Time Come?," *Proceedings: SDIO/DARPA Workshop on Laser Propulsion*, CONF-860778, Vol. 2, p. 12.
[200]McAleer, p. 11. This is one of Arthur Kantrowitz's ideas. Kantrowitz is the father of laser propulsion.
[201] Kare, p. 64.
[202]Very high energy laser beams tend to fry the air as they pass through it, converting the gas of the atmosphere into a fourth state of matter—ionized plasma. This produces undesireable effects on the transmission of the beam.
[203]Krier, et al., p. 20.
[204]T.A. Heppenheimer, "Free Electron Lasers," *Popular Science*, (December, 1987), p. 63.
[205]C. Kumar, et al., "Strategic Defense and Directed Energy Weapons," *Scientific American*, Vol. 257, No. 3, (September, 1987), p. 42.
[206]Phillip J. Klass, "Scientists Focus on FEL Technology to Counter ICBMs," *Aviation Week and Space Technology*, (August 18, 1986), p. 41.
[207] Klass, p. 40.
[208]"Induction-Type Laser Faces Critical Tests," *Aviation Week and Space Technology*, (August 18, 1986), p. 55.
[209]The electrons in atoms inhabit only closely defined 'orbits'. The distance of these obits from the nucleus determines the energy levels of the electrons in them. Since these orbits are fixed, and electrons can only exist in them, electrons changing orbits absorb and produce energy only in discreet quantities, like the steps in a stair case. It is this discreet production of energy, the production of 'quanta' of energy, that gives the science of 'quantum mechanics' its name. These energy quanta and the corresponding wavelengths of the light they produce is what characterizes the laser beams produced from various substances.
[210]Klass, p. 40.
[211]Kantrowitz, p. 4. Kantrowitz cites a focal limit of 833 km. for a wavelength of 100,000 angstrom, and a target diameter of one meter. At a wavelength of 5000 Å with a mirror radius of five meters and a target radius of one meter, the laser can remain focused at a distance of up to 33,000 km.
[212]"Free Electron Lasers: Stanford University Prepares to Test New Storage-Ring FEL," *Aviation Week & Space Technology*, (August 18,

1986), p. 77. The production of beams of much shorter wavelengths and correspondingly higher energies are now possible, and will certainly be within our grasp by the time we actually build the Bifrost Bridge. Lasers in the x-ray portion of the electromagnetic spectrum, 500 å, can remain much more tightly focused, and pass throught the atmosphere much more easily. Beams of this wavelength can be created if the energy of the electron beam is boosted into the giga-electron-volt range.

[213]Rice, p. 3-91.

[214]Rice. p. 3-93. Energy lost to atmospheric drag is 2% of total kinetic energy.

[215]*Air Command and Staff College Space Handbook*, Lt. Col. Curtis D. Cochran, et al. eds., (Maxwell Airforce Base, 1985), p. 13-6.

[216]Rice, p. 3-33.

[217]*Ibid.*

[218]Depending on atmospheric response to the laser beams, it may be necessary to utilize reflecting mirrors to transfer the beam to the launch capsule. These mirrors would, of course, have to be in space. They would therefore have to be in low earth orbit, at altitudes of a few hundred kilometers. As the launch capsule heeled over into horizontal flight and traveled down range, the angle at which the ground-based laser would have to penetrate the atmosphere would rapidly increase, necessitating punching the beam through a relatively thicker cross section of atmosphere. To avoid this the laser could be handed over to a satellite based reflector for part of the laser burn. Launches would have to be timed so that capsules would enter their horizontal burn phase just as a reflecting satellite was approaching in its orbit. To maintain a multiple launch capability would necessitate a number of reflecting satellites. An alternative would be a single reflecting satellite with a much larger mirror in geosynchronous orbit. The laser beam could be fired vertically to the stationary reflector, and from there it could be beamed to the launch capsule, thereby minimizing atmospheric interference.

[219]Rice, p. 3-105.

[220]*Ibid.*

[221]Rice, p. 5-21.

[222]*Ibid.*

[223]Beth Dickey, "Catch the Wave," *Final Frontier*, (Nov./Dec.1990), p. 12. Waveriders will be of enormous importance throughout our colonization of the solar system. By using their high lift to drag ratios, we can skim the atmospheres of planets and gain tremendous velocity boosts from gravity assists. Such techniques will enable us to traverse the solar system at high speeds with minimal expenditures of energy. By incorporating waveriders into our earliest plans, we will master this vital technology at the outset of our endeavors in space.

[224]"Shockwave Surfers," *Popular Mechanics*, Dec., 1990, p. 16.

[225]Lunan, p. 26.

---

[226]John O. Bickel, ed., *Tunnel Engineering Handbook*, (New York, 1982), pp. 244-245. Bickel rates costs as $500/ft. for 12 ft. tunnel.
[227]Bickel, p. 275.
[228]Multiple tunnel sections will necessitate additional surface adits. These, can, however, be temporary. The machines bore themselves into the earth at an angle, until they reach tunnel depth where they begin cutting their section of the tunnel. The entrance bore is then sealed, and any disturbance to the surface is repaired.
[229]James S. Hanrahan and David Bushnell, *Space Biology: The Human Factors in Space Flight*, (New York, 1961), p. 99. At 30 *g* the acceleration will last 17 seconds. These high *g* launch capsules will be equipped with liquid acceleration seats, in which the space cadets are totally immersed in a fluid with a high specific gravity. Fluid countermeasures like these have increased experimental animal's tolerance to *g* forces by a factor of 10.
[230]Space Station Freedom is like the incredible shrinking man. It keeps getting smaller. By the time you read this it is likely to have vanished altogether.
[231]The Astrodome has a clear span of 195.7 meters, and rises to a maximum of 63.4 meters above the playing field. The dome covers 160,000 square feet of interior space, and encloses a total volume of 60 million cubic feet—1.7 million cubic meters.
[232]Iain Nicolson, *The Road to the Stars*, (New York, 1978), p. 162.
[233]In Aquarius we are planning on 117 square meters (1260 sq. ft.) per person, if we multiply that by an average overhead clearance of say 4 meters, that means each person in Aquarius would have around 470 cubic meters of space for all purposes.
[234]1,100 cubic meters is the volume of a dwelling with 4,855 sq. ft. of floor space.
[235]With a population of one hundred thousand, this works out to 1130 cubic meters of space per person.
[236]Throughout the colony, but particularly in the large open spaces, large currents of air will be maintained. These gentle breezes will waft any airborne debris like leaves, feathers, and other detritus onto continuously cleaned belt-screens.
[237]By using two layers of fabric, the bubbles can be made self supporting. The fabric is impermeable, and air is pumped into the space between the two layers. The layers are attached to each other at regular intervals, giving it a quilted appearance. The air pressure in the space between the fabric layers causes the sphere to inflate. The inner layer of fabric restrains the gas, keeping the bubble inflated without having to remain closed.
[238]The Holodeck is a special room aboard the Starship Enterprise (*Star Trek: the Next Generation*), where an artificial reality is created by a computer. The Holodeck transforms energy into a special form of matter,

using technology derived from the 'transporter beam'. It is doubtful we will ever develop a real technology to match this fanciful concept. In the meantime, high resolution, liquid crystal screens, and real-time supercomputer animation will provide an approximation that should be every bit as much fun.

[239]The rate of rotation actually varies somewhat. This is necessary to adjust for the variations introduced by the orbit of Asgard around the Earth, and the orbit of the Earth around the Sun.

[240]Asgard is oriented with the Earth, so that Asgard's north corresponds to Earth's.

[241]It would be difficult to find large supplies of nitrogen outside of Earth's atmosphere. It is not a large component of lunar soil, for example.

[242]Physical stresses induced by atmospheric pressures can be intense. In Gerard O'Neill's *Island One* space colony, atmospheric pressure is to be around 6 psi, or 4.6 tons per square meter. The rest of the load on his cylinder, from soil, structures, people and equipment amounts to only 2.85 tons/$M^2$.

[243]Inspiration takes place when downward movement of the diaphragm creates a zone of low pressure in the lungs. At pressures under 2.2 psi, the diaphragm can no longer generate a pressure differential large enough to fill the lungs. However, if mechanical means were available for inflating the lungs, adequate oxygenation could still be achieved at even lower pressures.

[244]Emergency respirators would require a demand valve system, not unlike that on conventional SCUBA regulators, which will deliver a controlled flow of oxygen when a slight pressure reduction is developed through respiratory effort.

[245]In memoriam: Virgil I. "Gus" Grissom, Edward H. White II, and Roger B. Chaffee.

[246]Hanrahan and Bushnell, p. 156.

[247]Stine, *Handbook,* p. 126.

[248]Charles E. Cobb, Jr., "Living With Radiation," *National Geographic*, (April, 1989), p. 418.

[249]Stine, *Handbook* , p. 126.

[250]G. Harry Stine, "Hardening Humans," *ANALOG Science Fiction/Science Fact*, (Dec. 12, 1987), p. 127.

[251]Alexander Besher, "New Fashion Statement Radiation Resistant—Super Supplements," *Rocky Mountain News*, (Dec. 7, 1990), p. 82.

[252]Paul Birch, "Radiation Shields for Ships and Settlements," *Journal of the British Interplanetary Society*, Vol. 35, (November, 1982), p. 515. Note that doubling the shield density cuts radiation exposure by a factor of ten. This means that small increases in shield thicknesses can lead to large reductions in exposure.

---

[253]S.L Huston, et al. "Radiation Shielding for Lunar Bases Using Lunar Concrete," 43rd Congress of the IAF, IAF-92-0339, p. 4. At a shield density of 45 g/cm$^2$, water is 25% more effective at absorbing galactic cosmic rays than lunar regolith. This is due to water's high hydrogen content and correspondingly low molecular weight. Water's low average atomic number results in the production of fewer secondary radiation particles.
[254]One thousand grams per square centimeter is equivalent to the density of the Earth's own atmosphere, a shield that has guarded us well these past three billion years.
[255]Gold is a good electrical conductor. The sphere is apt to experience a build up of electrostatic charge on its surface as it interacts with ions in the solar wind. This is a problem common to most conductive structures in space. The dimensions of the problem and means for dealing with it will require more detailed study.
[256]Atomic oxygen is produced when molecular oxygen, $O_2$, is broken down in the upper atmosphere. Atomic oxygen can be a serious problem in low earth orbit; at Asgard's altitude—36,000 km.—it is of minor concern
[257]Harold V. Thurman, *Introduction to Oceanography* (Columbus, 1975), p.150.
[258]John E. Roymont, *Plankton and Productivity in the Oceans* (New York, 1963), p. 199. Fifty percent of the energy in sunlight is infra red.
[259]10,000 people each using 1000 watts of power will generate 820 million BTU of waste heat per day. The heat capacity of water is 62 BTU/ft3/° F. A 5.6° difference gives the shield a heat capacity of 350 BTU/ft3. To absorb the waste heat generated, the shield needs a volume of around thee million cubic feet.
[260]Oxygen atoms are 8 times more massive than hydrogen.
[261]Note that 'meteor' is a term that properly applies only to incandescent bodies that are passing through the earth's atmosphere at high speed. Once a meteor hits the ground it becomes a 'meteor<u>ite</u>', while it is still in space it is a meteor<u>oid</u>.
[262]M. Vertregt, *Principles of Astronautics*, (New York, 1965), p. 286.
[263]Thomas Damon, *Introduction to Space: The Science of Spaceflight*, (Malobar, 1989), p. 57.
[264]Editors of Heron House, *The Odds on Virtually Everything*, (New York, 1980). Odds of being hit by lightning are one in 607,000. Odds of being killed in a fatal car crash are one in 3,178. You are a thousand times more likely to be killed while driving to the grocery store than to be hit by a meteor while walking in space.
[265]*Ibid.*
[266]Eric C. Hannah, *Meteoroid and Cosmic Ray Protection in Space Habitats*, Lecture given at Princeton University Space Conference, (May 7, 1975), p. 11.

---

[267]Meteors are likely to break up in or skip off the atmosphere, so the chances are not really the same. The point is that a small object like Asgard is no more likely to be hit by a large meteor than is a relatively large target like a city on Earth.
[268]Barbara Wood-Kaczmar, "The Junkyard in the Sky," *New Scientist*, (13 Oct. 1990), p. 37.
[269]Greg Freiherr, "The Orbiting Junkyard," *Final Frontier*, (Nov./Dec., 1990), p. 42.
[270]'China Syndrome' is a term used in nuclear disaster scenarios. It refers to a nuclear melt-down, in which a pile of uranium melts the earth beneath itself and sinks through the lower density materials of the crust on its way to China, hence the name.
[271]Wood-Kaczmar, p. 39.
[272]Since the leaks will be tiny, the patches can be very small. Something the size of a quarter should be more than adequate. A simple pre-glued patch of the same silicone material used in the rest of the dome should work fine. The glue will be developed in Aquarius through genetic engineering techniques, duplicating the biological glue used by barnacles and other crustaceans to attach themselves to substrates. These biological glues are able to form strong bonds even when wet.
[273]The plants in the hanging gardens are mostly for aesthetic pleasure. They perform some air purification functions, and provide a backup to the redundant algae systems, but are not crucial to the main process loop.
[274]The Volvox is not a diatom. But "Space flagellates" has a distinctly twisted S&M ring to it. "Space diatoms" sounds better and conjurs to mind the phosforescent blue greens that, together with burnished gold, are Asgard's colors.
[275]Chapman. p. 230.
[276]Kenneth L. Jones, Louis W. Shainberg, and Curtis O. Byer, *Foods, Diet, and Nutrition*, 2nd Ed. (San Francisco, 1975), p. 40. A person with 15% bodyfat, weighing 175 pounds, requires 2496 calories per day to meet needs of resting metabolsim.
[277]Jones, et al., p. 45.
[278]Gerald Parkinson, "New Techniques May Squeeze More Chemicals From Algae," *Chemical Engineering*, (May 11, 1987), p. 19.
[279]*Ibid.*
[280]Required volume for colony = 600,000 liters. Requires two thousand kilometers of tubing twenty centimeters in diameter. Tubing is wound around a water filled core one meter in diamter. Required length of core = 5300 meters. Circumference of ecosphere = 1885 M. Three wraps of core around ecosphere's equator is sufficient. Six wraps needed to provide for sufficient residence time of algae in sunlit hemisphere.

---

[281]E. Gustan and T. Vinopal, "Controlled Ecological Life Support System: A Transportation Analysis," NASA-CR-166420, NASA-Ames Research Center, 1982, p. 366.
[282]David Dooling, "Closed Loop Life Support System," *Spaceflight*, Vol. 14, (April, 1972), p. 134.
[283]R.D. McElroy, "CELSS (Closed Ecological Life Support Systems) and Regenerative Life Support for Manned Missions to Mars," *Manned Mars Mission Working Group Papers*, NASA, NTIS N87-17749, p. 370.
[284]David Armstrong, et al., "Natural Flavors Produced by Biotechnological Processing," *Flavor Chemistry Trends and Developments*, Roy Terranishi, ed.,,(Washington, 1989), p. 112.
[285]L. S. Bassett, et al., "Prevention of Disuse Osteoporosis in the Rat by Means of Pulsing Electromagnetic Fields," *Electrical Properties of Bone and Cartilage: Experimental Effects and Clinical Applications*, Carl T. Brighton, Jonathan Black, and Solomon Pollack, eds., (New York, 1979), p. 317.
[286]Lunan, p. 38.
[287]Richard Harding, *Survival in Space*, (New York, 1989), p. 171.
[288]Anthony Delitto and Andrew J. Robinson, "Electrical Stimulation of Muscle: Techniques and Applications," *Clinical Electrophysiology*, (Baltimore, 1989), p. 159.
[289]"Stimulating Atheletes," *Discover*, (March, 1989), p. 14.
[290]Plays The Incredible Hulk on TV.
[291]Olympic high diver.
[292]Lady body-building champion.
[293]Actress and exercise guru.
[294]Solomon Pollack, et al., "Microelectrode Studies of Stress Potentials in Bone," *Electrical Properties of Bone and Cartilage: Experimental Effects and Clinical Applications*, (New York, 1979), p. 80.
[295]Carl T. Brighton, "Bioelectrical Effects on Bone and Cartilage," *Clinical Orthopaedics and Related Research*, Vol. 124, (1977), p. 2.
[296]Bassett, p. 317. Free-ranging rats had a bone strength of 9.5 lbs., immobilized and untreated rats had a bone strength of less than 1 lb., immobilized rats that were treated with pulsing electromagnetic fields maintained a bone strength of 8 lbs.
[297]I live not far from a Hot Springs resort in the Rocky Mountains. In the winter it is delightful to go there and soak your bones while snow falls steaming around you. A common sight at the hot pool are geriatrics from all walks of life. It is easy to recognize the old-timer cowboys. Aside from their bandy bow-legs, bent from a life-time on horseback, they can be recognized by the craggy rawhide of their faces, as brown and wrinkled as sun cured leather, which in fact it is. The faces of these old men are so craggy, they seem to have actually absorbed the canyons, gullies, and arroyos of this rough country. And yet, when one of these living fossiles

stands up, out of the water, he reveals a body that is the startling pure white of freshly carved ivory, as smooth and flawless as a baby's butt. You look at the face and then at the skin below it and you can't help but wonder where in the hell that face has been that the body didn't go. The answer is simple: that weather-beaten old face has been out in the mountain sunshine every day for 75 or 80 years, but the skin on his chest has scarcely ever seen the light of day.

[298]Gen. J.D. Ripper, (Burpelson A.F.B., 1964). *Dr. Strangelove Or: How I Learned to Stop Worrying and Love the Bomb.*

[299]Aluminum ammonium sulphate, alum, is used in modern water treatment systems to precipitate out particulates.

[300]B.N. Ames, R. Magaw, and L.S. Gold, "Ranking Possible Carcinogenic Hazards," *Science*, No. 236, April 17, 1987, p. 272.

[301]Exact duration of the day/night cycle would depend on the particular orbit chosen. The higher the orbit, the longer the orbital period. Shuttles typically orbit in 90 minutes, geosynchronous satellites orbit in 24 hours, the Moon orbits in a month.

[302]O'Neill favored L5 for his "Island One" space colony.

[303]Arthur C. Clarke, "Extraterrestrial Relays," *Wireless World*, Oct. 1945, p. 306.

[304]All geostationary satellites are geosynchronous, but not all geosynchronous satellites are geostationary. Geosynchronous satellites in orbits inclined to the plane of the equator appear to move up and down in latitude, traversing a figure 8 pattern around a fixed line of longitude.

[305]Arthur C. Clarke, *Ascent to Orbit: A Scientific Autobiography*, (New York, 1984), p. 62.

[306]KOMA, Oklahoma City, @1520 kHZ.

[307]Clemens P. Work, "Wiring the Global Village," *U.S. News & World Report*, (February 26, 1990), p. 44.

[308]Edward Fennessy, *JBIS*, V. 32, No. 10, (October 1979), p. 376. Phones per capita growing at rate of 6%/yr., number of calls growing at 18.5%/yr.

[309]Barry Fox, "Portable Phones Ring the Globe by Satellite," *New Scientist*, (7 July 1990), p. 30. Plans already exist for global cellular phone networks operating with low power satellites using phased-array technology in low earth orbit.

[310]Jonathan Beard, "Balloon in Space Takes the Heat off Spacecraft," *New Scientist,* 14 Oct. 1989, p. 35.

[311]Robert A. Heinlein.

[312]Paul Webb, "The Space Activity Suit: An Elastic Leotard for Extra-Vehicular Activity," *Aerospace Medicine*, Vol 39, (April, 1967), p. 378.

[313]*Ibid.*

[314]Normal atmospheric pressure at sea level on earth is just over one kilogram per square centimeter (14.7 psi.); this pressure will support a column of mercury 760 millimeters high. Accordingly, atmospheric

pressure is often given in millimeters, meaning millimeters of mercury, or mmHg, with 760 being earth normal, at sea level.

[315]John Grossmann, "The Blue Collar Spacesuit," *Air & Space*, (Oct./Nov. 1989), p. 64.

[316]The volume inside a balloon suit is only a few cubic feet. Even a small tear will cause it to deflate in seconds. Unlike Asgard, which has a volume of hundreds of millions of cubic feet, and would take hours to deflate, through even a fairly large hole.

[317]James E. and Alcestis R. Oberg, *Pioneering Space*, (New York, 1987), p. 101.

[318]Grossmann, p. 63.

[319]Actually the bolts blow the airlock off the pod, and the air pressure in the pod blows him out.

[320]Webb, p. 378.

[321]*Ibid.*

[322]*Ibid*

[323]*Ibid.*

[324]*Ibid.*

[325]Twenty meters of clear tubing with an inside diameter of two centimeters wound around 1.6 meters of core tubing with an outside diameter of 6 cm.

[326]K. Eric Drexler, *Engines of Creation* (New York, 1987), pp. 90-92.

[327]The inner belt is composed mostly of protons with energies on the order of 30 Mev, while the outer belt is made up mostly of electrons with energies of around 1.5 Mev.

[328]Harding, p. 78.

[329]Tensile strength of a typical industrial grade steel with moderate carbon content = 200,000 psi, tungsten = 600,000 psi.; density of lead = 11.35 g/cc, forged tungsten = 20+ g/cc; melting point of carbon = 3550° C., melting point of tungsten = 3,410° C. (6,170° F.); boiling point of tungsten is the highest of any element = 5927 ° C. (10,700° F.).

[330]Glass containing lead is often refered to as "lead crystal". Glass of this type can have densities as high as 8 g/cc., and is used extinsively as radiation shielding in the nuclear industry. Such glass also has a very high index of refraction which will be very important in a spherical helmet.

[331]Borosilicate glass has a very low coefficient of expansion. This will be an important property in a space helmet which will routinely be exposed to temperature extremes. Borosilicate glass is used extensively in laboratories under trade names like Pyrex.

[332]Micrometeoroids large enough to penetrate .5 mm of aluminum will hit a "large" space station about once a year. Tungsten armor covering vital organs will be two millimeters thick, equivalent to 14 millimeters (1/2 inch) of aluminum. The smaller size of human targets and the tungsten

armor's 28 fold increase in resistance to penetration will make fatalaties from meteoroids extremely rare.

[333]To put it in some perspective, consider that the distance to the nearest star is 4.3 light years away. Our sun is only 8 light minutes away. If you traveled 99.9999965% of the way to the next star, you would then be at the same distance from it as we are from our sun

[334]Everything inside the heliopause—the point where the solar wind meets interstellar gas—is roughly speaking, inside the sun's atmosphere. The heliopause is 100 AU from the sun.

[335]Peter E. Glaser, "The Satellite Solar Power Station," *Princeton Conference on Space Manufacturing Facilities*, (Princeton, 1975).

[336]A small hemispheric mirror will focus light onto a single intense spot. Larger spherical mirrors experience 'spherical abberation'. This means that the light is distributed along the focal axis. While such aberration makes large spherical mirrors useless as optical instruments for telescopes, they nonetheless work well for the concentraion of solar energy.

[337]The area along the focal axis will be saturated with an intense concentration of light. Through this focal zone will run a coil of piping carrying water. The water in the coil boils creating high pressure steam which spins a turbine. The turbine turns a generator, producing electric power. The steam is then circulated to the outside of the balloon, behind the reflective hemisphere. There it passes though the pipes of a radiator with a large surface area. Behind, the balloon, where no light can penetrate, the temperature will be extremely low, around 150° below zero C. The steam radiates its waste heat into space, and recondenses. The water is then recycled back through the heating coil. Electricity is conducted out of the generator and along a cable, to the space colony. The cable also holds the baloon in place against the gentle push of the solar wind.

[338]American total demand is 10 kw per capita, continuous. Domestic needs are 1 kw, continuous.

[339]Assumes 25 megawatts for internal consumption and 75 megawatts for communications, industry, maneuvering lasers, etc. Mirror's reflectivity = 90%, conversion efficiency = 40%. .486 kw/M2. Note that effective collection area is the area of a disk, not the entire surface area of the reflective hemisphere.

[340]The six collector bubbles will fly in formation with the main colony, each occupying one vertex of a hexagon. Three bubbles will be anchored to each pole, and each bubble will be anchored to two others, around the perimeter of the hexagon, giving the whole formation stability. The formation must move as a unit since it needs to slowly precess, always keeping the same face toward the sun, as the Earth passes through its yearly orbit. To this end, the connecting lines will be filled with

pressurized gas, so that the lines can transfer compressive as well as tensile forces.
[341]Solar wind pressure, the most significant of these forces, is negligible. Pressure of sunlight at earth's orbit is 5 E -6/ $M^2$. Each solar bubble has an effective surface area of 34,000 $M^2$, making the total solar pressure on each bubble only .17 newtons. Total mass of the power bubble is 1500 kg., mass of power plant is 166,000 kg. Acceleration = 1 E -6 M/$sec^2$.
[342]Sir Thomas Malory, *Le Morte D'Arthur.*
[343]Avallon, Avalon, Avollon, Afallon, Avylyon Avilion, Avelion. All derived from the Welsh *afal*, for "apple".
[344]*Genesis*, 6-19.
[345]Richard Grieve, "Impact Cratering on Earth," *Scientific American*, (April, 1990), p. 66.
[346]Rick Gore, "Extinctions," *National Geographic*, (June 1989), p. 664.
[347] Kenneth Weaver, "Meteorites: Invaders From Space," *National Geographic*, (September, 1986), p. 406.
[348]Shell Silverstein. *Ride the Perfect Wave.* "And cities fell beneath its swell, And mountains turned to mud."
[349]Clark R. Chapman and David Morrison, "Chicken Little Was Right," *Discover*, (May, 1991), p. 40. Relatively smaller meteorites, up to a mile in diameter, strike the Earth about once every 300,000 years. These are not planet killers, and are unlikely to exterminate the human race, but pack a wallop nonetheless. Such a meteorite would hit with the explosive energy of five million atomic bombs—ten times the force of the world's combined nuclear arsenal.
[350]Jeff Hecht, "Will We Catch a Falling Star?," *New Scientist*, (7 Sept. 1991), p. 48. This startling calculation was made by Clark Chapman of the Planetary Science Institute. His estimate for the chance of dying in a meteor strike is one in 6,000; the chance of dying in an air crash is one in 20,000. Chapman bases his calculation on the assumption that a meteor large enough to kill everyone on the planet hits every 100,000 years—a very pessimistic view.
[351]*Ibid.*
[352]Duncan Lunan, *Man and the Planets*, (Bath, 1983), p. 186.
[353]Bruce G. Blair and Henry W. Kendall, "Accidental Nuclear War," *Scientific American*, (December 1990), p. 53.
[354]Launch capacity of the Bifrost Bridge is 360 tonnes per day, 75,000 tonnes per year, assuming 208 launch days per year. 88 years to launch 6.6 X $10^6$ tonnes.
[355]The chemical formula for water is $H_2O$, two hydrogen atoms and one oxygen atom. However, the ratio of two to one does not hold for the masses inolved, since oxygen is 8 times as massive as hydrogen.
[356]According to O'Neill, minimum equipment for colonization of Asteroids is two tonnes/colonist. Equipment allotment for Millennial

colonists should be substantially less, due to reliance on tensile technology and dedication to material simplicity.

[357]99% $O_2$, @ 3 psi.

[358]100,000 @ 70 kg. ea.

[359]100 MW @ 10 kg/kw.

[360]Specific heat of lunar soil taken as .2 gm. calories/gm, equal to specific heat of powdered cement. Mean temp. of soil taken as 150° C., target temp. of 1500°C., 200 gcal./kg. X 1350°C. = 270,000 gcal./kg. @ 60% O2, it requires 1.66 kg. soil/kg. O2 produced. 1.66 X 270 kg. cal. = 450 kgcal.

[361]Alfred A. Levinson and S. Ross Taylor, *Moon Rocks and Minerals: Scientific Results of the Study of the Apollo 11 Lunar Samples with Preliminary Data on Apollo 12 Samples*, (New York, 1971), p. 102.

[362]David R. Criswell, "Lunar Materials for Construction of Space Manufacturing Facilities," *Princeton Conference on Space Manufacturing Facilities*, (Princeton, 1975), pp. 15-19.

[363]Frederick Golden, *Colonies in Space: The Next Giant Step*, (New York, 1977).

[364]These are figures for escape velocity, not orbital velocity. At orbital velocity an object may leave the surface and orbit a planet, but it is still essentially attached by its gravitational umbilicus.

[365]Brian O'Leary, "Mining the Apollo and Amor Asteroids," *Science*, Vol. 197, (22 July 1977), p. 364.

[366]At 70% efficiency.

[367]Robert Bustin, "Determination of Hydrogen Abundance in Selected Lunar Soils," NASA-N87-30239, (Oct. 31, 1987).

[368]B. Zellner, "Asteroid Taxonomy and the Distribution of the Compositional Types," *Asteroids,* Tom Gehrels, ed., (Tucson, 1979).

[369]Adrian Berry, *The Next Ten Thousand Years*, p. 59.

[370]O'Leary, p. 364.

[371]Gerard K. O'Neill, *The High Frontier: Human Colonies in Space*, (New York, 1977), p. 58.

[372]Harry K. Savage, *The Rock That Burns*, (Boulder, 1967), p. v. Oil shale is actually a misnomer, since this mineral in fact contains no oil whatsoever. Oil shale contains a rich organic substance properly known as 'kerogen', that when properly treated by the application of heat and high pressure hydrogen can be induced to produce an oil that is "similar to petroleum".

[373]Concentration of nitrogen in lunar soil is 100 parts per million. Atomic weight of nitrogen is 14.

[374]O'Leary, p.364.

[375]Clifford J. Cunningham, *Introduction to Asteroids*, (Richmond, 1988), p. 55. I hope the reader will forgive a minor obsession with the organic content of asteroids in general and kerogen in particular. I am the product

---

of three generations of oil shalers. Had my mother not put her foot down, my father would have named me "Kerogen Blue Savage." To carry my family's ancestral preoccupation into space with me is a sweet synchrony.

[376]W. Scott Meddaugh, et al. "Variations in Organic Geochemistry and Petrography with Depth of Burial, Green River Formation Oil Shales Piceance Creek Basin, Colorado", *17th Oil Shale Symposium Proceedings*, (Golden, August 1984), p. 162.

[377]Lunan, p. 189.

[378]*Ibid.*

[379]Cunningham, p. 116. Major strikes average out at one every 26 million years.

[380]O'Leary, p. 365.

[381]Hecht, p. 48.

[382]Lunan, p. 193.

[383]Hecht, p. 52.

[384]Cunningham, p. 139.

[385]Escape velocity = $\sqrt{2gr}$, g = .00128 M/sec/sec, r = 4000M. = $\sqrt{100}$ = 10 meters per second.

[386]O'Leary, p. 363. Another reason for using the AA asteroids is that they receive five times as much solar energy as do the main belt asteroids.

[387]*Ibid.*

[388]The mass launcher needs to be as close to the lunar equator as possible. The Moon doesn't spin rapidly on its axis the way the Earth does, so there is no boost to be gained by an eastward launch. An equatorial outpost will nevertheless minimize the amount of energy required to send payloads up to an orbital station at one of the Lagrange points.

[389]*Geologic Atlas of the Moon*, U.S. Geological Survey, (Reston, 1975).

[390]Arthur C. Clarke, *Ascent to Orbit*, (New York, 1984). The concept of a lunar mass launcher was first put forward by Arthur C. Clarke in the 1950s.

[391]Bear in mind that mass is not a function of gravity. Just because a ton of liquid oxygen on the Moon weighs only one sixth of of what it weighs on Earth does not mean that it is six times easier to accelerate. Its resistance to acceleration remains the same. However, launch to orbital velocity on the Moon is much lower than it is on Earth, so the launch can be made with a much shorter acceleration run, and much less energy. Payloads launched from the Moon do not require as much laser propellant as capsules launched from Earth. There is no atmospheric barrier to punch through, and orbital velocity can be easily attained with the launch tube alone. Therefore, more of the launch mass can be in the form of payload. Some laser propellant will be required, however, to make the journey to GEO and to rendezvous with Asgard.

[392]A mirror diameter of 15 meters should allow us to focus a green laser out to $L_1$, a distance of 58,000 km.

---

[393]Heppenheimer, *Colonies In Space*, p. 62.
[394]L1 and L2 are not as stable as L4 and L5. Solar mirrors positioned in halo orbits around these L points will require dynamic positioning. The impulses involved, however, are small and can probably be imparted by using the pressure of sunlight.
[395]The orbit must be oriented perpendicular to the Moon's equator.
[396]If the bubble membrane were installed as a simple hemisphere, anchored to the foundation ring, it would have to be held down against the entire force of the atmosphere— 3 psi X total area of the dome. Such uplifting forces would be enormous, particularly in the larger ecospheres, and would require extensive and expensive countermeasures to negate.
[397]2500/6
[398]Two meters beneath the lunar surface the temperature never rises above -20° C. (-30° F.).
[399]Unlike the earth, where plate tectonics are a fact of life, the Moon is geologically dead. The crust is entirely stable and solidified, so moonquakes simply do not happen. The only seismic disturbances we can expect are from the shockwaves of large meteor impacts. However, large meteor falls are no more likely on the Moon than they are on Earth. Engineering a building to withstand the shock of a major meteor strike would be considered ludicrous paranoia here, and is no less so there.
[400]This assumes, of course that trees are limited in size by their ability to support their own weight, and to lift water to their highest leaves through evapotranspiration—both functions being directly limited by gravity. Even if this is not the case, and the size of trees is genetically predetermined, independent of gravity, we can alter that genetic characteristic and still achieve the growth of giants.
[401]Donald H. Menzel, *Astronomy*, (New York) p. 109.
[402]The entire crater, including the walls, is just over 90 km. across.
[403]Ian Ridpath, *The Illustrated Encyclopedia of Astronomy and Space*, Revised Ed., (New York, 1979), p. 134.
[404]In Aquarius people live quite comfortably in an area of less than 100 square meters (1000 sq. ft.). With the same space allotment, 700 people could live comfortably spread out over the crater bottom. Actually, the colonists would probably erect a smaller version of an Aquarian surface structure which could accommodate many more people.
[405]The underground rooms can also serve as a storm shelter for protection from solar flares.
[406]For a good description of the outcome of a war with the Moon, see *The Moon is a Harsh Mistress* by Robert A. Heinlein.
[407]From an idea by Keith Spangle.
[408]Assuming an acceleration of .3*g*.—about equivalent to what you might experience on Disneyland's Space Mountain.

[409]Lorien is the wooded realm of the Elven folk in the Tokein Trilogy, *Lord of the Rings*. "Loth" is an Elven prefix meaning "blossom".
[410]Sustained growth rate of 2%, from a base of 6 billion in 2000 A.D.
[411]Norman Myers, *Gaia: An Atlas of Planet Management*, (New York, 1991).
[412]Deimos and Phobos (Fear and Panic) are the horses who pull Mars' chariot into battle.
[413]When Mars is on the same side of the sun as the Earth the distance is minimized, whereas, when Mars is on the opposite side the distance is at its maximum.
[414]Anywhere inside the solar system, that is. Escaping the gravitational empire of the sun is another matter.
[415]Arthur C. Clarke, "An Elementary Mathematical Approach to Astronautics", *Ascent to Orbit,* p. 21. These figures are actually for Deimos.
[416]This is only possible because Mars has an atmosphere. Phobos orbits Mars at 3.06 kps, and any object you threw off Phobos would retain almost this entire velocity. Since orbital velocity around Mars is 2.2 kps, your rock would simply continue to orbit. But, since Mars has an atmosphere, it could eventually approach Mars close enough to be decelerated and fall to the surface.
[417]Isaac Asimov, *Mars the Red Planet*, (New York, 1977), p. 173.
[418]Deimos is smaller, 15 km. in diameter, and more distant, 14,600 km. from Mars.
[419]Passengers and cargo inbound to the Martian surface will travel in wave riders, and so will require little fuel for landing.
[420]This does not count energy losses to ineffeciencies at any point in the system, due to the Earth's thick atmosphere, the energy penalty for loads from the Earth are apt to be even steeper than reflected here.
[421]Arthur C. Clarke, "An Elementary Mathematical Approach to Astronautics," *Ascent to Orbit,* p. 21.
[422]Based on electicity cost of 5¢/kwh.
[423]At continuous 1*g*, a ship will travel around 37 million km. per day. Accelerate for 24 hours and decelerate for 24. Fly straight across from Earth to Mars.
[424]Arthur C. Clarke, *Interplanetary Flight: Introduction to Astronautics*, (New York, 1960).
[425]Buzz Aldrin, "The Mars Transit System," *Air & Space*, (Oct./Nov. 1990), pp. 40-47. Aldrin—second human ever to set foot on another world—bases his Mars Transit System on "cyclers": reusable spacecraft that make continuous gravity-assisted transits between the Earth and Mars. The system proposed here is identical to that suggested by Aldrin, except that the cyclers are replaced by Valhalla-class ecospheres.
[426]Aldrin, p. 42.

[427]The orbital dynamics for each transfer habitat will differ, depending on the relative positions of Earth and Mars. Some habitats might follow relatively complex orbits, passing by Venus on their way to Mars, for example.
[428]Eugen Sanger, *Space Flight: Countdown to the Future*, (New York, 1965), p. 241. Deuterium contains ten million times as much energy per kg. as chemical fuels like coal. One kg. of deuterium is equal to 10,000 tons of coal.
[429]One kilogram calorie will raise the temperature of one litre of water one degree C.
[430]Helium3 is not needed for straight deuterium fusion, or for deuterium tritium fusion, but if we had straight deuterium fusion we would have a terrestrial source of helium3 and fusion would be easier.
[431]Iain Nicolson, *The Road to the Stars*, (New York, 1978), p, 59.
[432]Lunan, p. 57.
[433].1 g = .98 M/sec/sec X 24 hrs = 84 kps.
[434]Roughly, the multiple of the exhaust velocity translates into the power of the mass ratio. This means that rockets can achieve only a few times their own exhaust velocity. To illustrate, consider the mass ratio of a rocket with an exhaust velocity of 3 kps trying to attain a velocity of 30 kps: its mass ratio would be 1: $10^{10}$, one to ten billion. So to accelerate a single ton of payload to 32 kps would take over ten billion tons of fuel- an impossible requirement.
[435]David Dooling, Jr., "Controlled Thermonuclear Fusion for Space Propulsion," *Spaceflight*, (January, 1972), p. 26. Fusion Temp. $10^7$ °K.
[436]*Ibid.*
[437]*Shaded Relief Map of the Tharsis Quadrangle of Mars*, M 5M 15/112 R, I-926 (MC-9), 1975, Dept. of the Interior, USGS.
[438]Mercury is too close. Water on that planet's surface would instantly vaporize. The surface of Venus is too hot for liquid water, but this is due to the atmosphere, not the proximity to the sun.
[439]Nomads roaming among the comets of the Oort cloud may be the exception to prove this rule.
[440]M.M. Averner and R. D. Macelroy, *On the Habitability of Mars: An Approach to Planetary Ecosynthesis*, NASA SP-414, (1976), p. 37.
[441]Averner and Macelroy, p. 44.
[442]Surveyor 3 was an unmanned spacecraft which made a soft landing on the Moon in 1967. Astronauts from Apollo 12 landed within a few hundred meters of the defunct instrument and brought parts of it back to Earth for study.
[443]Averner and Macelroy, p. 61.
[444]James Lovelock, *The Ages of Gaia*, (New York, 1988).
[445]*Topographic Map of Mars*, 1:25,000,000 Topographic Series, M 25M 3 RMC, U.S. Geological Survey, Department of the Interior, (1976). Mars,

of course, has no sea level—at least not yet. Therefore the zero altitude level has been somewhat arbitrarily assigned by earth-bound areographers as that altitude corresponding to an atmospheric pressure of 6.1 millibars.
[446]Carbon dioxide equals only .03% of the volume of Earth's atmosphere, weighing just 1.7 trillion tons.
[447]John P. Holdren, "Energy in Transition," *Scientific American*, (September, 1990), p. 160.
[448]Averner & Macelroy, p. 65.
[449]225 years.
[450]Dust must have an albedo of .25.
[451]Averner & Macelroy, p. 66.
[452]Averner & Macelroy, p. 8.
[453]Averner & Macelroy, p. 65.
[454]Lake Superior has a volume of 2900 cubic miles.
[455]Thomas Damon, *Introduction to Space: The Science of Spaceflight*, (Malabor, 1989), p. 188.
[456]David G. Stephenson, "Comets and Interstellar Travel," *JBIS*, Vol. 36, (May, 1983), p. 210
[457]Lunan, p. 251.
[458]Fred L. Whipple, *The Mystery of Comets* (Washington, 1985), p. 74.
[459]"Mother of Short Comets," *Discover*, (February, 1991), p. 8. There is considerable controversy surrounding the origin of these comets. It is unlikely that they have all been captured by the gravitational influence of the giant planets. There is a proposed second belt of comets, the Kuiper Belt, surmised to exist just outside the orbit of Neptune, thought to be the source of these comets. Alternatively, it may be that all these comets are just chunks that have sloughed off of Chiron, a giant comet, 240 km. dia., that orbits between Saturn and Uranus.
[460]Note that the Martian year is 687 days, in that time six comets will cross Mars' orbit.
[461]David A. Seargent, *Comets: Vagabonds of Space*, (Garden City, 1982), p. 23.
[462]Gregory L. Matloff and Kelly Parks, "Interstellar Gravity Assist Propulsion: A Correction and New Application," *JBIS*, Vol. 41, (November, 1988), p. 523
[463]David G. Stephenson, "Comets and Interstellar Travel," *JBIS*, (May, 1983), p. 211.
[464]Holdren, p. 160.
[465]Ian Ridpath, *The Illustrated Encyclopedia of Astronomy and Space*, Revised Ed., (New York, 1979), p. 54. And, Roy A. Gallant, *National Geographic Picture Atlas of Our Universe*, (Washington, 1986), p. 151. The meteorite which created the Barringer Crater in Arizona weighed 250,000 tons and blew out 400 million tons of rock—a ratio of 1600:1.
[466]Averner & Macelroy, p. 65.

[467]Averner & Macelroy, p. 66.
[468]This may take quite some time. Release of frozen carbon dioxide from perma-frost may be particularly slow.
[469]Averner & Macelroy, p. 17.
[470]James F. Kasting, Christopher P. McKay, and Owen B. Toon, "Making Mars Habitable," *Nature*, Aug. 8, 1991, p.489.
[471]*Ibid.*
[472]Thomas A. Mutch, et al., *The Geology of Mars*, (Princeton, 1976), p. 307.
[473]Nigel Henbest, "Vast Oceans may have Covered Surface of the Red Planet," *New Scientist*, (24 August 1991), p. 19.
[474]At first, domes on Mars must still incorporate a water shield to filter ultraviolet light and protect inhabitants from cosmic radiation. Later, though, as the atmosphere thickens, a layer of ozone will form in the upper atmosphere, and the atmospheric mass will absorb cosmic rays.
[475]From Edgar Rice Burroughs, as emphasized by Carl Sagan in *The Cosmic Connection*, (New York, 1973). p. 102.
[476]Freeman Dyson, "The Greening of the Galaxy," *Disturbing the Universe*, (New York, 1979), p. 236.
[477]Neil McAleer, *The Cosmic Mind-Boggling Book,* (New York, 1982), p. 9.
[478]"Solar Energy," *McGraw Hill Encyclopedia of Science and Technology*, 1966 ed., Vol. 12, pp. 466-467. Total planetary insolation = $6.4 \times 10^{20}$ kgcal./yr. Total energy utilized by terrestrial and marine plants = $3.6 \times 10^{17}$ kgcal./yr.
[479]"From women's eyes this doctrine I derive:
They sparkle still the right Promethean fire;
They are the books, the arts, the academes,
That show, contain, and nourish all the world.
William Shakespeare, *Love's Labour's Lost.*
[480]Nathan Keyfitz, "The Growing Human Population," *Scientific American*, (September, 1989), p. 119. Annual global growth rate is 1.74%.
[481]Ben R. Finney and Eric M. Jones, *Interstellar Migration and the Human Experience*, (Berkely, 1985), p. 11.
[482]2.12%
[483]33 doublings.
[484]*The Brain—is wider than the Sky—*
*For—put them side by side—*
*The one the other will contain*
*With ease—and You—beside.*
Emily Dickinson, No. 632 [1862]
[485]Water shields are a very primitive solution to the problems of radiation exposure. Two centuries from now we will undoubtedly have perfected active radiation shields. This will reduce the water requirements to barely

more than the metabolic and organic needs of the people, plants and other life forms. This might only average a thousand kg. per person. Scale factor reductions in water requirements would reult in a commensurate increase in the carrying capacity of this region of space. For our purposes, however, let's assume the Asgard average requirement; this way we will be making ample allowance for leakage, propellant expenditures, mining inefficiencies, and other losses. Once the water from the near Earth asteroids has all been absorbed, more can be produced by importing hydrogen from Earth. This is expensive, but not out of the question. Of course, any exports from the Earth must be strictly limited. Demands on the order of those being contemplated here could quickly strip Earth of vital commodities.

[486]This assumes close to 30 billion in the Martian system: 16 billion around Phobos, and 3 billion on Deimos, with 10 billion on Mars itself. The Moon-Asgardian complexes will support 5 to 10 billion, with hydrogen imports from Earth, and water at a premium from the asteroid belt. That leaves 10 to 15 billion people on Earth.

[487]This figure does not count terrestrial oceans, ice caps and glaciers, or ground water, but it does include soil and atmospheric moisture. Assumes CC water content of 20%, which is probably high.

[488]Millennial ecospheres are closed and self-sustaining in the same sense as the Earth's own planetary ecosphere. Important commodities like water are not expended; they simply pass around the closed-cycle loops indefinitely. Therefore, one of the residents of a free space habitat may require a water supply of 60 tons, but that is a life time allottment. The only losses to the system are due to leakage. With advanced materials and careful management this resource cost can be minimized.

[489]During the California gold rush of 1849 the rail head was at St. Joseph, Missouri.

[490]Eric Drexler, "Deep Space Material Sources," *Princeton Universtiy Conference on Space Manufacturing Facilities*, (Princeton, 1975).

[491]Jim MacNeill, "Strategies for Sustainable Economic Development," *Scientific American,* (September, 1989), p. 156. Achieving such a global average requires rapid rates of growth in the developing world. During the '60s and '70s such rates were attained. In the '80s however, economic growth in some regions, like Sub-Saharan Africa virtually ceased.

[492]*Sykes has set up a small solar-powered mass launcher which has run continuously for about twelve years. During that time the launcher has expelled enough rock to despin the asteroid from its intial rotational rate of 4.5 hours, down to a stately 18 hours. The Robinson's will continue to operate this stabilizing mass launcher for several more years, until the spin rate is optimized at 24 hours.*

[493]Gary Stix, "Desktop Artisans," *Scientific American*, April, 1992, pp. 141-142.

[494]Total area of 31,400 $M^2$. Living area of 113.5 $M^2$ is standard on Aquarius. Total area available for 276 persons.
[495]In this process, people of advanced age gradually replace their failing biological systems with bio-mechanical ones.
[496]Half of one percent of one percent.
[497]Freeman Dyson, "The World, the Flesh, and the Devil," *Communication with Extraterrestrial Intelligence (CETI)*, C. Sagan, ed., (MIT Press: Cambridge), 1973. As reprinited in *Comet*, p. 353. Freeman Dyson has proposed the growth of gigantic trees in free space, rooted in comets. According to Dyson such trees could be genetically engineered to thrive in a vacuum environment. The proposal here is substantially less ambitious. The trees on the asteroids are sheltered within the membrane of an ecosphere. Presumably the same capacity for growth in the low gravity environment would hold.
[498]Spain, France, Belgium, Holland, Germany, Italy, Austria, Switzerland, Poland, Checkoslovakia, Hungary, and Yugoslavia combined.
[499]Edgar Rice Burroughs. In addition to his more famous works about Tarzan and Mars, Edgar Rice Burroughs wrote a series of books about a primordial world populated by dinosaurs inside the hollow Earth. He called this world Pellucidar.
[500]Clark R. Chapman, "Asteroids," *The New Solar System*, J. Kelly Beatty, Brian O'Leary, and Andrew Chaikin, Eds., (Cambridge, 1981), p. 100.
[501]Cunningham, p. 74.
[502]Cunningham, p. 27.
[503]*Ibid.*
[504]With an ecosphere diameter of 28 kilometers, internal volume = 11.5 X $10^{12}$ $M^3/1000$ = 11.5 X $10^9$ people. If even half the mass of the asteroid remains in place there is still ample room for five billion people. The mass of water on an asteroid this size is 2.5 trillion tons. The water shield would not need to be a full five meters thick, since people actually live inside the asteroid, completely shielded from radiation. If it were a five meters, however, it would weigh 12 billion tons. This is less than half of one percent of the available water supply.
[505]Observations of asteroids smaller than a kilometer in diameter in the main belt have not actually been made. This number is a ball-park extrapolation based on a phone conversation with Cunningham.
[506]T. Ransome, "Lagrange Points and Applications," *Spaceflight*, Vol. 12, No. 12, (December, 1970), p. 489. Passage between asteroids that are not in the same orbital plane is more costly. For each 2° change in the orbital plane, an additional $\Delta v$ of 1 kps is necessary.
[507]Chapman, p. 106.

---

[508]Mean distance of asteroids from sun is 2.9 A.U., which is very close to the value of 2.8, predicted by Bode's Law. Sunlight obeys the inverse square law, so for each additional A.U. from the sun, the solar constant will be cut to $1/d^2$: 325 watts/$M^2$ at 2 A.U., and 144 watts/$M^2$ at 3 A.U.
[509]Edward Mazria, *The Passive Solar Energy Book*, (Emmaus, 1979), p.10.
[510]Mazria, p. 14. At an angle of incidence of 45° the radiation received on a surface is equal to 70% of that received at an angle of incidence of 0°.
[511]The intensity of the sun in space near Earth is equivalent to that of 21,000 candles held one foot away from a piece of paper.
[512]James B. Pollack, "Atmospheres of the Terrestrial Planets", *The New Solar System*, pp. 57 & 59.
[513]John P. Holdren, "Energy in Transition," *Scientific American*, (September, 1990), p. 162. Per capita energy demands in the future are predicted to top out at around three kilowatts per person. Americans today typically consume about 7.5 kilowatts each.
[514]K2 is the unlikely name of the Earth's second highest mountain.
[515]In the U.S. in 1986 there were five million scientists and engineers out of a total workforce of 109 million (not counting unemployed). In the next Millennium the proportion of sceintists and engineers is apt to rise from its present 5% to a much higher proportion of the population.
[516]Average weight for men 5'9" = 170 lbs. (77 kg.); women 5'5" = 134 lbs. (61 kg.). Avg. human = 152 lbs. (69 kg.) X .6 = 91 lbs. (41.3 kg.) water.
[517]This assumes that the economies of scale offered by larger ecospheres—especially with regard to radiation shield masses per capita—are resistable. The average ecosphere size, at least in terms of shielded volume, could easily grow to ten kilometers. Each time an ecosphere's diameter is increased by a factor of ten, the required shield mass per person falls by a factor of 100. At an average ecosphere diameter of ten kilometers, water requirements for shielding will fall from 300 to just 3 cubic meters per person. All other water requirements, e.g., for humidity, plants, food production, etc., are considered to be included in the shield mass calculations, since water molecules will absorb radiation regardless of place or function.
[518]98% of Earth's water is held in the ocean basins. When all water supplies, including ice caps is included the total equals 1,326,530,000 cubic kilometers.
[519]The asteroid belt has enough water for five million billion people—5 X $10^{15}$. Jupiter's two largest satellites, Ganymede and Callisto, are both more than half water. Considering that neither of these bodies is much smaller than the planet Mars, one gets an appreciation for the amount of water involved. Together, Ganymede and Callisto contain 150 million trillion tons—15 X $10^{19}$of water. Nor is that the bottom of the solar system's water barrel. Titan, the largest moon of Saturn is almost as large

as Ganymede and has about the same composition. In addition to Titan, Saturn has 16 smaller moons which are mostly ice. Altogether, the Saturnian pool holds 86 million trillion tons of water. The twin gas giants Uranus and Neptune, together have seven major satellites, all of which are about half made of ice. The last significant body of water in the outer solar system is Pluto, which, together with its stygian companion Charon, is almost entirely ice. Altogether the small bodies of the solar system hold a reserve of water equal to three hundred million trillion tons.

[520]Lunan, p. 251

[521]This would be true for objects of small mass, not in orbit around Jupiter, like the Trojan asteroids. More massive objects orbiting Jupiter would require larger $\Delta$vs.

[522]Personal communication with Mr. P. Blass, November 1, 1993.

[523]*Ibid.*

[524]Note that the caption for Plate 13 is in error. The gravitational "sling-shot" maneuver described in the caption actually requires more energy than a direct launch from Callisto: 7.445 kps vs. 4.12. *Ibid.*

[525]Andrew Ingersoll, "Jupiter and Saturn," *The New Solar System*, p. 118. The atmosphere of the sun contains 1000 times more water than Jupiter's atmosphere appears to. Jupiter and the sun should consist of the same elements in roughly equal proportions. Therefore, it is suspected that there may be much more water, in the form of droplets or ice particles deeper in Jupiter's atmosphere.

[526]Not counting the Oort cloud.

[527]Arthur C. Clarke, "The Dynamics of Space-Flight," *Ascent to Orbit*, p. 131. To determine the depth of a planet's gravity well, multiply its value of $g$ by its radius. In the case of Jupiter: 2.65 X 71,900.

[528]The distance from the sun to the Earth is called the 'astronomical unit' (A.U.), 150 million kilometers.

[529]Sagan and Druyan, p. 197.

[530]*Ibid.* If the average diameter of a comet is four kilometers, then its mass is approximately 33.5 billion tons. One hundred trillion comets of this size will weigh 550 times as much as the Earth. This is somewhat more than the aggreagate mass of all the planets in the solar system combined. While the mass of the Oort cloud is unknown, this estimate may be quite high. The consensus of estimates falls more in the neighborhood of between one and five times the mass of the Earth.

[531]Jeff Hecht and Nigel Henbest, "Ice Dwarfs at the Edge of the Solar System," *New Scientist*, (13 July 1991), p. 24.

[532]According to Alan Stern, of the University of Colorado in Boulder, the number of larger comets is not known with any degree of precision, and the actual number may vary from estimates by two or three orders of magnitude. If theories of cometary formation and gravitational interaction with the rest of the solar system are correct, then some fraction of the large comets have been hurled out of the Oort cloud into interstellar space.

According to Stern, the total population of large comets remaining in the cloud, and their aggregate mass can only be guessed at. Based on a phone conversation with Alan Stern, July 22, 1991.

533Oort himself estimated the total number of comets in the cloud at 100 billion with an aggregate mass equal only to ten percent of the mass of the Earth.

534David G. Stephenson, "Comets and Interstellar Travel," *JBIS*, Vol. 36, (May, 1983), p.212.

535David A. Seargent, *Comets: Vagabonds of Space*, (Garden City, 1982), p. 23.

536Stephenson, p. 212.

537Gregory L. Matloff and Kelly Parks, "Interstellar Gravity Assist Propulsion: A Correction and New Application," *JBIS*, Vol. 41, (November, 1988), p. 523.

538*Ibid.*

539*Ibid.*

540Includes use of fossil fuels for transportation, electricity, industrial steam, etc.

541The sun's total annual power output is around 3.5 X $10^{27}$ kwh/yr.

542This could be alleviated by placing the collectors in polar solar orbits, but this only helps for a little while. Such orbits must cross the plane of the Sun's equator, and at large scales, they too will create the same problem.

543This same technique is used in proposed solar sails. Solar sails need only to counter the gravitational attraction of the sun. This is easily accomplished with thin sheets of mylar plastic in space near Earth. Since both the intensity of sunlight and gravitaional attraction increase by the inverse square of the distance to the sun, what holds in space near Earth will hold in space closer to the sun. A black material will have to be half the weight of a completely reflective material. Power bubbles would work in this application. Although they reflect light onto a collector, this cancels only half the "lift" imparted by sunlight. Private communication, Mr. P. Blass, November 1, 1993.

544The solar corona reaches out about 15 million kilometers from the photosphere. If the solar collectors are positioned 20 or 30 million kilometers from the sun, they will be safe from coronal disturbances. At these distances the solar constant would be around 10 kilowatts per square meter.

545Since the sun presents a shining disk, occulting any fraction of the disk will result in a proportional decrease in light received. Therefore, when viewed from the asteroid belt, Mars, or Earth, the collectors must obscure none of the Sun's face.

546The material in Mercury could form a shell centimeters thick around the entire sun with a radius of about half of Mercury's orbital distance.

[547]Freeman J. Dyson, "Search for Artificial Stellar Sources of Infrared Radiation," Science, Vol. 131, (3 June 1960), p. 1667.
[548]"I could be bounded in a nutshell and count myself a king of infinite space..."—Shakespeare, *Hamlet*, II,ii.
[549]Astronomers on Earth and even the Moon will probably not appreciate this change in the night sky. By 2500 A.D., however, the cutting edge of optical astronomy will be out in the Trojan asteroids, far from the light contamination of Solaria. In an asteroid's microgravity, or floating in free space in the gravitational islands of the Jovian Lagrange points, focusing mirrors thousands of meters across can be built. Armed with such instruments, astronomers working in the unsullied darkness of the outer solar system can probe deeply into the mysteries of the universe.
[550]Population would exceed 1 X $10^{23}$ some time between 3300 and 3350 A.D.
[551]The volume of Jupiter is 1400 times that of the Earth.
[552]Clark R. Chapman, "Asteroids," *The New Solar System*, p. 97.
[553]Real town. Northeast of Ft. Wayne, near the Indiana and Ohio border.
[554]Peter Russell, *The Global Brain: Speculations on the Evolutionary Leap to Planetary Consciousness*, (Los Angeles, 1983), p. 56.
[555]1.2 million km.
[556]Eric M. Jones and Ben R. Finney, "Fastships and Nomads," *Interstellar Migration and the Human Experience*, Finney and Jones, eds., (Berkeley, 1985), p. 96.
[557]Stephenson, p. 211
[558]Finney and Jones, "Fastships and Nomads," p. 97. Assumes a power supply of one megawatt for each colonist. This is very high, equal to fifty times the energy consumption of the average American today. It may be justified, however, since these people will dwell in interstellar space with nothing but starlight to brighten their dark little worlds.
[559]Stephenson. p. 212.
[560]Distance to Alpha Centauri is 272,000 A.U. Presumably there is a gap of around 72,000 A.U. between the Oort clouds of the two systems.
[561]Sagan & Druyan, *Comet*, p. 355.
[562]Dyson, *Disturbing the Universe*, p. 236.
[563]"*A Route of Evanescence*
*With a revolving Wheel —*
*A Resonance of Emerald—*
*A Rush of Cochineal—...*"
Emily Dickinson, No. 1463, [c. 1879].
[564]Dyson, *Disturbing the Universe*, p. 213. For poetic effect we can imagine a galaxy of green stars; in reality, as is usually the case, the situation will be more complex. Solaria would appear to be green only if viewed edge-on with respect to the plane of the ecliptic. A distant observer, seeing Solaria from either pole would detect not a green star, but

an infra-red source. The large solar collector caps will completely occlude visible light. They will absorb sunlight and heat up in the process. At their equilibrium operating temperatrue they will glow with infra-red energy with a wavelength of approximately ten microns, 100,000 å.
[565] George Abell, *Exploration of the Universe*, (New York, 1973), p. 271. Stars are classified in part according to the color of their photosphere. This classification is called the color index. Hot stars are at the head of the index and appear blue, while cooler stars, at the tail of the index, are red. When stars are plotted on a graph with the color index or corresponding temperatures plotted on the horizontal axis and stellar absolute visual magnitudes plotted on the vertical axis the result is the H-R diagram, showing stars clumpted into groupings, like the main sequence. Though star temperatures span the spectrum, from 50,000° K. for O type blue-giants to 3,000° for M type red-dwarves; stars ranging in temperature from 8,000 to 5,000° K. of the A and F types, neither appear to be, nor are they classified as "green"; A types are considered blue and F types are blue-white; G types are white to yellow; K types are orange to red; and M types are red.
[566]Dyson, *Disturbing the Universe*, "The Greening of the Galaxy," pp. 225-238. Dyson, of course, meant this in a much less literal sense. Whether a star like the sun would actually appear green to the naked eye from interstellar distances is speculative and may fall in the category of artistic license. The important point is that the spread of animate matter will leave some unmistakable mark on the galaxy. I suggest we go ahead and do it, then we can find out for sure what it will really look like.
[567]2X exhaust velocity is the practical limit for each stage of a step-rocket, so using triplet helium at high efficiency a rocket could accelerate by 60 kps in each stage. This would require 500 stages to reach 30,000 kps, each subsequent stage being 20 times more massive than the first stage. Assuming a payload of 1000 tons, the first stage would weigh in at 20,000 tons, the second stage at 400,000 tons, the third stage at eight million tons, etc. Since the mass increases exponentially, each additional step adds a zero to the total mass. So at the end of the exercise the final mass of the rocket is a number with five hundred zeros!
[568]Clarke, *Astronautics*, p. 64.
[569]Jerry Pournelle, "That Buck Rogers Stuff," *Galaxy: The Best of My Years*, James Baen, ed., (New York, 1980), p. 284.
[570]Eugene F. Mallove and Gregory L. Matloff, *The Starflight Handbook*, (New York, 1989). p. x.
[571]Mallove and Matloff, *The Starflight Handbook*, p. 66. Freeman Dyson described such a world ship, which could be propelled by tens of millions of thermonuclear explosions.
[572]Columbus sailed three tiny caravels to the New World 500 years ago.
[573]Frank E. Stewart, *Basic Units in Physics*, (Flushing, 1957), p.363.

[574]Mallove and Matloff, *The Starflight Handbook*, p. 252. A Saturn V launch released $1X10^{13}$ Joules.
[575]The bombs dropped on Hiroshima had an explosive power of around 20 kilotons of TNT. Energy content of TNT = 700 cal/gm. 20,000 tons TNT = $5.85 X 10^{13}$ Joules.
[576]Nigel Henbest, *The Exploding Universe*, (New York, 1979), p. 189.
[577]Robert L. Forward, "Antimatter Revealed," *Omni*, (November, 1979), p. 46.
[578]"Radio Frequency FELs May Win Role in SDI," *Aviation Week and Space Technology*, (August, 1986), p. 507.
[579]Devon G. Crowe, "Laser Induced Pair Production As a Matter-Anti-Matter Source," *JBIS*, Vol. 36, (November, 1983), p. 507.
[580]This will require lasers operating at very short wavelengths.
[581]Crowe, p. 507.
[582]*Ibid.* For starship fuel it will be necessary, however, to produce proton/anti-proton pairs.
[583]Robert L. Forward, "Feasibility of Interstellar Travel: A Review," *JBIS*, Vol. 39, (September, 1986), p. 381.
[584]Eugen Sanger, *Space Flight*, p. 255. Sanger's "photon rocket" has a theoretical exhaust velocity of 300,000 kps—light speed.
[585]Forward, "Antimatter Revealed," p. 48.
[586]*Ibid.*
[587]*Hi-Yo Silver. Away*!
[588]Reaching light speed is a real-life version of Zeno's Paradox. Zeno asserted that Achilles could not outrun a tortise because he must first cover half the distance between them, during which time the tortise would have moved, then he must cover half the remaining distance, etc.
[589]@ .5*c* ¥ = 115%
[590]@ .9*c* ¥ = 230%
[591]@ .99 *c* ¥ = 710%
[592]Other great apes had also progressed. Orangutans had become philosophers, while gorillas had evolved into warriors. The chimps were scientists
[593]This is not to say that the apparent velocity of the light wave will change from your frame of reference. It won't; the waves will still appear to propagate at light speed. The clock you are observing, however, will appear to slow.
[594]*Voyage Through the Universe: Between the Stars*, (New York, 1990), p. 16.
[595]One cubic centimeter of gas at sea level pressure contains $3 X10^{19}$ molecules.
[596]Sanger, p. 278.
[597]*Ibid.*

---

[598]Thomas Damon, *Introduction to Space: The Science of Spaceflight*, (Malobar, 1989), p. 410.
[599]Nigel Henbest, "Inside Science: Dust in Space," *New Scientist*, (18 May 1991), p. 1.
[600]Martyn J. Fogg, "The Feasibility of Intergalactic Colonization and its Relevance to SETI," *JBIS*, Vol. 41, (November, 1988), p. 493.
[601]Forward, "Antimatter Revealed," p. 48. @ 100 kg. anti-matter per 10 tons of payload. To attain speeds of 30,000 kps, anti-matter can be diluted with water. Such an interstellar freighter will have a payload mass of 100,000 tons. It will take 400,000 tons of water, but only 1000 tons of anti-matter to make the voyage.
[602]This assumes perfect efficiency in the means used to accelerate the payload. No real system can be 100% efficient. A more realistic figure would be 250 million kwh/kg.
[603] Anti-matter contains, at best, only half of the energy required to produce it, but half the energy of propulsion comes from the annihilation of normal matter, so the energy conversion at high efficiencies is virtally one to one. Actual values will nevertheless have to be two to three times higher due to the inefficiencies involved in converting energy into thrust.
[604]This figure is even higher when the relativistic effects of *gamma* are taken into account, in which case it requires 31 times more energy to accelerate a given mass to 50% of the speed of light.
[605]Accelerating one kg. to .5 *c* requires at least 4 billion kwh. Decelerating at the end of the voyage takes an equivalent amount of power. Eight billion kwh @ 5¢/kwh = $400 million. @ 50% efficiency = $800 million/kg.
[606]Time dilation is a phenomenon of relativistic motion at speeds approaching the speed of light. At the speed of light the progression of time comes to a complete stop. As a ship approaches the speed of light, time aboard the ship—relative to an outside observer—appears to slow. Of course, aboard the moving ship, everything seems to progress normally. At 99% of the speed of light, travel time, after the acceleration phase, would be cut to a seventh. At 99.99% of the speed of light it would be one seventieth.
[607]Fifty tons of anti-matter would allow 830 tons per passenger, including ship structure, fuel and payload.
[608]A single kilogram of matter contains 25 billion kwh of energy. It would take at the very least twice that much energy to produce a kilogram of anti-matter.
[609]John H. Gibbons, Peter D. Blair and Holly L. Gwin, "Strategies for Energy Use," *Scientific American, Special Issue: Managing Planet Earth*, (September, 1989), p. 136. World energy usage, from all sources in 1988 was 318 X $10^{18}$ Joules = 88 X $10^{12}$ Kwh

---

[610]Anti-matter costs \$2.5 trillion per ton. Cargo ship requires 1000 tons of anti-matter at a cost of \$2.5 $X10^{15}$. Cost of ship and cargo are estimated as being equal to fuel cost. Total cargo ship cost = \$5 $X10^{15}$. Transport ship requires 2500 tons of anti-matter at cost of \$6.25 $X10^{15}$. Transport ship and crew costs estimated at half of fuel costs = \$3 $X10^{15}$. Total expedition costs = \$14.25 $X10^{15}$.
[611]At \$200 per capita, star colonization budget would equal \$20,000 trillion per year. To people earning \$5 million per year, an expenditure of \$200 is equivalent to the expenditure of \$1 to people earning \$25,000 per year.
[612]GSSP (Gross Solar System Product) in year 2500 is likely to exceed $\$10^{20}$/yr.
[613]*Statistical Abstract of the United States-1988*, Table 900.
[614]Rather than sending out lone expeditions of 50 people every year, the Solarians will send combined expeditions of a few hundred people every few years.
[615] *Star Trek II: The Wrath of Khan*, (Paramount Pictures, 1982). The "Genesis Effect" is a force, that once unleashed, causes the spontaneous conversion of inanimate matter into animate forms. It is a creature of science fiction. We, the forces of life, as we spread through the galaxy, will be the Genesis Effect made real.
[616]J.B. Birdsell, "Biological Dimensions of Small Human Founding Populations," *Interstellar Migration and the Human Experience*, p. 111.
[617]*Ibid*., p. 114.
[618]*Ibid*., p. 118.
[619]At 10% of the speed of light, this will save a couple of weeks. More significantly, the Rigilian system will be five light months closer to us in 500 years when these journeys begin.
[620]The distance between the Sun and Neptune is 30 A.U. At the outer limit of its eccentric orbit, Pluto is as far as 50 A.U. from the Sun. Pluto has a very eccentric orbit that carries it within the orbit of Neptune. It is presently inside of Neptune's orbit and will remain so for a number of years. Neptune is now, therefore, the most distant planet in the Solar system.
[621]As in all gravitational systems, the two stars actually orbit a common point between them.
[622]Rigil Kent A is 1.1 times as massive as the sun, and has a radius 1.23 times the sun's. Stars are classified according to spectral type, color, which varies from ultra-hot blue stars to relatively cool red ones. The spectral class is designated by a letter: O, B, A, F, G, K, or M. O stars are blue and M stars are red. There is also a number designation indicating the stars absolute magnitude—its true brightness when viewed from a specific distance of 10 parsecs ( 32.6 LY)

---

[623]It is theorized that the asteroid belt did not aggregate because it was torn between the gravitational influences of the sun and Jupiter. A similar gravitational disequilibrium might prevail throughout a multiple star system like Rigil Kent.
[624]Michael Rowan-Robinson. *Our Universe: An Armchair Guide.* (New York, 1990), p. 21.
[625]Recent mathematical analyses have concluded that stable planetary orbits are possible out to distances of 2.5 A.U around either star. Poul Anderson, personal communication, 10 March 1994.
[626]Ian Ridpath, ed., *The Illustrated Encyclopedia of Astronomy and Space,* Revised Edition, (New York, 1979), p. 59. 15%-single star systems, 46%-binary star systems, 39%-multiple star systems averaging 3.5 stars each.
[627]By the year 3000, the freighters launched in 2750 will have travelled 25 light years. Ships launched before 2750 are all assumed to have had destinations within 25 light years.
[628]Abell, p. 298. The parsec—parallax second—is equal to 3.2616 LY. It is the distance at which a star would show a parallax shift of one second when viewed from opposite sides of the Earth's orbit around the sun. The density of stars in our region of galactic space is approximately one per ten cubic parsecs. The average distance between stars is around 7 LY.
[629]Mallove & Matloff, *The Starflight Handbook*, pp. 236-237.
[630]Rowan-Robinson, p. 26.
[631]Nigel Henbest, *The Exploding Universe*, (New York, 1979), p. 135.
[632]Eds. of Time-Life Books, *Voyage Through the Universe: Stars*, (Alexandria, 1988), p. 63.
[633]During the 500 years of the Rigilian's early development, Solaria will have contiued to send out seed ships.
[634]Distance from Sun to Neptune is $4.5 \times 10^9$ km.
[635]Reaction forces, unless countered, will tend to move the mass-driver out of position. Counter forces can be applied by ejecting some mass out the back end of the launch tube. Eventually, star-pod launches could be made in opposite directions, thereby keeping the individual mass-drivers in place.
[636]Mallove and Matloff, *The Starflight Handbook,* p. 144. This distance will minimize interference from the sun's gravity well.
[637]Assumes a pair of continuous tubes with diameters of 20 meters, and a mass of 20 to 200 kg., per running meter. This would equate to a wall density of between 10 and 100 grams/cc. Tube length of $4.5 \times 10^{14}$ meters.
[638]At this speed *gamma* is 700 million. Apparent transit time equal to .2 seconds.
[639]Energy and time issues aside, it is doubtful that much higher velocities will be desireable because of the innordinate tube lengths imposed by limiting acceleration forces.

[640]Enrico Fermi—first man to construct a nuclear reactor—was one of the 20th Century's great wise men. He was not, however, infallible. In 1945, before the detonation of the first atomic bomb at Alamogordo, New Mexico, he gave odds on a bet that the explosion would ignite the atmosphere and vaporize the entire SW quadrant of the United States. Luckily, he lost this bet.
[641]Robert Sheaffer, "Are We Alone After All?," *Spaceflight*, (Nov.-Dec., 1980), p. 335. Space arks, traveling at only 1% of the speed of light, crossing an average of 100 light years between stars, and stopping for 1000 years before proceeding on to the next star system would expand into the galaxy at one light year per century.
[642]This allows the first five billion years of the galaxy's history for the first intelligent species to evolve.
[643]Carl Sagan, *Cosmos*, (New York, 1980), p. 301.
[644]Strieber has written several books, most notably *Communion*, in which he purports to have been captured by aliens who performed horrible experiments on his brain. One can only conclude that the experiments were not a great success.
[645]Peter Adams, *Moon, Mars and Meteorites*, (London, 1977), p. 24. The surface of the Moon is not entirely free of erosive forces. The surface is continually bombarded by meteoritic dust and swept by the solar wind. Even so, footprints are apt to persist for millions of years, and larger artifacts are virtually permanent. Rates of erosion on the lunar surface average around a millimeter per million years. After a billion years on the lunar surface a meter of the desecent stage will have been worn away.
[646]The ancients flattered themselves. "Man fears time, but time fears the Pyramids," they used to say. As far as Father Time is concerned, the Pyramids don't even rate goose-bumps. The Pyramids will be entirely wheathered away in a few scores of thousands of years.
[647]The surface of the Moon has been scrutinized in excruciating detail, observed and photographed down to a resolution of a few feet. The Ranger series of probes photographed the Lunar surface to a resolution of 3 feet, and the later Lunar Orbiters mapped the entire surface of the Moon, with particular attention paid to potential landing sites.
[648]An alien civilization would not necessarily incorporate chlorophyll, of course. However, as Dyson suggsets, any K2 culture will tend to degrade the energy coming from its home star, thereby creating some sort of characteristic signature.
[649]At the outer surface of the radio shell is Marconi's transmission of the letter *S*, made in 1901.
[650]Carl Sagan and Frank Drake, "The Search for Extraterrestrial Intelligence," *Scientific American Exploring Space*, Special Issue, Vol. 2, No. 1, (1990), p. 155.
[651]Quasi-stellar objects, or as in this case quasi-stellar radio sources.

---

[652]Radio Raheem is a character in Spike Lee's film *Do the Right Thing*. Raheem is a big black kid who carries a gigantic 'ghetto blaster', on which he plays rap music at deafening volumes.
[653]A gamma ray observatory now in orbit has revealed mysterious sources of gamma radiation emanating from all over the sky. Could this be the signature of an alien race? No one knows.
[654]Occam's razor is a philosophical principle that dictates that the simplest explanation is almost always the best.
[655]A K2 civilization could exist out of radio range. But that would mean it had sprung up only in the past 100,000 years. Since this amounts to only a minuscule fraction of the age of the galaxy, it seems an improbable coincidence.
[656]The Borg is an implacable alien race that periodically terrorizes the galaxy in *Star Trek: The Next Generation*. Ewoks are animated teddy bears that populate the Moon of Endor in George Lucas's *Star Wars* universe. A populated galaxy would be subject to the harsh laws of Darwinian selection. We are more likely to be contacted by aggressive predatory species, than benign ones.
[657]The Vogons are a noxious species populating Douglas Adams' classic *Hitchhiker's Guide to the Galaxy*. When the Vogons came through our Solar system they demolished the Earth to make way for an interstellar by-pass. Terrible cock-up, the mice were furious.
[658]Sagan, *Cosmos*, p. 301.
[659]Harlow Shapley, *The View From a Distant Star: Man's Future in the Universe*, (New York, 1963), p. 62.
[660]Sagan, *Cosmos*, pp. 30-31.
[661]Sheaffer, p. 337.
[662]John Pfeiffer, *The Cell*, (New York, 1964), p. 61.
[663]Michael H. Hart, "Atmospheric Evolution, the Drake Equation, and DNA: Sparse Life in an Infinite Universe," *Extraterrestrials—Where Are They?*, (New York, 1982), p. 160. Life may have first been based on RNA, a nucleic acid in which there is a single strand of molecules, rather than the double strand in DNA.
[664]The googol is $10^{100}$.
[665]Edward Argyle, "Chance and the Origin of Life," *Extraterrestrials—Where Are They?*, p. 102. That is a mass equating to 2 $X10^{15}$ tons, out of an oceanic mass of 1.3 $X10^{18}$ tons, or .15%.
[666]The formation of a self-replicating molecule is only the tip of the iceberg. Such an occurrance presumes a whole set of preconditions. It presumes the existence of a stable long-lived star, and the basic nuclear reactions which sustain its fusion; it presumes the creation of elements like carbon, essential to our form of life; it presumes the delicate arrangement of binding electrons which allows the formation of complex

molecules, etc. In short, this first gigantic 'coincidence' is really the end result of a whole series of coincidences.

[667]Richard D. Meisner, "The Compatability of the Universe to Complex Order: Paradigms and Speculations," *JBIS*, Vol. 39, p. 124.

[668]George Greenstein, *The Symbiotic Universe: Life and the Cosmos in Unity*, (New York, 1988), p. 65.

[669]Zeyphod is a character from *The Hitchhiker's Guide to the Galaxy* by Douglas Adams. Zeyphod has two heads and is the "Coolest cat in the known universe." At one point, characters are subjected to a mind-melting torture in which they are exposed to the unmitigated grandeur of the universe. Zeyphod emerges from this encounter unscathed, because it turns out that the universe was created especially for him.

[670]Zeyphod often wore an ultra-cool pair of sunglasses that would get darker as the level of danger increased. In panic situations the glasses would go entirely black, thereby sparing their wearer from having to cope.

[671]Paraphrased. "... has changed everything save our modes of thinking ..." In reference to the atomic bomb.

[672]The unsurpassed description of a human pan-galactic civilization is contained in Isaac Asimov's *Foundation Trilogy: Foundation, Foundation and Empire,* and *Second Foundation.*

[673]Calcification of the American democratic process now confirms this.

[674]For a good grounding in Chaos theory read *Chaos: Making a New Science*, by James Gleick, (New York: Penguin Books), 1987; and *Turbulent Mirror* by John Briggs and F. David Peat, (New York: Harper & Row), 1989.

[675]This example originally used the Apartchik in Moscow, but they have gone the way of all flesh. Now China is undergoing a quiet but inexhorable revolution as free market economics infiltrate the communes. Consequently, one can buy bacon in Beijing, but it is despite the communist appartatus, not because of it.

[676]Alvin Toffler, *The Third Wave*, (New York, 1980), p. 48.

[677]Dendrites do not actually touch each other, but are separated by a gap, the synapse, across which signals are conveyed by chemicals called neurotransmitters.

[678]This principle of brain behavior is well illustrated by the work presently being done on 'neuronal nets' for artificial intelligence in computers. In these networks a 'neuron' is set up to receive inputs from several others. Depending on the input from the other neurons in its net, the individual cell may or may not send a signal on down the line. This enables a neural network to actually learn from past mistakes.

[679]Russell, p. 195.

[680]77,777 is the square root of 6,049,261,729.

[681]"Anlage" is a synonym for "foundation"; it also means the beginning stage in the development of an organ or embryo.

---

[682]David Chaum, "Achieving Electronic Privacy," *Scientific American*, August, 1992, pp. 96-101.
[683]The colony will need 42 million kilowatt hours of electric power per year. This will require one 5 megawatt OTEC. A 5 megawatt OTEC will require an uptake of 210,000 gallons of cold deep sea water per minute.
[684]The Waikoloa in Hawaii cost $360 million and accommodates up to 2500 guests. Eden's resort will be about the same size and cost. The average cost per night for a room at such a resort is over $250. With room, meals, activities, and gaming, guests will spend an average of $375 a day each. At 75% occupancy this will produce $250 million a year in gross revenues for the colony. Tourism is one of the most powerful economic forces on the planet. For example, in 1982 the State of Hawaii had a gross domestic product of $11.5 billion. Of this amount, $3.6 billion, or 31% was brought in by outside tourists. In fact, tourism as an industry is a microcosm of organic economics. In essence, people save up their surplus production and then go somewhere else and spend it. Tourism provides a direct conduit for the transference of surplus economic power to the Foundation.
[685]At the beginning of the Eden phase, the Foundation's laser will have a power output of 200 megawatts ($200 million). For the first two years of this period, all of this power will be focused on the construction of the Eden colony and resort—a 400 megawatt proposition. After Eden is up and running, the Eden resort will pump an additional net 100 megawatts into the laser. Staffing in Eden, through the extensive use of cybergenic systems will be about half of the level at equivalent resorts. In Eden the ratio of staff to guests will be one to one. The 2500 staffers, all Foundation Colonists, will account for $25 million of the $250 million gross generated by Eden's guests. Assuming a 50% margin on other factors consumed by the guests, this will leave a net balance of $100 million for the Foundation. With this additional burst of power, the Foundation's laser will be pumping with a voltage of 300 megawatts.
[686]Includes all oceanic waters between the Tropics of Cancer and Capricorn, an area of 50 million square miles. This larger area is taken as the resource base, since the waters circulate and will enter the 20 million square mile OTEC zone along the equator.
[687]Arthur W. Hagen, *Thermal Energy From the Sea* (Park Ridge, 1975) p. 44. Warm-water intake of 100 MW (gross) OTEC = 1.604 X $10^7$ lbs./min.
[688]Robert A. Muller & Theodore M. Oberlander, *Physical Geography Today*, (New York, 1978), pp. 60 & 67. The solar constant at the top of the atmosphere is 1.94 Langleys per minute. A Langley is 1 calorie per square centimeter. The year -round average for solar radiation received at the surface at the equator in the Indian Ocean is around 500 Langleys/day. One square foot = 929 square centimeters, so the average solar flux is 464,500 calories/day/sq. ft. X .004 = 1860 BTU/sq. ft.

---

[689]Norman Myers, *Gaia, An Atlas of Planet Management*, (New York, 1984), p. 112.
[690]Breaking strength of Kevlar 29 = 400,000 lbs./in$^{2}$.
[691]Jonathan M. Ross & William A. Wood, "OTEC Mooring System Development: Recent Accomplishments," *NOAA Technical Report: OTES-4*, (Rockville, October, 1981), p. 11. Five and a half inch dia. steel cable has breaking strength of 2,567,000 lbs. = 108,000 lbs/sq.in. 97 in dia = 7390 sq. in X 108,000 = 798 X $10^{6}$ lbs.
[692]Roymont, p. 155
[693]Roymont, p. 176.
[694]Roymont, p. 29.
[695]O.A. Roels, et al., "Potential Mariculture Yield of Floating Sea Thermal Power Plants: Part I - General Statement," *Energy From the Oceans - Fact or Fantasy*?, CONF-751235-2, (Raleigh, Jan. 27-28, 1976), p. 16.
[696]G.E. Fogg, *Algal Cultures and Phytoplankton Ecology*. (Madison, 1965), p. 20.
[697]Roymont, p. 195.
[698]Fogg, p. 93.
[699]Switzer, p. 103.
[700]Sanger, p.167. Note that at these extreme temperatures water disassociates into an ionized plasma with a molecular weight of 2.
[701]Arthur C. Clarke, *Man and Space*, (New York, 1964), p. 164. Assumes an orbit at 115 miles above the surface.
[702]Curtis D. Cochran, ed., *Space Handbook*, U.S. Air Force Air Command and Staff College, (Maxwell Air Force Base, 1985), p. 13-7.
[703]If V/v = 1, then mass ratio = 2.718. If, however, V/v = 2, the final velocity is twice the exhaust velocity, then the mass ratio must be such that the natural logarithm of the mass ratio is not 1 but 2. In this instance that means that the mass ratio must be 7.4. Even assuming that the structural mass of the rocket remains the same, the amount of fuel required increases by almost 4 times.
[704]Bob Parkinson, "Superfuels," *Spaceflight*, Vol. 18, No. 10, (October, 197), p. 399.
[705]"Producing Metallic Hydrogen," *Spaceflight*, May 1977, p. 175.
[706]Rice, p. 6-17.
[707]Dr. Eyssa Yehia, University of Wisconsin Applied Superconducting Center, Madison, Wisconsin. Phone converstation of July 9, 1990.
[708]The launch tube is only 85% efficient, so 15% of the energy input must be eliminated as waste heat. Making the conservative assumption that all of this waste heat would be in supercooled components, means disposing of 10,500 kwh of waste heat after every launch—up to 31,500 kwh/hr. A kilogram of Liquid Nitrogen ($LN_2$) can absorb .055 kwh of waste heat. To keep the launch tube cool will require about 572 tons of

---

$LN_2$ per launch. As much as 40,000 tons of liquid nitrogen might be required for cooling purposes each day. Capital costs for liquid nitrogen production plants are on the order of $10,000 per ton of capacity. The $LN_2$ plant will therefore cost around $400 million. Storage tanks for a three day reserve of liquid nitrogen together with distribution piping, valves and controls will cost 1.5 times the cost of the plant itself, or an additional $600 million.

[709]The HPGs, which are massive, have to be supported in place. The support structures will weigh about the same amount as the HPGs themselves, or around 19 million kg. The support structures will cost about $2/kg installed, or $38 million. The inductors and switches needed to control the flow of current to the HPGs will cost about a third of what the generators themselves cost, or around $40 million. Installed they will cost about $120 million.

The launch tube will be of aluminum, with an exterior protective container of plastic. The aluminum tube will be about 4 cm. thick, and the plastic will be about the same. Both aluminum and plastic both cost about $2.20/kg. (Rice, p. 6-23) The aluminum pipe will weigh about 21 kg./cm. so the entire launch tube will weigh about 170 million kilograms and will cost $375 million. The outer plastic pipe will weigh and hence cost about half that amount, or $187.5 million. Installed, the launch tube, equipped with HPGs, will cost a total of around $1.5 billion.

[710]*Proceedings: SDIO/DARPA Workshop on Laser Propulsion*, CONF-860778, Vol. 1, "Executive Summary", p 6.

[711]*Ibid.*

[712]This is not to say that this lonley form of existence is to be recommended, it is only to illustrate that once an ecosphere is established, it can continue to operate at virtually no cost.

[713]4.3% in 1987. *Statistical Abstract of the United States: 1988*, Table 678, "Personal Income and its Disposition: 1970 to 1986."

[714]*Statistical Abstract of the United States: 1988*, Table 727, "Household Net Worth—Percent Distribution by Selected Characteristics: 1984." and Table 728, "Household Net Worth—Ownership Rates and Median Value of Holdings: 1984."

[715]Gerard K. O'Neill, "The Colonization of Space", *Physics Today*, (September, 1974), p. 33.

[716]J.E. Gordon, *The Science of Structures and Materials*, (New York, 1988), p. 190.

[717]Thomas Herzog, *Pneumatic Structures*, (New York, 1976), p. 18. Assumes 400,000 psi reinforcing cables with a maximum diameter of 15 centimeters and an ultimate tensile strength of 11 million pounds. A safety margin, of factor five. And disregards the tensile strength of the membrane. Membrane tensions in a sphere are expressed as $T = P \times r/2$ where T = tension in pounds, P = pressure in psi, and r is the radius in inches.

[718]The shield thickness in very large ecospheres can be substantially reduced, thanks to the mass of the internal atmosphere. At a pressure of 3 psi, air has a density of 235 $g/M^3$. Twenty kilometers of air at this density will absorb as much radiation as five meters of water. Therefore, the inner realm of such an ecosphere would be radiation shielded, even without a water layer.
[719]These are particles with a high enough energy to penetrate a shield mass of 1 $gram/cm^2$.
[720]Eric C. Hannah, "Radiation Protection for Space Colonies," *Journal of the British Interplanetary Society*, Vol. 30, (August, 1977), p. 310.
[721]Hanrahan, p. 157.
[722]K.A. Trukhanov, "Active Shielding of Spacecraft", U.S. AirForce, AD-742410, p.5.
[723]Eric C. Hannah, "Meteroid and Cosmic Ray Protection in Space Habitats," *The Princeton University Conference on Space Manufacturing Facilities*, (Princeton, 1975), p. 17.
[724]Total energy input from the sun is 5.4 million quad per year. Assuming all the energy is perfectly clean—like hydrogen or solar electric, producing no $CO_2$ or other pollutants—the theoretical limit on the amount of energy a terrestrial civilization can produce within its own atmosphere is 1% of the incident solar radiation. In the case of the earth this would be 54,000 quad, equal to 18 times the maximum we will probably ever need to use.
[725]This is based on a 1000 year turn-over rate for deep ocean waters, making only 1000th of the total energy available in any given year.
[726]Gross global product of 12 trillion. Population of 4.5 billion.
[727]With a space allottment of over 1000 cubic meters per person.

# Index/Glossary

*2001 A Space Odyssey* 161,180, 220
abalone 54, 59
Acacia trees 219
**Accretion disk** - When the Solar system originally formed, materials condensed out of a cloud of interstellar gas. As gravity pulled the cloud in on itself, a disk of material formed around the nascent proto-star. The planets eventually formed within this cloud. 301
**Acid rain** - Precipitation in any form that has been acidified by sulfur or other atmospheric pollutants. 26
Acidalia Planitia 257
adenine 352
Africa 89, 217, 218
**Age of Aquarius** - The ancient astrologers divided the broad span of human history into separate epochs. According to the astrologists, the Age of Aquarius will begin in the year 2000, and will usher in a golden age for mankind. 19
**AIDS** - Acquired Immune Deficiency Syndrome. Viral infection that attacks the immune system—universally fatal. No cure is presently known. 25
air car 216
air cycles 216
airship drawing 53
airships 122
albedo 251
algae 34, 42, 46-53, 58, 248
  adaptability 46
  annual production of 30
  benthic 54
  biomass 46
  closed loop system 155-156
  $CO_2$ sponge 30
  nutrient composition 155
  protein supplement 45
  radiation antidote 142
  reproduction 46
  synthetic food 59, 159-160
  terraforming with 249
  waste recycling 158
**Algin** - Long chain organic polymer derived from brown algae. 56
**Ali Baba** - A character from the roster of *Tales of the Arabian Nights*. Ali Baba discovered a magical cave, the portal of which could only be opened by the magic words "Open Simsim"—the Arabic word for "sesame". 18
Alnico 282
**Alpha Centauri** - Nearest star system, at a distance of 4.3 light years. Also known as Rigil Kent. 314, 330-334
Altair 338
alum 168
aluminum 106, 197, 207, 213, 282
**Amalthea** - The mythical goat who nursed Zeus. One of the goat's horns became the cornucopia. 298
amino acids 48
ammonia 43, 56, 296
**Ancient Mariner, Rime of** - Epic poem by Samuel Taylor Coleridge. 90